AUTOMATIC CONTROL, MECHATRONICS AND INDUSTRIAL ENGINEERING

PROCEEDINGS OF THE INTERNATIONAL CONFERENCE ON AUTOMATIC CONTROL, MECHATRONICS AND INDUSTRIAL ENGINEERING (ACMIE 2018), OCTOBER 29–31, 2018, SUZHOU, CHINA

Automatic Control, Mechatronics and Industrial Engineering

Editors

Yigang He

School of Electrical Engineering, Wuhan University, Wuhan City, Hubei Provice, P.R. China

Xue Qing

School of Mechanical Engineering, Beijing Institute of Technology, Haidian District, Beijing, P.R. China

CRC Press
Taylor & Francis Group
Boca Raton London New York

CRC Press is an imprint of the
Taylor & Francis Group, an **informa** business

A BALKEMA BOOK

CRC Press
Taylor & Francis Group
6000 Broken Sound Parkway NW, Suite 300
Boca Raton, FL 33487-2742

First issued in paperback 2020

© 2019 by Taylor & Francis Group, LLC
CRC Press is an imprint of the Taylor & Francis Group, an Informa business

No claim to original U.S. Government works

Typeset by V Publishing Solutions Pvt Ltd., Chennai, India

ISBN-13: 978-1-138-60427-8 (hbk)
ISBN-13: 978-0-367-73147-2 (pbk)

**Visit the Taylor & Francis Web site at
http://www.taylorandfrancis.com**

**and the CRC Press Web site at
http://www.crcpress.com**

Table of contents

Mechanical engineering

Mechatronics

Pattern recognition

Automatic Control, Mechatronics and Industrial Engineering – He & Qing (Eds)
© 2019 Taylor & Francis Group, London, ISBN 978-1-138-60427-8

Preface/foreword

The proceedings of the International Conference on Automatic Control, Mechatronics and Industrial Engineering (ACMIE 2018) is a compilation of the papers presented at the conference held in Suzhou, China, 29–31 October, 2018. ACMIE 2018 consists of keynote speeches, invited speeches, oral presentations and poster presentations. It attracted many scientists from all around the world to attend, providing ideal forums for them to exchange the most up-to-date and authoritative knowledge from both industrial and academic worlds. The proceedings collects not only the presentations given by the renowned world-class scientists, but also novel results in the field of Mechatronics, Automation and Industrial Engineering from different regions of the World.

We would like to thank all participants, especially the invited speakers, for their efforts and contributions to the success of the conference. We also thank the reviewers for their valuable comments on the conference proceedings.

ACMIE2018 Committee

Organizing committee

WUHAN UNIVERSITY

Scientific Committee

GENERAL CHAIR

Prof. Tao Lin, *Wuhan University, China*

COMMITTEE MEMBERS:

Prof. Yigang He, *Hefei University of Technology, China*
Prof. Qing Xue, *Beijing Institute of Technology, China*
Prof. E. Amirhossein Eslampanah, *Machine Sazi Arak (MSA) Company, Iran*
Dr. D. Devika, *Saveetha University, India*
Dr. M. Mohan Reddy, *Curtin University, Malaysia*
Dr. Daschievici Luiza, *"Dunarea de Jos" University of Galati, Romania*
Dr. Mukund Nilakantan Janardhanan, *Aalborg University, Denmark*
Prof. Andrey V. Brazhnikov, *Siberian Federal University, Russia*
Prof. Jin Su Jeong, *Technical University of Madrid, Spain*
Dr. Nasser Shahsavari Pour, *Vali-e-Asr University, Iran*
Dr. Anindita Ganguly, *Jadavpur University, India*
Prof. Vo Ngoc Phu, *Dui Tan University, Vietnam*
Dr. Delin Chu, *National University of Singapore, Singapore*
Prof. Tzyh Jong Tarn, *Washington University in St. Louis, USA*
Dr. Eric T.T. Wong, *The Hong Kong Polytechnic University, Hong Kong (China)*
Dr. Saeed Hamood Ahmed Mohammed Alsamhi, *Tsinghua University, China*
Dr. Hamid Yaghoubi, *Iran Maglev Technology (IMT), Iran*
Dr. Ali Safaei, *Universiti Sains Malaysia, Malaysia*
Dr. Luis Manuel Sanchez Ruiz, *Universitat Politècnica de Valencia, Spain*
Prof. Ting Yang, *Tianjin University, China*
Prof. Wen-Tsai Sung, *National Chin-Yi University of Technology, Chinese Taipei*
Dr. Yau Hon Keung, *University of Kent at Canterbury, UK*

Automation

Development and application of creep test remote-monitoring system

Z.L. An, W.X. Wang, Z.Y. Wang, T.H. Chen & N. Wang
Faculty of Railway Transportation, Shanghai Institute of Technology, Shanghai, China

ABSTRACT: In this work, to solve the information island existing in the Distributed Control System (DCS) of the creep laboratory, set up a network-monitoring system which could integrate multiple information to deal with the information construction problems. Based on a TCP/IP RS485 communication protocol and data-mining technology, a system with field and remote monitoring software in charge of monitoring and control rights is built. Up to 256 creep and endurance testing machines can be remotely monitored by means of a mobile client or web browser. The system provides an enterprise management system, which makes the enterprise and test information highly integrated. The performance results show that the system has many advantages, such as stability, reliability and real-time monitoring.

Keywords: information island, Creep test, remote monitoring, data-mining

1 INTRODUCTION

In recent years, due to the advantages of fast transmission rate, low cost and high reliability, Industrial Ethernet has been widely used to monitor the test for measuring properties of metallic materials at high temperature (Ning, 2015; Zhang, 2016; Chen, 2016; Wang, 2016; Sun, 2016). At present, creep testing machines such as INSTRON, MTS, and Shimadzu have successively developed remote network monitoring systems, while domestic Creep testing machines still use a Distributed Control System (DCS) (INSTRON Co, 1993; Ni, 2007; Wang, Z, 2006). With the grouping of the steel, boiler and other industries, its organizational structure and product types are showing diversity. At the same time, the geographical locations of the laboratories are scattered, which increases the difficulty of information construction (Zhu, 2018). It is therefore necessary to optimize the original application system and develop a highly integrated remote monitoring system to improve social benefits and meet industrial needs.

The paper uses an RD2-3 creep and endurance testing machine as a prototype, based on network and serial communication protocol and data-mining technology, to build a set of Intranet and Internet hybrid network monitoring systems. We developed the Creep Test Remote-Monitoring System, which uses max language programming in Visual Basic, JAVA, improving remote query efficiency and portability. It can automatically acquire and monitor the test data, such as temperature, stress, strain and break time, of up to 256 electronic creep testers. In addition, Considering the timeliness of the alarm, the system is equipped with a GSM/GPRS module. Hence the alarm signals such as break and overheating will be successively sent to the user's mobile phone, iPad and other devices in the form of short messages and application-notifications.

2 SYSTEM HARDWARE STRUCTURE

As shown in Figure 1, the Creep Test Remote-Monitoring System of the paper adopts a three-tier structure, comprising the field layer, service layer and client layer. The field layer mainly includes sensors for temperature, displacement and pressure, an analog-to-digital conversion module, MR13, PLC, 32-channel PCI-1611U multi-serial communication card, IPC

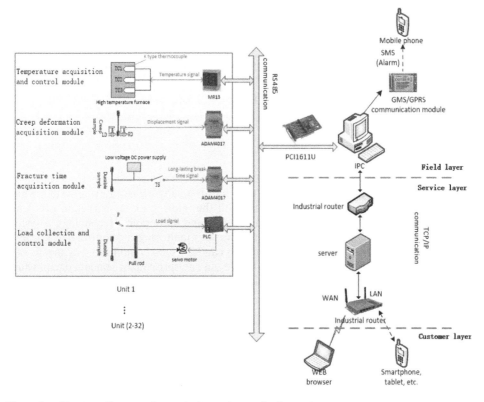

Figure 1.   Structure diagram of creep test remote monitoring system.

and GSM/GPRS communication module. Based on the RS485 data communication bus, the system uses the address differentiation technology to realize the automatic acquisition and parameter control of the temperature, strain, load and break time of 64 RD2-3 testing machines. The IPC is connected to the GSM/GPRS communication module through the RS232 port and directly sends the alarm signals, such as break and overheating, to the user's mobile phone. The four acquisition modules are as follows:

1. Temperature acquisition and control module. The three-stage independent temperature control is used to reduce the temperature interference of the interval. The IPC sends a read/write command to the MR13 through the RS485 bus to realize the independent acquisition and control of the temperature during the test (An et al., 2007);
2. Creep deformation acquisition module. The CBW digital displacement meter is used to convert the deformation displacement signal into a DC signal, which will be converted into a digital signal by the ADAM4017 module and then transmitted to the IPC to realize automatic acquisition and recording of creep deformation (An et al., 2007);
3. Fracture time acquisition module. The broken electrical signal is converted into a digital signal by the ADAM4017 module and then transmitted to the industrial computer to realize automatic recording and alarm of the long-term test break time (An et al., 2007);
4. Load collection and control module. While the system retains the lever loading mode of the long-term test, the Siemens S7-200PLC is used to control the servo motor to rotate the rod to achieve precise loading. The communication between the PLC and the pressure sensor can be realized by using the Modbus library provided by Siemens (Huang & Wang, 2018). The IPC further reads the load data or sends the load control parameters through the PCI-1611U.

The service layer uses the Advantech EKI-2528 industrial router to form a network between the server and four IPC, set a fixed IP for each IPC, and use a TCP/IP communication protocol to monitor up to 256 creep testers on the server at the same time. This provides a remote monitoring service to solve the problem of information construction caused by the dispersed locations of the field systems.

The client layer includes handheld devices and network terminals. The test status, data and enterprise information are put out to the network by the service layer. Users can directly monitor test on smart phones, tablet computers and other devices to make up for the lack of remote monitoring services.

3 SYSTEM SOFTWARE STRUCTURE

As shown in Figure 2, the field monitoring system was developed in the field layer's IPC using VB language, and the remote monitoring problem was lacking for the domestic testing machine DCS system. The remote monitoring system was developed on the server using java, which improved the applicability and portability. Multi-threaded and optimized queue sorting prevents program blocking, allowing the server to handle both access methods in an orderly manner: lightweight web and client applications. In order to cater to the diversification brought on by enterprise grouping, the system is equipped with an enterprise information management platform to facilitate enterprise structure adjustment and product grasp.

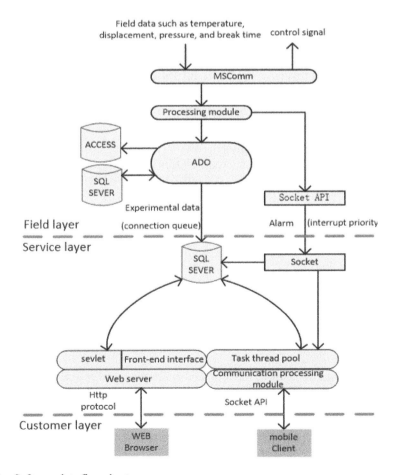

Figure 2.   Software data flow chart.

3.1 Field layer

In the field control layer, the DCS user software system was developed based on the Win XP operating system using VB6.0. The MSComm control is used as the serial communication tool of the application program to allow the upper computer to read and write data from the serial port to the lower computer. In this paper, we use the link technology of OLE object to get the control handle of the ADO data object to complete VB and database connection, in order to store the test data in the background database. The ACCESS database is used as the original collection database. The IPC periodically sends a connection request to the server, and the SQL database data is queued and replicated to the server through the connection queue, data consistency, continuity and integrity. The following describes the implementation of the key parts:

1. GSM/GPRS communication module. Considering that the alarm is only covered by the server message push, it will be overwritten by other push messages of the user's mobile client. Therefore, the GSM/GPRS wireless transmission technology is directly sent to the user's mobile terminal in the form of short message.
2. ActiveX Data Objects (ADO). The ADO provides remote operation of the database. The IPC connects to the server's SQL SEVER database through the server's IP, port number, database port number, and account password to copy the data to the server database. The DCS system has a good command flow application with ACCESS connection. Therefore, the communication code between the industrial computer and the server SQL SEVER is given below:

```
Dim AdoConnTool As New ADODB.Connection
Dim AdoRsTool As New ADODB.Recordset
Dim strAdoConnTool As String
strAdoConnTool = ""
strAdoConnTool = "Provider = SQLOLEDB.1;Persist Security Info = False;User ID = sa;
Password = sa123;Initial Catalog = new;Data Source = 210.35.96.142;"
'SQL server account password and IP address
strSQL = "........"
'Data selection
AdoConnTool.Open strAdoConnTool
'AdoRsTool.Open strSQL, AdoConnTool, adOpenStatic, adLockReadOnly
AdoRsTool.Open strSQL, AdoConnTool, adOpenForwardOnly
AdoRsTool!username
Dim strTotal As String
strTotal = ""
If AdoRsTool.RecordCount <> 0 Then
MsgBox "connection succeeded"
......'Data replication'
End If
AdoRsTool.Close
AdoConnTool.Close
End Sub
```

3.2 Service layer

This system is based on database technology, and builds the server of the two frameworks of B/S and C/S in the server. The monitoring adopts the B/S architecture as the main scheme and the C/S as the auxiliary scheme to ensure the efficient and stable operation of the system. In the B/S framework, Apache+Tomcat is used as the web application server and the standard World Wide Web port 80 is configured as the web terminal request port. The Servlet written by java is responsible for interactively browsing and modifying data. The eclipse is used to develop jsp dynamic pages to realize user login, life prediction, real-time data curve, historical curve and other functions.

In the C/S framework, the server is configured with a custom TCP port greater than 1024, and the Socket Server is programming in java for the client connection to achieve data transmission and timely push of alarm signals. The framework includes three modules as follows:

1. Data processing module: used to complete data processing and thread control. The task thread pool is used to provide threads for each connected client. When idle occurs in the collection, the queue task is started, otherwise it waits.
2. Processing module: used to complete client connection and XML file encapsulation and parsing. Connects to the client by comparing the sockets. When connecting, it uses the socket connection pool method to ensure the connection of multiple clients by initializing several long connections and adding identification bits.

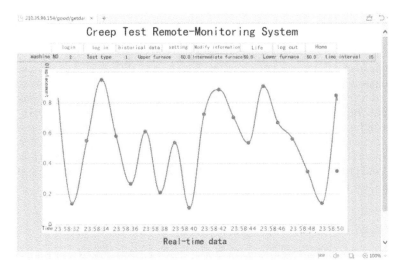

Figure 3.   Web client interface.

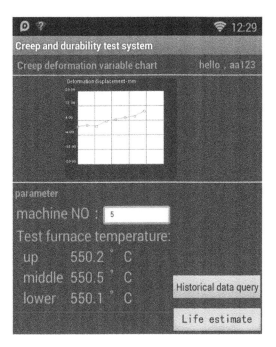

Figure 4.   Mobile client interface.

7

3. Communication module. Through the data transmission channel, the data information is sent to the server side in an XML data format in an asynchronous manner to implement data interaction.

3.3 *Customer layer*

The client layer is mainly used for interface display and human-computer interaction. The web browser is only responsible for front-end page display. The mobile client uses the eclipse tool to develop the mobile application, which achieves user login, life prediction, real-time data curve, history curve, break alarm and other functions. The interface is shown in Figures 3 and 4.

4 CONCLUSIONS

This paper builds a remote monitoring system based on TCP/IP, RS485 communication protocol and data-mining technology. The system is divided into the field layer, the service layer and the customer layer. The on-site and remote software are in charge of monitoring and control rights. Both can exchange data through ActiveX Data Objects (ADO), automatic acquisition and remote monitoring of temperature, stress, strain and break time data for up to 256 electronic creep testers. The system provides remote monitoring and enterprise management services for creep testing, which makes the enterprise and test information highly integrated, and provides a means for enterprise information construction. The system operation results show that the system runs stably and the test data is reliable.

REFERENCES

An, Z., Xuan, F. & Tu, S. (2007). Measurement and control software system of creep test machine based on VB, Excel and Access. *Chinese Testing Technology, 3*, 84–87.
Chen, S. (2016). Development of fieldbus standards and industrial ethernet technology. *Telecom Power Technology, 33*(04), 150–151.
Huang, H. & Wang, J. (2018). The application of PLC control system in electrical automation equipment. *Enterprise Technology and Development, 2*, 189–190.
INSTRON Co. (*1993*). Metal heat treatment [R]. A new material testing machine technology conference in Beijing.
Ni, X. (2007). Several technical studies in MTS test. *Proceedings of the 7th National MTS Materials Testing Conference.* MTS Materials Testing Cooperation Committee Chinese Society of Theoretical and Applied Mechanics.
Ning, J., Gao, F. & Liu, J. (2015). Analysis of the key technologies of industrial Ethernet. Communication world, *2*, 21–22.
Wang, Z (2006). The latest development of Shimadzu electronic universal material testing machine in Japan [J]. *Physical Testing,* 06, 62.
Sun, Z., Yu, M. & Zhu, X. (2016). Design of power management system based on industrial Ethernet. Electronic Design Engineering, *24*(21), 150–152.
Wang, J. & Li, J. (2016). Design of communication network based on industrial ethernet + Fieldbus. Communication World, *14*, 118–119.
Zhang, B. (2016). On the technical characteristics and application of industrial Ethernet. *Electronic Technology and Software Engineering, 10*, 10.
Zhu, S. (2018). Information system integration and data integration strategy. *Technology Economic Market, 3*, 13–14.

Automatic Control, Mechatronics and Industrial Engineering – He & Qing (Eds)
© *2019 Taylor & Francis Group, London, ISBN 978-1-138-60427-8*

A new method for integrated navigation of hypersonic cruising aircraft using non-Keplerian orbits

H.L. Li, D.W. Wu, B. Zhang, B.F. Yang & Y.H. Zhao
Information and Navigation College, Air Force Engineering University, Shanxi Xi'an, China

ABSTRACT: Although deeply coupled Strapdown Inertial Navigation System/Global Positioning System (SINS/GPS) technology has been considered the foremost navigation system in Hypersonic Cruise Vehicles (HCVs) by many research institutions, the wireless signal from the Global Navigation Satellite System (GNSS) is almost completely shielded by the real-gas effect in the hypersonic state, which results in an obvious decline in precision of the deeply coupled INS/GPS. Integrated INS/GNSS/CNS(Celestial Navigation System)navigation systems were adopted for HCVs. However, this increased the complexity of the system and led to the main filters tending to diverge in the information fusion process. Thus, a new method for INS/GNSS/CNS integrated navigation in hypersonic cruising aircraft based on non-Keplerian orbits is presented in this paper. By using the flight characteristics of the HCV, without adding any sensor, the non-Keplerian orbit mode of the HCV is considered to be the fourth virtual sensor; its exact non-Keplerian orbital model and an optimal data fusion model of a SINS/GNSS/CNS integrated navigation system constitute a deeply coupled mode, and the orbital model modifies the integrated navigation system and the integrated navigation system feeds information back to the orbital model. Experimental results indicate that the new method has much better precision with respect to position and velocity than the traditional method; the main filters still achieve effective constringency when they are subject to interference; the precision and reliability of the INS/GNSS/CNS integrated navigation of the hypersonic cruising aircraft are improved. The next step of this paper is to continue the research on the nonlinear filtering algorithm of the integrated heterogeneous navigation sensor.

Keywords: Hypersonic cruising aircraft, integrated navigation, non-Keplerian orbits, information fusion, deeply coupled mode

1 INTRODUCTION

A Hypersonic Cruise Vehicle (HCV) is an aircraft which can fly for a long time continuously at hypersonic speeds (5–25 Ma) in near space. The HCV has three superlative characteristics, which are superfast velocity of sound, super high altitude, and super long flying time. It centralizes new technologies in all aeronautic and aerospace disciplines, and represents future development directions in aeronautic and aerospace fields. It has become an important stratagem for development by today's great military powers of the world (Jiang et al., 2015). However, the navigation system functionality is a challenge in the HCV, and thus it has become a research hotspot (Gao et al., 2014; Li et al., 2014).

The integration of the Inertial Navigation System (INS), the Celestial Navigation System (CNS) and the Global Navigation Satellite System (GNSS) is regarded as being the optimal navigation system for an HCV (Peng,2014; Kaiser et al., 2017; Serrani & Bolender, 2014). However, the traditional INS/CNS/GNSS adopts a federal filter that is a decentralized filtering method, which has two data processes consisting of many local filters and a main filter. The INS's drift is respectively revised by the CNS and the GNSS. The system state equation is based for the most part on the INS errors, which become the system state vectors.

In actual application, however, the INS/GNSS/CNS integrated navigation system uses three dissimilar sensors to obtain observational information. The parameters concerned in the integrated navigation system are the position and velocity information, but only the platform misalignment angle is observed in the CNS, which does not provide the position and velocity information for the filter directly. For the same state, only two sensors (INS/GNSS or INS/CNS) are used. In the information fusion, the information from each sub-filter is different. Therefore, the output information from two sub-filters if the federal Kalman filtering is straightly used is an isomerous in the main filter, which leads to the squared error matrix estimator of the main filter being diverged (Yang et al., 2013). Meanwhile, the satellite navigation equipment is vulnerable to interference or shielding during extended-time and long-distance flights (He et al., 2014). All of these problems make the traditional INS/CNS/GNSS integrated navigation system inefficient for navigation on extended-time and long-distance HCV flights.

To eliminate the main filter distortion in traditional integrated navigation for INS/GNSS/ GNS, the research focused on the characteristics of the HCV's non-Keplerian orbits. The CNS system gathers the position and velocity information using a non-Keplerian orbit model, and then fuses that information in the filter. Only the state model of the CNS is revised using the non-Keplerian orbits of the HCV in the literature (Heiligers & McInnes, 2014), which has been expanded to the INS/GNSS/CNS integrated navigation in this paper. We expand the main filter to achieve optimal information fusion, which not only satisfies the requirement for filter stability, but also enhances the navigation accuracy and the comprehensive performance of the system.

2 THE PRINCIPLE OF INS/GNSS/CNS INTEGRATED NAVIGATION IN HCV NON-KEPLERIAN ORBITS

In the INS/GNSS/CNS integrated navigation system of an HCV,SINS is used as the basic system for the integrated navigation system because of its advantages in comprehensive navigation parameters, good maneuvering target tracing, timely continuous output, good concealment, and powerful anti-jamming performance. GNSS can be compatible with systems such as BeiDou, GPS and GLONASS. To improve fault tolerance and anti-jamming performance, each system can be a backup system. CNS adopts the COMS APS star sensor, which has good radiation toleration and a large dynamic range, to output the real-time vector position with high accuracy. In that regard, the parameters concerned for the integrated navigation are the position and velocity information, though accurate observation of the platform offset angle in the CNS system does not provide the position and velocity information for the filter directly. Therefore, by adopting a soft correction method, a system equation

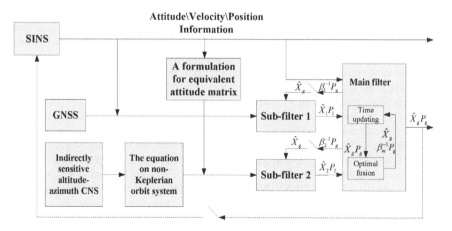

Figure 1. Schematic diagram of the integrated navigation system.

between the platform drift angle and the errors of position and velocity in the INS are established. The platform offset information is gathered by the globally estimated outputs of the federal filter and the CNS observation, while the error estimation for position and velocity is calculated via the non-Keplerian orbits' system equation. The errors for the position and velocity information are highly accurate, and from that, the CNS platform observation represents true real-time values and the global estimation output has the highest accuracy. Sub-filter 1 is composed of the position and velocity information coming from the system correction and the INS, while sub-filter 2 is composed of the position and velocity information derived from the INS and the GNSS observation. The outputs of the two sub-filters are then fused in the main filter. In addition, the INS can output information by itself. The principle of INS/CNS/GNSS integrated navigation is depicted in Figure 1.

3 A NEW METHOD FOR INS/GNSS/CNS INTEGRATED NAVIGATION FOR HCV NON-KEPLERIAN ORBITS

3.1 The model for HCV non-Keplerian orbits

HCV belongs to the non-Keplerian orbits through derivation. Moreover, the model and applied scope between Keplerian and non-Keplerian orbits are given inNazari et al. (2014). Under the action of multiple forces, these forces can be synthesized into four forces, $\mathbf{C_x}$, $\mathbf{C_y}$, $\mathbf{F_G}$ and $\mathbf{\bar{F}_s}$, as depicted in Figure 2. $\mathbf{C_x}$ is the control vector on the HCV, $\mathbf{C_y}$ is the aerodynamic vector on the HCV, $\mathbf{\bar{F}_s}$ is the thrust vector on the HCV, and $\mathbf{F_G}$ is the gravity vector on the HCV.

According to Newtonian mechanics, the resultant force $\mathbf{F_H}$ on the HCV is the vector sum of forces $\mathbf{C_x}$, $\mathbf{C_y}$, $\mathbf{F_G}$ and $\mathbf{F_s}$. Assuming that r is the vector of the earth, $\bar{r}$ is the unit vector of r, G is gravitational constant, M is earth's mass, and m is HCV's quality, the formulations $\mu = GM$ and $F_G = G\frac{Mm}{r^2}$ are substituted into the formulation of non-Keplerian orbits derived in Equation 1, because of $\mathbf{F_H} = m\mathbf{f_h}$:

$$\frac{d^2\mathbf{r}}{dt^2} = \frac{1}{m}(\mathbf{C_x} + \mathbf{C_y} + \mathbf{F_S} + \mathbf{F_G} - \bar{r}F_G)$$ (1)

Using the relation $\mathbf{F'_G} = \mathbf{F_G} - \bar{r}F_G$, Equation 1 can be rewritten as:

$$\frac{d^2\mathbf{r}}{dt^2} = \frac{1}{m}(\mathbf{C_x} + \mathbf{C_y} + \mathbf{F_S} + \mathbf{F'_G})$$ (2)

The return coordinate system rotates at an angular velocity ω_e, which is relative to the inertial coordinate system; the point (0,0,0) is defined as the center of mass of the earth; the x-axis is permanently fixed in a direction relative to the celestial sphere; the z-axis points

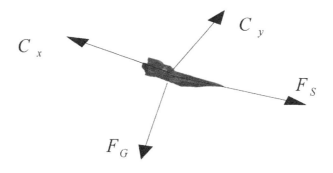

Re-entry Stage

Figure 2. Stress on the HCV.

toward the North Pole in a coincidence with the instantaneous earth rotational axis, and the y-axis is in the direction of the right-angle coordinate system.

Based on the vector derivation, $m\frac{d^2\mathbf{r}}{dt^2} = m\frac{\delta^2\mathbf{r}}{\delta t^2} + 2m\omega_e \times \frac{\delta\mathbf{r}}{\delta t} + m\omega_e \times (\omega_e \times \mathbf{r})$ can be substituted in Equation 2 as:

$$m\frac{\delta^2\mathbf{r}}{\delta t^2} = \mathbf{F_S} + \mathbf{C_Y} + \mathbf{F_G'} + \mathbf{C_X} - m\mathbf{a}_e - m\mathbf{a}_k \qquad (3)$$

To develop the dynamic and kinematic equations in the return coordinate system, the vectors in Equation 3 should be projected to the return coordinate system.

Let state vector $\mathbf{X} = [x\ y\ z\ v_x\ v_y\ v_z]^T$, and state model noise $\mathbf{W} = [w_x\ w_y\ w_z\ w_{vx}\ w_{vy}\ w_{vz}]^T$; the state equation of the aircraft movement is given by:

$$\dot{\mathbf{X}}(t) = \mathbf{f}(X(t),t) + \mathbf{W}(t) \qquad (4)$$

3.2 Integrated navigation system model for hypersonic cruising aircraft based on non-Keplerian orbits

According to the source of system error, the strap-down inertial navigation system error and the star sensor installation error will be included in the state of the integrated navigation system, including the mathematical platform attitude error of the strap-down inertial navigation system ϕ_E, ϕ_N, ϕ_U, the velocity error δv_E, δv_N, δv_U, the location error δL, $\delta\lambda$, δh, the gyro random constant drift ε_{bx}, ε_{by}, ε_{bz}, the accelerometer random constant error ∇_{bx}, ∇_{by}, ∇_{bz}, and the star sensor installation error δA_x, δA_y, δA_z, and the state vector of the integrated navigation system $\mathbf{X}$ is given as follows:

$$\mathbf{X} = [\phi_E,\ \phi_N,\ \phi_U,\ \delta v_E,\ \delta v_N,\ \delta v_U,\ \delta L,\ \delta\lambda,\ \delta h,\ \varepsilon_{bx},\ \varepsilon_{by},\ \varepsilon_{bz},\ \nabla_{bx},\ \nabla_{by},\ \nabla_{bz},\ \delta A_x,\ \delta A_y,\ \delta A_z]^T$$

According to the system error model and state vector $\mathbf{X}$, we can get the state equation for inertial/satellite/integrated navigation as follows:

$$\dot{\mathbf{X}}(t) = \mathbf{F} \cdot \mathbf{X}(t) + \mathbf{G} \cdot \mathbf{W}(t) \qquad (5)$$

where $\mathbf{F}$ is the system state matrix, $\mathbf{G}$ is the system noise drive array, and $\mathbf{W}(t)$ is the white noise. Moreover, w_{gx}, w_{gy}, w_{gz} is the gyroscope white noise, and w_{ax}, w_{ax}, w_{ax} is the accelerometer white noise (Miao & Shi, 2014).

3.3 The fusion algorithm for the Keplerian orbits' integrated navigation in hypersonic aircraft

Because the operating cycle is long for an HCV, and usually in a hypersonic maneuvering state, the influence of the engine's thrust is an uncertain factor in the system state model. Measurement residuals are no longer regarded as the white-noise sequence whose mean value is zero, such that the aircraft is navigated automatically. There is a limited capacity for the federated Kalman filtering method to whittle down this uncertain factor (Peng et al., 2014). To decrease the filter's estimated error in the hypersonic maneuvering process for an HCV, and to improve the filter's stability, an improved Kalman filtering algorithm designed for this work was fitted to decrease the negative influence of the state model's uncertainty, thereby improving the performance in navigating automatically during the HCV orbital maneuvering period.

3.3.1 Prediction and judgment

According to the state estimated value $\hat{x}_{t-1}$ and the error covariance matrix P_{t-1}, which are given at the last moment, and the model (Equation 5) which describes the aircraft's state of motion, predicting the present moment's state variable $\hat{x}_{t/t-1}$ and homologous error covariance $P_{t/t-1}$ is as follows:

$$\hat{x}_{t/t-1} = \hat{x}_{t-1} + f(\hat{x}_{t-1})T \tag{6}$$

$$P_{t/t-1} = F_t P_{t-1} F_t^T + Q_t \tag{7}$$

$F_t = I + T \frac{\partial f(x)}{\partial x}\Big|_{x=\hat{x}_{t-1}}$ is the Jacobian matrix of the state equation, Q_t, R_t represent the covariance matrix, and T represents the filter period.

$$\sum P_{t/t-1} = \begin{cases} (P_{t/t-1}^{-1} - \gamma^2 L_t^T L_t)^{-1}, & \overline{P}_{y,t} > \alpha P_{y,t} \\ P_{t/t-1}, & \overline{P}_{y,t} < \alpha P_{y,t} \end{cases} \tag{8}$$

According to Equation 8, to estimate whether there is a need to reset the prediction error matrix, the adjustable filter parameter $\alpha > 1$, and L represents the adjustable parameter matrix:

$$L_t = \gamma \left(P_{t/t-1}^{-1} - \lambda_t^{-1} P_{t/t-1}^{-1}\right)^{1/2} \tag{9}$$

$$\lambda = \frac{tr(\overline{P}_{y,t})}{tr(P_{y,t})} \tag{10}$$

$$P_{y,t} = H_t P_{t/t-1} H_t^T + R_t \tag{11}$$

$$\overline{P}_{y,t} = \begin{cases} \Delta y_t \Delta y_t^T & t = 0 \\ \dfrac{\rho \overline{P}_{y,t-1} + \Delta y_t \Delta y_t^T}{\rho + 1}, & t > 0 \end{cases} \tag{12}$$

where $tr(\nabla \cdot)$ represents the trace of the matrix, $\Delta y_t = y_t - H_t x_{t/t-1}$ is the residual error, and ρ represents the forgetting factor.

3.3.2 *Revise and update*
In revising the estimation value $\hat{x}_{t-1}$ according to the observation quantity y_t, we then get the present state variable's prediction value $\hat{x}_t$:

$$\hat{x}_t = \hat{x}_{t/t-1} + K_t(y_t - H\hat{x}_{t/t-1}) \tag{13}$$

$$K_t = \sum_{t/t-1} H_t^T \left(\sum_{t/t-1} H_t^T + R_t\right)^{-1} \tag{14}$$

$$P_t = \left(\sum_{t/t-1}^{-1} + H_t^T R_t^{-1} H_t\right)^{-1} \tag{15}$$

where $H_t = \frac{\partial h(x)}{\partial x}\Big|_{x=\hat{x}_{t/t-1}}$ represents the observation equation's Jacobian matrix, and K_t represents the filter gain. The arithmetic ensures the estimation error is lower than the specified attenuation level of γ.

$$\frac{\|L_t \tilde{x}_t\|^2}{\|w_t\|^2 + \|\Delta_t\|^2 + \|v_t\|^2} \leq \gamma^2 \tag{16}$$

In Equation 16, $\|x\|$ is the 2-bound norm of x, $\tilde{x}_t = x_t - \hat{x}_t$ represents the estimation error, and the high-order Taylor series expansion of Δ_t is:

$$\Delta_t = f(x_t) - f(\hat{x}_{t/t-1}) - F_t \hat{x}_{t/t-1} \tag{17}$$

As for the HCV autonomous navigation system, if the HCV maneuvers on the orbits at hypersonic speed, then the state equation conflicts with the truth, so the estimating error

13

becomes greater and leads to the estimating residual error Δy_t becoming greater. If it rises to a certain degree and leads to $\bar{P}_{y,t}$ exceeding the threshold $\alpha P_{y,t}$, the real estimating residual variance exceeds the theoretical value, and then the prediction error variance matrix needs to be reset. The method for this is to increase the value of the prediction error variance matrix, which is conducive to optimizing the filter gain matrix K_t, and improving the revising function of the estimating information and the position precision of the hypersonic orbits, and it also improves the filter's stability in tracking performance.

4 EXPERIMENT AND RESULT ANALYSIS

With regard to the hypersonic vehicle as a simulation object, the aircraft's parameters adopt the data from the US experimental aerospace aircraft called X-33 (Xiong et al., 2014), whose mass is 900 kg, angle of attack is 20°, and angle of tracking yaw is 8.5°. Following the simulation calculation, the initial velocity is 7,000 m/s, the height is 100 km, the flight time is 600 s, and the sampling period of SINS is 0.1 s, and 2 s for Kalman filtering. The selection for the initial filter values are as follows: position error $\delta L = \delta \lambda = 3$ m, $\delta h = 10$ m, velocity error 0.2 m/s, attitude angle error $\varphi_E = \varphi_N = 5$", $\delta h = 10$ m. The gyro constant drift and gyro random drift are 0.1°/h, the gyro white-noise drift is 0.01°/h; the accelerometer random drift is 10^{-4} g, and the accelerometer white-noise drift is 10^{-5} g. The GPS receiver's velocity error is 0.5 m/s, and the position error is

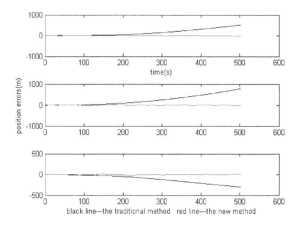

Figure 3. Comparative curves for position errors.

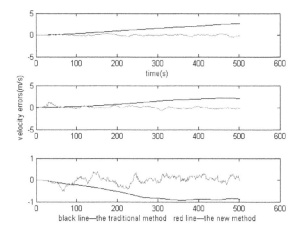

Figure 4. Comparative curves for velocity errors.

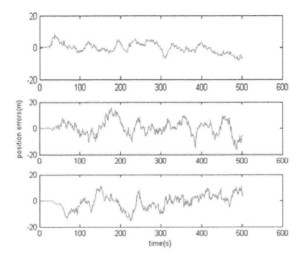

Figure 5. Position error curves for the new method.

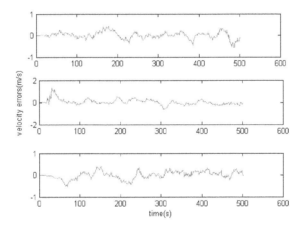

Figure 6. Velocity error curves for the new method.

10 m. The observation error of the celestial navigation system is 3". Set the malfunction on the GPS for 200–250 s for time, and bring in an outlier of 100 m for the position, with 10 m/s for the velocity. Set the malfunction on the GPS for 300–350 s for time, and bring in an estimated outlier of 5 minutes of arc. The results of the experiment are shown in Figures 3, 4, 5 and 6.

Figures 3 and 4 show that the new method used in this paper is more accurate than the traditional method; the filter can achieve effective convergence, and the traditional method starts to diffuse after 200 s, where the velocity error becomes more obvious. Figures 5 and 6 show that the new method can achieve high-precision navigation for a hypersonic vehicle, and the filter can effectively converge when encountering interference in the 300–350 s period, indicating that the non-Keplerian orbit parameters enhances the stabilization of the filter as the fourth sensor information.

5 CONCLUSIONS

In this paper, a new method for INS/GNSS/CNS integrated navigation for hypersonic vehicles is proposed and the state model of the system is built using non-Keplerian orbital characteristics of the hypersonic vehicle. Based upon not increasing the number of sensors, the aircraft's internal resources are fully utilized, and the orbital parameters of the aircraft are used as the information for a fourth virtual sensor. First, the position and velocity information and the information from the CNS system are obtained, and then the information symmetry of the two sub-filters is given. The non-Keplerian orbits' state equation is used as *a priori* information. Finally, the main filter proceeds to optimal information fusion. The simulation results show that the new method is better than the traditional method. Because the fourth virtual sensor information is used for filtering existing information, even though the integrated navigation system of a given subsystem may be disturbed, the filter can still effectively converge and maintain high precision.

Under the existing conditions, the dynamic equations for a near-space hypersonic vehicle based on non-Keplerian orbits are necessary to provide the control force and the control torque. This is related to the aerodynamic layout and mission of the aircraft, and the solution process of this paper is simplified. This aspect will be further improved in future research.

ACKNOWLEDGMENTS

This study was conducted under Funds for the National Natural Science
Foundation of China:
61603412
Shaanxi Province Natural
Science Foundation Research
Project: 2017 JQ6027

REFERENCES

Gao, Z., Cao, T., Lin, J. & Qian, M. (2014). Observer-based H-infinity tracking control design for a linearized hypersonic vehicle model with external disturbance. In *Proceedings of 2014 IEEE Chinese Guidance, Navigation and Control Conference* (pp. 1649–1653).

He, F., Zhou, S.S., Hu, X.G., Zhou, J., Liu, L., Guo, R., … Wu, S. (2014). Satellite-station time synchronization information based real-time orbit error monitoring and correction of navigation satellite in Beidou system. *Science China, 57*(7), 1395–1403.

Heiligers, J. & McInnes, C.R. (2014). Cylindrically and spherically constrained families of non-Keplerian orbits. In *2013 AAS/AIAA Astrodynamics Specialist Conference, Astrodynamics 2013* (pp. 134–141).

Jiang, W., Li, Y. & Rizos, C. (2015). Optimal data fusion algorithm for navigation using triple integration of PPP-GNSS, INS, and terrestrial ranging system. *IEEE Sensors Journal, 15*(10), 5634–5644.

Kaiser, J., Martinelli, A. & Fontana, F. (2017). Simultaneous state initialization and gyroscope bias calibration in visual inertial aided navigation. *IEEE Robotics and Automation Letters, 2*(1), 18–25.

Li, S., Jiang, X. & Liu, Y. (2014). Innovative Mars entry integrated navigation using modified multiple model adaptive estimation. *Aerospace Science and Technology, 39*, 403–413.

Miao, L. & Shi, J. (2014). Model-based robust estimation and fault detection for MEMS-INS/GPS integrated navigation systems. *Chinese Journal of Aeronautics, 27*(4), 947–954.

Nazari, M., Butcher, E. & Mesbahi, A. (2014). On control of spacecraft relative motion in the case of an elliptic Keplerian chief. In *2013 AAS/AIAA Astrodynamics Specialist Conference, Astrodynamics 2013* (pp. 101–119).

Peng, H. (2014). The key technology research on error modeling and compensation of the INS for the HCV. Nanjing University of Aeronautics and Astronautics.

Peng, H., Zhi, X., Wang, R., Liu, J.Y. & Zhang, C. (2014). A new dynamic calibration method for IMU deterministic errors of the INS on the hypersonic cruise vehicles. *Aerospace Science and Technology, 32*(1), 121–130.

Serrani, A. & Bolender, M.A. (2014). Nonlinear adaptive reconfigurable controller for a generic 6-DOF hypersonic vehicle model. American Control Conference (ACC). Portland, Oregon, USA. June 4–6 (pp. 1384–1389).

Xiong, K., Zhang, H. & Liu, L. (2008). Adaptive robust extended Kalman filter for nonlinear stochastic systems. *IET Control Theory & Applications, 2*(3), 239–250.

Yang, B., Hu, J. & Liu, X. (2013). A study on simulation method of starlight transmission in hypersonic conditions. *Aerospace Science and Technology, 29*(1), 155–164.

Automatic Control, Mechatronics and Industrial Engineering – He & Qing (Eds)
© 2019 Taylor & Francis Group, London, ISBN 978-1-138-60427-8

Novel utilization of machine vision technology in a pattern matching operation and blob analysis using an image processing inspector for automation and robotics applications

S. Sharma
CSIR-Central Leather Research Institute, Regional Center for Extension and Development, Jalandhar, Punjab, India

S. Shalab
DAV University, Jalandhar, Punjab, India

S. Mithilesh
IKG Punjab Technical University, Jalandhar, Punjab, India

M. Sundaram
Department of EEE, PSG College of Technology, Coimbatore, Tamil Nadu, India

T. Jishnu
Department of Robotics and Automation, PSG College of Technology, Tamil Nadu, India

ABSTRACT: Machine vision technologies have been the focus of research in the fields of industrial robotics and automation for more than two decades now. Over the years, these technologies have revolutionized the manufacturing and inspection processes industries with the help of sophisticated vision-based robotics systems that can gage the location of a desired object in a 3-dimesional space, ascertain its orientation relative to its surroundings and perform the task accordingly. A basic machine vision system has a single stationary camera mounted over the workspace to enable image acquisition. With this form of active sensing, the position, orientation, identity and condition of each part can be obtained. This high level information can be used to plan the robotic motion, such as determining how to grasp a part and how to avoid collisions with obstacles.

The objective of this paper is to perform a pattern matching operation and blob analysis on an image processing inspector. An overhead CCD camera performs the task of image acquisition.

1 INTRODUCTION

The study and development of robot mechanisms can be traced back to the mid-1940s, when master slave manipulators were designed and fabricated at Oak Ridge and the Argonne National Laboratories for the purpose of handling radioactive materials. The first commercial computer controlled robot was introduced in the late 1950s by Unimation Inc., and a number of industrial experimental devices suit during the next 15 years (Bhanu, 2013). A robot can be defined as (Chin & Dyer, 1986).

'A Robot is defined as reprogrammable device designed to both manipulate and transport parts, tools or specified manufacturing implements through variable programmed motion for the performance of specific manufacturing task.' (British Robot Association, 1984).

As is the case with humans, vision capabilities endow a robot with a sophisticated sensing mechanism that allows the machine to respond to its environment in an intelligent and flexible manner. The use of vision and other sensing elements are motivated by the continuing need to increase the flexibility and scope of the application of the robotic system (Chauhan & Surgenor, 2015). While proximity, touch and force sensing play a significant role in the improvement of the performance of robots, vision is considered to be the most powerful of a robot's sensory capabilities. Robot vision is defined as a process of extracting, characterizing and interpreting information from images of three dimensional worlds (Li, 1996). This process is commonly known as computer vision or machine vision and comprises six principle areas: sensing, preprocessing, segmentation, description, recognition, and interpretation.

To achieve the required results, the View Flex 2.8 version is used as the machine vision software, with the Matrox Inspector version 3.0 being used as the image processing tool. The View Flex is provided with an Intel camera that grabs the image. Before capturing the image and performing any analysis, the environment is to be calibrated using the Calibration icon in the View Flex. The robot used is SCORBOT ER-4u, which is equipped with SCORBASE software and a PID controller. The sorting is performed for three matchstick boxes using the pattern matching feature of the View Flex (Nagrale & Bagde, 2013).

2 THEORETICAL BACKGROUND

The View Flex interactive vision software is based on the Matrox Inspector Image Processing Engine. The system also includes a USB digital color camera that provides both still and video images. The advanced features of the View Flex allow users to implement scientific and industrial applications. View Flex software offers an extensive set of optimized functions for image processing and enhancement, blob analysis, gauging and measurement, and pattern matching. The system supports applications such as precision measurement, flaw detection and assembly inspection to enable compliance with exacting quality requirements.

The Matrox Inspector 3.0 is an independent hardware application designed to work interactively with images for image capture, storage and processing applications. Inspector is a 32 bit Windows-based package, giving the full potential and ease of use of a graphical interface. It is capable of running on any system that runs Windows 95/98 or Windows NT 4.0.

2.1 *Image processing capabilities*

Inspector includes a variety of point-to-point, statistical, spatial filtering, Fourier transform and morphological operations. These operations can be used to smooth, accentuate or qualify images. Inspector supports geometric operations, including interpolated resizing, rotation, and distortion correction, as well as polar to rectangular unwrapping. Processing can be restricted to rectangular or non-rectangular Region of Interest (ROI). Processing can also be restricted to a selected component of a color image.

2.2 *Pattern matching capabilities*

This enables the finding of a pattern (referred to as a model) in an image, with sub pixel accuracy. Pattern matching can help solve machine vision problems, such as alignment and the inspection of objects.

2.3 *Pattern matching*

1. Automatic or manual model creation.
2. Edit model, rotate model, make/load mask, change hotspot values.
3. Apply 'don't care' pixel mask to model.
4. Preprocess model to optimize search speed (circular over scan).
5. Modify default search parameters: model center, search area, positional accuracy, number of matches, acceptance threshold, speed, and spacing.
6. Search results: model position and angle, number of model matches in target, match scores.

Blob analysis capabilities

Inspector's blob analysis capabilities allow the finding and measurement of connected regions of pixels (normally called blobs) within a gray scale image. Once these regions are identified, features for these regions can be automatically calculated, then the regions that are not of interest can be discarded, and the remaining regions classified according to the values of the features. The identification is done by the process of segmentation. The applications of blob analysis are:

- Presence detection
- Inspection of mechanical parts
- Tissue and material analysis
- Sequence analysis

The values of different blob features can be calculated and the image can be analyzed.

3 PROBLEM STATEMENT

The following experiments are to be conducted with the machine vision software:

I. To perform pattern matching both with and without the application of the filter (all objects the same) in View Flex software using Matrox Inspector as the image processing tool. Also to see the effect of changing the acceptance level.

   Description of the objects used: Three similar matchstick boxes were used for the purpose of analysis. The boxes were of a similar color, size and brand.

II. To perform blob analysis on the image and compare the variations in area, compactness and roughness in the View Flex machine vision software in conjunction w.ith Matrox Inspector.

   Description of the objects used: The object is a mild steel bar prepared on a lathe machine with a taper turning area and a thread cutting area.

The above features of these parts are to be calculated and compared.

4 EXPERIMENTAL PROCEDURE

1. Execute the home routine and calibrate the robot with vision software.
2. Define ROI and click the pattern matching icon. A model window appears.
3. Click the Search button to search for the three occurrences of the model in the image.
4. Change the acceptance level to 60% and 40% in the model window and apply the Search.
5. Apply a smooth filter using the image processing filters/smooth filters. The following dialog box appears and applies smooth filters.
6. Apply the pattern matching process again with a 70% acceptance level. A measurement table is displayed listing the match score.
7. Apply blob analysis: A blob window appears. Select the compactness and roughness feature from the feature tab. Apply segmentation so that the regions to be identified as blobs are in some logical pixels state foreground state. Apply the count and the result window appears.

5 RESULTS AND OBSERVATIONS

For pattern analysis: The search window giving the occurrences of the three boxes is shown in Figure 1. The measurement table showing the results of the pattern matching is shown in Figure 2. The results have been summarized in Tables 1 and 2.

The following observations were made while performing pattern matching.

The products were successfully sorted by View Flex software. Small discrepancies were taken care of by the software. It is noted that:

- Decreasing the acceptance level increased the number of matches.
- With the application of smooth filters:

i. The percentage acceptance increased.
ii. Objects were sorted at once and at one acceptance level.
iii. The match score was increased.

Figure 1. Objects selected.

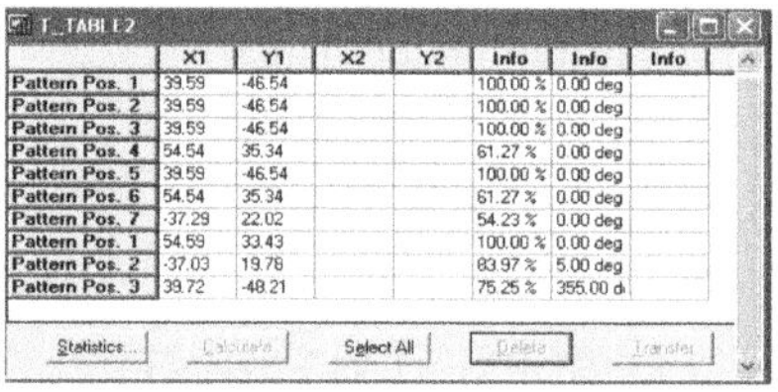

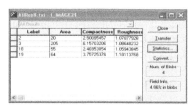

Figure 2. Measurement table.

Table 1. Pattern pos. 1 to pattern pos. 7: Without applying filters.

Pattern pos. no.	Acceptance level	No. of objects selected
1 & 2	70%	1
3 & 4	60%	2
5, 6 & 7	40%	3

Table 2. Pattern pos. 1 to pattern pos. 3: With smooth filters.

Pattern pos. no.	Acceptance level	No. of objects selected
1, 2 & 3	70%	3

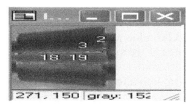

Figure 3. Blobs for taper turning zone.

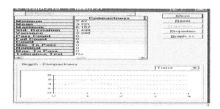

Figure 4. Result window for taper turning.

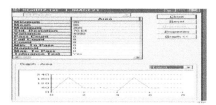

Figure 5. Statistics and trend chart for area for taper turning.

Figure 6. Statistics and trend chart for compactness for taper turning.

For blob analysis: The following are the results of blob analysis on the two areas. The results table clearly gives the values of the feature selected and its trend charts, which are used for making inferences.

Area 1: Taper Turning Zone.

After applying blob analysis, four blobs are clearly detected in Figure 3.

The results table shown in Figure 4 gives the values of area, compactness and roughness. Using the Statistics tab from the result window, the statistics of all of the features can be studied. The trends and statistics of the three features are clearly seen in Figures 5, 6 and 7.

Area 2: Thread Cutting Zone

As above, the following results are given below for the thread cutting zone.

These statistics can be summarized in Tables 3 and 4.

The following points can be concluded from the above experiment:

i. There is a greater variation in the values of compactness and area for the taper turning zone.

ii. The roughness trend for both zones is very similar, but the thread cutting zone is rougher than the taper turning zone.

This type of analysis will be helpful for identifying faulty parts by comparing blob features calculated with the set of expected feature values.

Figure 7. Statistics and trend chart for roughness for taper turning.

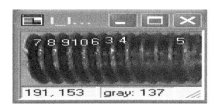

Figure 8. Blobs for thread cutting.

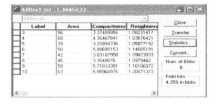

Figure 9. Results table for thread cutting.

Figure 10. Statistics and trend chart for compactness for thread cutting.

Figure 11. Statistics and trend chart for area for thread cutting.

Figure 12. Statistics and trend chart for roughness for thread cutting.

Table 3. Features for taper turning.

Taper turning	Minimum	Maximum	Mean	Standard deviation	Variance	Peak to valley value
Roughness	1.059	1.181	1.101	0.04705	0.002213	0.122
Compactness	2.47	6.157	3.721	1.499	2.248	3.687
Area	20	205	86	70.64	4990	185

Table 4. Features for thread cutting.

Thread cutting	Minimum	Maximum	Mean	Standard deviation	Variance	Peak to valley value
Roughness	1.037	1.201	1.11	0.0510	0.002662	0.164
Compactness	3.209	5.897	4.437	1.1132	1.282	2.688
Area	39	68	52.75	9.744	94.44	29

6 CONCLUSIONS

Machine vision and robotics are multidisciplinary in the sense that these fields use knowledge taken from traditional engineering and computer programming for different parts of the process. The image processing with Matrox Inspector was successfully achieved. The pattern matching operation and the blob analysis made with the inspector were also successful. Pattern matching helped to sort objects of both the same and different types. The effect of applying a filter to the image and changing the acceptance level was studied in the correct way. Blob analysis helped to compare the roughness and compactness of two surfaces, which can then be given to the robot for use in different applications. Blob analysis helped to find out the areas that are not of interest, classify them and then discard them.

7 SCOPE OF FUTURE WORK

The pattern matching analysis can be extended for PCB board for determining whether or not the correct numbers of diodes are placed and whether they are structurally intact. FFT analysis is also proposed in order to identify consistent spatial patterns in an image, which can be caused by systematic noise. Blob analysis can be applied to compare the features of surfaces made by different cutting and forming processes. The vision and the robotics system can be used for complex pick and place operations with the utilization of blob analysis and pattern matching.

REFERENCES

Bhanu, P. (2013). Machine vision systems and image processing with applications. *Journal of Innovation in Computer Science & Engineering*, *2*(2), 1–4.

Chauhan, V. & Surgenor, B. (2015). A comparative study of machine vision based methods for fault detection in an automated assembly machine. *Procedia Manufacturing*, XXX, 1–13.

Chin, R.T. & Dyer, C.R. (1986). Model-based recognition in robot vision. *Computing Surveys*, *18*(1), 67–108.

Li, Y.F. (1996). A sensor based robot transition control strategy. *The International Journal of Robotics Research*, *15*(2), 128–136.

Nagrale, S.K. & Bagde, S.T. (2013). Application of image processing for development of automated inspection system. *International Journal of Computational Engineering Research*, *3*(3), 103–107.

Automatic Control, Mechatronics and Industrial Engineering – He & Qing (Eds)
© 2019 Taylor & Francis Group, London, ISBN 978-1-138-60427-8

Automatic driving system of plant protection robot based on differential GPS

X.M. Xia, C.S. Ai & Y.C. Du
School of Mechanical Engineering, University of Jinan, Shandong, China

ABSTRACT: This paper presents a system for the automatic driving of plant protection robots that also achieves accuracy and reliability. A method based on differential GPS obtains high-precision information on position and direction, and the A/B points are collected to obtain the AB line of the path planning. The preview point is calculated by using the straight-line tracking model, and the desired deflection angle is obtained. The electromagnetic proportional hydraulic valve is controlled by PWM modulation, through the opening of the valve to control the expansion and contraction of the hydraulic cylinder, and the wheel is rotated to achieve the purpose of adjusting the turning. The system realizes a control function that the plant protection robot can travel straight. The plant protection robot travels at a speed of 1.2 m/s, the maximum deviation is 5.3 cm, and the standard deviation is 1.4 cm. The test results show that the system can be used for the automatic driving of plant protection robots, and it also has reliability.

Keywords: plant protection robot, differential GPS, automatic driving, path planning, straight-line tracking model

1 INTRODUCTION

When working conditions are complicated and varied. For the field of work that lack reference, it is difficult to ensure the direction and spacing of the work by manual judgment, which often causes the track to be bent and overlapped. Autonomous driving technology therefore came into being. Autopilot technology is a relatively new trend. It is supported by information technology, and is used to implement a modern farming technology based on satellite positioning (Zhu et al., 2016). If a plant protection machine is equipped with an automatic driving system, it will improve the accuracy of operation, reduce repetitive operations, reduce staff workload, and improve production efficiency.

Automatic driving of agricultural machinery has been widely researched outside China. Thomas Bell of Stanford University installed four GPS receiving antennas on the top of a tractor cab and studied the application of agricultural vehicle auto-navigation technology by using CP-DGPS technology. The position and direction of the vehicle is estimated by measuring the angle of each GPS antenna relative to the satellite and the phase difference of the GPS signal (Ji & Zhou, 2014). This not only increases the GPS antennas and improves the cost, but also makes the control algorithm more difficult. Takai (Yang et al., 2015) used the real-time dynamic difference global positioning system and inertial measurement unit as the navigational sensor. The test result is that the error of the tractor is 1~3 cm at different speeds.

Chinese universities are also actively conducting research on autonomous driving of agricultural machinery. Ji and Liu (2009) have developed an automatic navigation and control system for agricultural vehicles based on CAN bus. The steering controller is designed by the adaptive PID control algorithm and the proportional parameters are adjusted to achieve automatic control of the agricultural machinery. However, the accuracy is not high—the lateral deviation is 32 cm, and the average deviation is 9 cm. Luo et al. (2009) developed an automatic navigation control system for tractors based on RTK-GPS. A Dongfanghong X-804 tractor was taken as the object of study. Based on the fusion of the kinematics model and the steering control model of the agricultural machinery, the state equation of the linear tracking is derived, and the navigation controller based on PID is designed. A control method was proposed, which is to turn across the line.

Plant protection robots acquire high-precision positioning databased on a differential GPS; two antennas are used to obtain direction of robot. The preview point and the expected deflection angle are calculated by using the straight-line tracking model. Combined with the electromagnetic proportional hydraulic valve to control the steering device, we propose a path-planning method based on the AB line in order to achieve the automatic driving of the robot.

2 RESEARCH PROGRAMME

The automatic driving of plant protection robots requires a three-part design: the vehicle body, vehicle-mounted system, and base station system. The vehicle body provides power and the steering mechanism. The vehicle-mounted system provides path planning, location and control decisions for automatic driving. The base station system provides differential information for automatic driving, so that high-precision information can be obtained. The overall structure of the GPS automatic driving navigation system is shown in Figure 1.

The GPS differential base station transmits the solved differential data to the receiver through the external radio, the obtained dual antenna GPS data are compared with the differential data from the receiver, and the information is sent to the Electronic Control Unit (ECU). The ECU controls the rotation of the steering wheel by controlling the hydraulic valve, thereby achieving the purpose of controlling the direction. The actual steering angle is collected by the angle sensor, and it is fed back to the ECU to complete the closed loop control.

2.1 *Vehicle body*

In this paper, the vehicle body adopts the electric forklift chassis. It can achieve front and rear steering. It provides different steering modes for the autopilot test. It is convenient to compare the advantages and disadvantages of automatic driving under different steering modes.

2.2 *The design of the vehicle system*

The vehicle system is a comprehensive system that integrates satellite receiving, high-precision positioning and directional control. The receiver adopts UN257 board, and it can be positioned and directed. The electromagnetic proportional hydraulic valve adopts PRM2-063Y11, which has the advantages of flexible movement, low impact and vibration, low noise and long service life.

2.2.1 *The principle of steering control*
The principle of automatic driving steering is as follows. The positioning and deviation angle are sent to the ECU by the GPS receiver. After the internal processing of the ECU, the offset

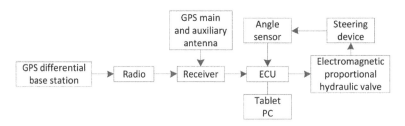

Figure 1.   Overall structure of GPS autopilot navigation system.

24

will be obtained. Combined with the deviation, the pre-pointing point will be calculated by the straight-line tracking model, and then the desired declination will be obtained. The ECU controls the opening of the electromagnetic proportional hydraulic valve through PWM modulation, then controls the expansion and contraction of the hydraulic cylinder, and turns the wheel to achieve the purpose of adjusting the steering. The angle sensor feeds the measured true angle to the ECU for closed loop control. This is shown in Figure 2.

2.2.2   *Hydraulic circuit*

The electromagnetic proportional hydraulic valve is installed near the reversing valve. Under the premise of not changing the original hydraulic steering system of the robot, three three-way connections are connected to the reversing valve, namely the left port, the right port and the oil return port; we design a parallel oil circuit with the original hydraulic steering system. When starting the automatic driving mode, the electromagnetic proportional hydraulic valve (the hydraulic valve in Figure 3) starts to work and the high-pressure oil enters the hydraulic valve from the P port and controls the left or right turn when in the left or right position. When the automatic driving mode is off, the manual adjustment mode will be initiated. At this time, the neutral position of the hydraulic valve starts working, the high-pressure oil is imported from port P, then passes through the output port to the port P of the reversing valve, the reversing valve starts to work and controls the steering by turning the steering wheel. This is shown in Figure 3.

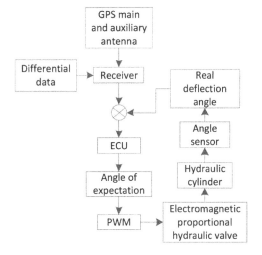

Figure 2.   Schematic diagram of steering.

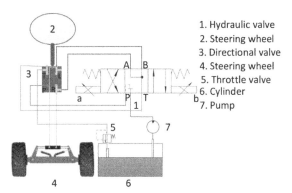

1. Hydraulic valve
2. Steering wheel
3. Directional valve
4. Steering wheel
5. Throttle valve
6. Cylinder
7. Pump

Figure 3.   Hydraulic circuit.

2.2.3 *The design of closed loop feedback*

Angle sensors provide the true corner measurements for automatic driving. The angle sensor is fixed to the bracket, and it is coaxial with the rotation centre of the left wheel. It is connected to the wheel through the connecting frame, so it rotates synchronously with the wheel to measure the true corner of the wheel. The true corner is fed back to the ECU to complete the closed loop control.

2.3 *Base station system*

The receiver has a built-in chip for interpreting the satellite message, and the satellite signal is processed into a fixed position deviation signal by a certain algorithm. The signal is transmitted through the radio. The automatic driving system receives differential signals through the vehicle radio, thereby obtaining high-precision positioning data.

3 THE TEST OF AUTOMATIC DRIVING

3.1 *Test process*

In our system, the high-precision RTK-GPS receiver is used to reduce errors, and it is also used to improve measurement efficiency and accuracy of later evaluation.

Before starting the power, it is necessary to establish a GPS-RTK receiver base station. Because tall buildings and trees will obscure the satellite signals, the base stations are generally installed at a high level to avoid satellite signal interference. After completion of construction, the power is turned on, and after the differential signal is transmitted stably (the indicator lights are flashing evenly), the next step can be started.

After the base station is set up, the vehicle parameters are first calibrated, because different parameters will affect the test data. Therefore, it is necessary to establish a vehicle body model to deal with various errors that are caused by the vehicle body, as shown in Table 1.

- Wheelbase: distance between the front and rear axles of the vehicle.
- Height of antenna: the height of the GPS antenna from the ground.
- Offset of antenna: the distance from the longitudinal center of the vehicle to the GPS antenna.
- Height of ECU: 1 cm above the bottom of the controller chassis.
- Lateral offset: the longitudinal center axis of the car body to the center of the ECU.
- Vertical offset: ECU to rear center distance.

After the vehicle parameters are calibrated, the left and right angle sensor left and right limit calibration is performed. In this paper, the left limit is 1,003, the right limit is 324, and the center is 659.

Before the test, we set the interval to 3 m in the path planning and we select a point on the ground as point A (south). We drive the robot along the straight line to the other end (about 60 m), selecting it as point B (north), and generate an AB line (north–south trend).

The automatic driving deviation of the AB line is measured. The plant protection robot has a forward speed of 1.2 m/s, and the frequency of GPS data update is 1 Hz. The experimental conditions are sunny. First, the front steering is measured from A to B: the plant protection robot is driven to the AB line, at point A, the automatic driving is activated, and

Table 1. Parameter calibration.

Name	Wheelbase	Height of antenna	Offset of antenna	Height of ECU	Lateral offset	Vertical offset
Numerical value (cm)	150	125	70.5	103.5	48	57

data is collected synchronously. The rear steering is then measured by collecting data from B to A.

3.2 *Analysis of test data*

After processing the collected data, we obtain the deviation and draw a normal distribution map, so that we can visually see the straight driving deviation of the automatic driving system.

We want to evaluate the performance index of autonomous driving. According to the driving characteristics, we evaluate the deviation between the actual driving route and the planning path. Therefore, we use standard deviation and extremum to describe the driving performance of the automated driving system.

3.2.1 *Data pre-processing*

In this study, we use the pre-processing method of experimental data to reduce the interference of the outliers generated by the test. It is difficult to ensure that the robot is zero distance from the path planning at the beginning of the test and we can only minimize this man-made error. Therefore, removing part of the data at the beginning of the test will help to ensure the accuracy of the test results. According to mathematical statistics research and analysis, combined with **MATLAB** software for graphical processing, we perform the corresponding error analysis.

The raw data of the GPS collected in the experiment uses the WGS-84 coordinate system. The WGS-84 coordinate system is a reference system based on the centroid of the Earth. In order to carry out the analysis of the test data, the evaluation results must be accurate and easy to understand and count. Therefore, when performing the evaluation, it is usually necessary to convert the test data from the Earth coordinates to the plane rectangular coordinates by projection.

At present, there are many projection conversion methods commonly used in surveying and mapping. Usually, the UTM projection and Gauss-Krüger projection are used for coordinate projection transformation. They are both isometric projections, and the projection method is basically the same. However, UTM has no distortion of the projection angle compared to the Gauss-Krüger projection. In addition, X and Y represent longitude and latitude (Zhou et al., 2013). The UTM projection is adoptedin the experimental data processing. UTM projection positive solution formula, longitude L, latitude B, This is the formula for calculating the rectangular coordinates of the plane from the geodetic coordinates (L, B):

$$y = F_E + k_0 N \left[A + \left(1 - T + C\right)\frac{A^3}{6} + \left(5 - 18T + T^2 + 72C - 58e'^2\right)\frac{A^2}{120} \right] \quad (1)$$

$$x = F_N + k_0 \{ M + N \tan B \left[\frac{A^2}{2} + \left(5 - T + 9C + 4C^2\right)\frac{A^2}{24} \right] + \left(61 - 58T + 600C - 330e'^2\right)\frac{A^6}{720} \quad (2)$$

Middle East latitude offset $F_E = 500,000$ m; north latitude offset $F_N = 0$ (northern hemisphere), $F_N = 10,000,000$ m; the scale factor of the UTM projection is $k_0 = 0.9996$, $T = \tan^2 B$;

$$C = e'^2 \cos^2 B; A = \left(L - L_0\right)\cos B; N = \frac{a}{\sqrt{1 - e^2 \sin^2 B}}$$

$$M = a\left[\left(1 - \frac{e^2}{4} - \frac{3e^4}{64} - \frac{5e^6}{256}\right)B - \left(\frac{3e^2}{8} + \frac{3e^4}{32} + \frac{45e^6}{1024}\right)\sin(2B) + \left(\frac{15e^4}{256} + \frac{45e^6}{1024}\right)\sin(4B)\right. \quad (3)$$
$$\left. - \left(\frac{35e^6}{3072}\right)\sin(6B)\right]$$

The above conversion formula is based on an ellipsoid, where a is the long semi-axis of the ellipsoid; b is the short semi-axis of the ellipsoid; e is the first eccentricity; e' is the second eccentricity; N is the radius of curvature of the ring; R is radius of the meridian circle; L_0 is the central longitude.

3.3 *Experimental data processing*

In this paper, a large volume of data from the automatic driving of the plant protection robot is collected and processed. The sample data include the front steering mode and the rear steering mode. In order to effectively reflect the sample of driving deviation, the sample data in both modes must be processed. In the following, two data samples are taken as an example for detailed description of the data processing. The error analysis is carried out according to the error distribution.

First of all, the paper studies the straight driving performance of the plant protection robot under automatic driving, so the collected measuring points are linear. The collected measuring points are converted into X/Y coordinates using the UTM projection, and then using the MATLAB mathematical tool, the straight-line equation is obtained by the least squares method of fitting the straight line (He et al., 2011). A normal distribution map is made according to the distribution of the error as shown in Figure 4.

The front steering and rear steering of the plant protection robot and their impact on the automatic driving deviation are analyzed. The blue dots in Figure 4a and 4b represent the traveling trace of the plant protection robot, and the red lines represent the fitted straight lines. It can be seen that the deviation of automatic driving is distributed on both sides of the line. For the front steering, Figure 4c represents the normal distribution of the deviation of automatic driving, where the maximum deviation is 5.3 cm, and the standard deviation $\sigma = 1.4$ cm. For the rear steering, Figure 4d represents the normal distribution of the deviation of automatic driving, where the maximum deviation is 10.9 cm, and the standard deviation $\sigma = 4.6$ cm. Among them, 4.6 is greater than 1.4. It shows that when the plant protection robot is front steering, the automatic driving deviation is more concentrated and the automatic driving accuracy is higher. This is because when front steering, the steering system only drives the front wheel to turn. The rotating instantaneous center of the robot is mostly

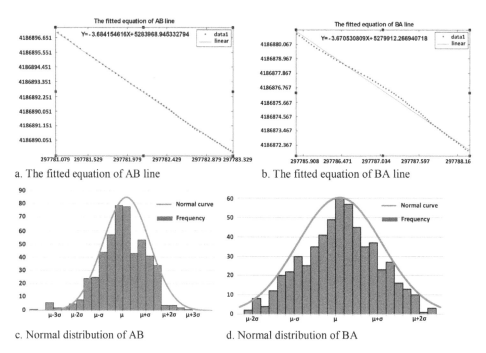

a. The fitted equation of AB line   b. The fitted equation of BA line

c. Normal distribution of AB   d. Normal distribution of BA

Figure 4. Data processing results.

near the axis of the rear axle, and the steering of the robot is consistent with the steering of the front wheel. When rear steering, the steering system only drives the rear wheel to turn. The rotating instantaneous center of the robot is mostly near the axis of the front axle, and the steering of the robot is opposite to the steering of the rear wheel. However, the automatic driving system measures the information is the position and direction of the robot, so the front steering is more consistent.

As shown in Figure 4, firstly, the performance of the plant protection robot based on differential GPS, combined with an electromagnetic proportional hydraulic valve to control the steering, and using an AB line to plan the path in the automatic driving mode, has high precision and stability along a straight line, and it can meet normal operations of plant protection. Second, under the conditions of this test, the front steering of the plant protection robot is significantly better than the rear steering.

4 CONCLUSION

The automatic driving system of a plant protection robot based on differential GPS, has the advantages of high accuracy and strong stability. The maximum deviation is 5.3 cm and the standard deviation is 1.4 cm. Position and direction information are based on differential GPS, and an AB line is used as the planning path to achieve the linear navigation of the plant protection robot. The preview point and the expected deflection angle are calculated by using the straight-line tracking model. The opening and direction of the electromagnetic proportional hydraulic valve are controlled by PWM modulation, controlling the expansion and contraction of the hydraulic cylinder, thus driving the wheel to rotate. The system achieves precise control of straight-line tracing and the tests show that the front steering of the plant protection robot is significantly better than the rear steering. This applies to plant protection robots that have hydraulic steering systems.

REFERENCES

He, B., Liu, G. & Ji, Y. (2011). Auto recognition of navigation path for harvest robot based on machine vision. *Computer and Computing Technologies in Agriculture IV, 344*, 138–148.

Ji, C. & Liu, G. (2009). Automatic navigation and control system of agricultural vehicle based on CAN bus. *Transactions of the Chinese Society for Agricultural Machinery, 40*, 28–32.

Ji, C. & Zhou, J. (2014). Current situation of navigation technologies for agricultural machinery. *Transactions of the Chinese Society for Agricultural Machinery, 45*(9), 44–54.

Luo, X., Zhang, Z., Zhao, Z., Chen, B., Hu, L. & Wu, X. (2009). Design of DGPS navigation control system for Dongfanghong X-804 tractor. *Transactions of the Chinese Society of Agricultural Engineering, 25*(11), 139–145.

Yang, L., Luo, T., Cheng, X., Li, J. & Song, Y. (2015). Universal autopilot system of tractor based on Raspberry Pi. *Transactions of the Chinese Society of Agricultural Engineering, 31*(21), 109–115.

Zhou, C., Fang, Z., Yu, C., Zhang, Y. & Gao, Y. (2013). UTM projection and Gauss-Krvger projection and their conversion. *Geology and Exploration, 5*, 0882–0889.

Zhu, Q., Gao, G. & Niu, W. (2016). The automatic navigation driving system of agricultural machinery. *Modern Agriculture, 5*, 65–67.

Automatic Control, Mechatronics and Industrial Engineering – He & Qing (Eds)
© 2019 Taylor & Francis Group, London, ISBN 978-1-138-60427-8

Key technologies of robot navigation based on machine vision: A review

B. Zhang
College of Mechanical and Electrical Engineering, Hohai University, Changzhou, China
College of Mechanical and Electrical Engineering, Sanjiang University, Nanjing, China

D.L. Zhu
College of Mechanical and Electrical Engineering, Hohai University, Changzhou, China

ABSTRACT: In recent years, navigation has been a hot topic in the field of robotics, and machine vision is the most common method of autonomous navigation. Vision-based robot navigation is first studied, based on the three corresponding areas of pattern recognition, pattern matching and 3D reconstruction. Then the current situation and prospects for various types of algorithms are discussed, which include the algorithms of obstacle detection, stereo matching and Simultaneous Localization And Mapping (SLAM). The advantages and disadvantages of different algorithms are summarized. Finally, the prospects for robot navigation technology are discussed, and the algorithm framework that may be useful for related research fields is summarized.

Keywords: Quadruped robot, obstacle detection, stereo matching, SLAM

1 INTRODUCTION

With the development of robot technology, the research and development of the quadruped robot has become increasingly focused on autonomy. This requires the functions of environment perception, terrain modeling and autonomous navigation, in addition to the basic operational function of the robot. Genuine autonomous navigation requires a quadruped robot to explore a completely unknown environment and establish an environmental map. As for the problems that accompany this process, the navigation can be defined in terms of three questions, namely, 'Where am I?', 'Where am I going?' and 'How do I get there?' The first and second questions can be answered by providing the robot with appropriate sensors, while the third question can be answered by using effective system navigation planning. Vision-based autonomous robot navigation can roughly be divided into two categories: systems that require prior knowledge of the environment in which they will be operating (in this case the systems require maps/folders), and systems that can see the environmental conditions in which they will navigate (Rahmani et al., 2015). Obviously, mapless navigation is much more complicated than navigation based on a map. Mapless navigation systems mostly involve technologies that use vision information from image segmentation, optical flow, and feature matching.

2 OBSTACLE DETECTION

Any object that forms an impediment in the path of a robot can be called an obstacle. In the process of mobile real-time autonomous robot movement and navigation, it is necessary to achieve obstacle avoidance. In recent years, the application of visual sensors to mobile robot navigation has attracted more and more attention. The main function of visual obstacle avoidance is to identify all kinds of scenes from visual information and determine the feasible operational region for the mobile robot.

2.1 Obstacle detection based on feature matching

Obstacle detection based on feature matching is the simplest method. It detects some special and obvious features of obstacles, such as edge, texture and shape. These methods perform image segmentation or pixel classification through color or edge information (Lowe, 2004).

2.1.1 Obstacle detection based on color

This method mainly uses the color difference between roads and obstacles to distinguish feasible areas from suspicious obstacles. In general, mobile robot vision obstacle avoidance systems mainly adopt the Hue-Intensity-Saturation (HIS) color system (She & Huang, 1994; Dima & Hebert, 2005; Batavia & Singh, 2001; Kunt, 1982). However, because the color of shadow, water-trace and leaves are not consistent with the color of a road, and may be mistakenly placed into the obstacle area, so we can only define the area as being a suspicious obstacle area by differentiating between the color information.

2.1.2 Obstacle detection by edge detection

Edge is the most basic feature of an image, which is the junction of the image region and another region, the place where the regional attributes are mutated, and the most concentrated part of the image information. Extracting edges can distinguish obstacles from backgrounds (Dhankhar & Sahu, 2013). There are three categories of image edge extraction methods. The first is based on some fixed local operation methods, such as differential and quasi-legality, which are classical edge extraction methods. The common differential operators include the Robert crossover operator, the Sobel operator, the Laplacian of Gaussian (LoG) operator and the Canny (1986) operator (Janani et al., 2012). The second category is the global extraction method based on the criterion of energy minimization, and the extraction of edges from the perspective of global optimization, such as relaxation and neural networks (Etemad & Chelappa, 1993). The third category has developed in recent years, such as the wavelet transform, mathematical morphology, and shape theory. These image edge extraction methods have good time-frequency local characteristics. A novel feature extraction method called robust sparse linear discriminant analysis is proposed by Wen et al. (2018), which can select the most discriminative features for discriminant analysis.

2.2 Obstacle detection based on motion

Motion detection involves the following algorithmic methods: background subtraction (Tougaard, 1989; Xiao et al., 2003); frame difference (Zhan et al., 2007; Chu et al., 2007); optical flow (Alard & Lupton, 1998; Horn & Schunck, 1981). Background subtraction is a method to detect moving objects by comparing the current frame and background reference image, and its performance depends on a background modeling technique. It is suitable for the target detection of a larger and significant motion deformation obstacle, but the effect is not ideal in the case of uneven or multi-bend roads. The optical-flow algorithm was originally proposed by Horn and Schunck (1981), and the principle of this algorithm uses the information of the surrounding environment at different times and estimates the optical flow through a sequence of images. The disadvantage is that when the contrast between the obstacle and the background is too small, or there is noise, shadow, transparency or occlusion, the calculated optical-flow distribution is not very reliable and accurate, and it has a poor real-time ability.

2.3 Obstacle detection based on stereo matching

The feature-matching algorithm or motion algorithm are monocular-vision obstacle-detection methods which are based on a single-frame image, and a single-frame image can easily lose depth information of a scene. However, stereo vision can obtain 3D information about objects (Sun et al., 2003; Kanade, 1994). The commonly used detection methods are mainly based on binocular vision and multi-vision. Binocular vision directly simulates the human vision system. The basic principles of binocular vision are: first, the scene images are obtained from two angles with two identical cameras; then, stereo matching is performed to generate a disparity map; finally, depth information about the obstacle is calculated according to the inside and outside parameters of the camera and the disparity map.

3 STEREO-MATCHING ALGORITHMS

The study of stereo matching began in the middle of the last century. At MIT, Robert extended 2D image analysis to 3D scene analysis, which marked the birth of computer stereo vision technology and developed rapidly into a new subject (Faugeras, 1993). At the end of the last century, the visual computing theory created by Marr had a great influence on the development of stereo vision (Sonka et al., 1993) and has given rise to a relatively complete system, from image acquisition to the visual surface reconstruction of scenery.

3.1 Deployment of vision sensors

There are many visual-sensor configuration methods for quadruped robots. These include monocular vision, binocular stereo vision, multilocular stereo vision and panoramic vision. In addition, the combination of vision sensors and laser or ultrasound sensors has been widely used. Among these, the most common configuration method in autonomous navigation is binocular vision. In this method, the features from the left and right cameras can be matched by the geometric constraint of the outer pole line, the depth information of the scene can be provided on the premise of the established frame rate, and the complete feature information is extracted to facilitate the feature initialization.

3.2 Stereo-matching primitives

To improve the accuracy of matching and reduce the ambiguity of matching points, we need to select the best matching primitives to represent images. The matching primitives used in most matching algorithms include image color, grayscale, phase, texture, gradient, and edge and corner information. In order to increase the robustness of the matching, these basic elements are usually used in combination. According to the differences of the primitives used, stereo matching can be divided into three parts: the matching algorithm based on the region (Veksler, 2003), the feature (Venkateswar & Chellappa, 1995) and the phase (Chan et al., 2006).

3.3 Stereo-matching strategy

The most commonly used matching algorithms are the global-optimal matching algorithm and the local-optimal matching algorithm (Tombari et al., 2008). To avoid the effect of local extrema, the global energy function is established in the global algorithm, which mainly includes the smoothing constraints and similar data items. Then the global-optimization method is used to minimize the global energy function, to get the globally optimal disparity distribution. The most commonly used global-optimal matching algorithms are graph cuts (Hong & Chen, 2004), belief propagation (Sun et al., 2005), neural networks, and genetic. The disparity obtained from global matching algorithms is more accurate, but is not suitable for real-time processing because of its low speed. The local matching algorithm directly solves the local energy optimal results on the respective matching regions. This algorithm has low complexity and fast computation speed, so it can meet real-time requirements. However, due to the lack of global constraints, the local algorithm is vulnerable to local noise, object deformation and occlusion. The local matching algorithms can be divided into three categories: adaptive-window stereo-matching algorithms, adaptive-weight stereo-matching algorithms and multi-window stereo-matching algorithms (Adhyapak et al., 2007).

4   SIMULTANEOUS LOCALIZATION AND MAPPING (SLAM)

Simultaneous Localization And Mapping (SLAM) is a prerequisite for realizing the real autonomy of mobile robots. Autonomous navigation requires robots to explore in a completely unknown environment, establish environmental maps, locate accurately, and implement planning tasks. For a long time, the research of mapping and positioning have been independent. The location research requires a priori maps guidance. In addition, the mapping research requires the robot's position and posture. Smith et al. (1988) proposed a probabilistic method to solve two aspects simultaneously, which uncovered the prelude to the study of SLAM. EKF-SLAM is a remarkable early SLAM which was designed by Paz et al. (2008). Based on Conditionally Independent Divide and Conquer, it was able to operate in larger environments, robustly and in real time. ORB-SLAM is a feature-based SLAM system that operates in real time, in small and large indoor and outdoor environments (Mur-Artal et al., 2015).

4.1   *Reconstruction of 3D environment*

3D information of scene acquisition and terrain construction by binocular vision is a prerequisite for a robot to walk autonomously in an unknown environment. The result of stereo matching is just some disparity information, and the 3D information of the scene is restored using a reprojection matrix of the disparity map. Using the internal and external parameters obtained by the camera calibration, the disparity map is converted to a depth map based on $z(x,y) = f \times b/d(x,y)$, in which f is the camera focal length, b is the baseline length, and $d(x,y)$ and $z(x,y)$ are the disparity and depth of pixel (x,y), respectively.

4.2   *Implementation mechanism of SLAM*

SLAM includes Visual Odometry (VO), optimization, loop closing, and mapping. VO estimates rough camera motion according to the information of adjacent images, and provides a good initial value for the back end. The back end of SLAM is optimization, which reduces the noise of SLAM. Loop closing is used to solve the problem of time drift.

SLAM is usually solved as an a posteriori probability estimation. SLAM can be divided into two major categories, filter and nonlinear optimization, involving techniques such as the Extended Kalman Filter (EKF), Particle Filters (PFs) and pose graphs.

The Kalman filter has the advantages of simplicity and astringency, but the linear hypothesis of the system model is not established in many cases. EKF is the generalization of the Kalman filter in nonlinear systems. However, because there are many more visual features, and the complexity of SLAM based on EKF is the square of the characteristic number, so it is difficult to adapt to a large-scale environment (Miro et al., 2005).

PFs are a sequential Monte Carlo filtering method, which use some samples (particles) set to approximate the a posteriori probability of SLAM. PFs can represent any robot probability model expressed by the Markov chain, and have the advantage of easy realization. However, the probability distribution of high dimensional space requires many particles. Rao-Blackwellized Particle Filters (RBPFs) (Miller & Campbell, 2007) are used to solve this problem: the approach divides the state space into an independent part and marginalizes one or more components of the part, providing an algorithm for simplifying the probability estimation.

5   CONCLUSION AND PROSPECTS

As this paper shows, much has already been accomplished in vision-based quadruped robot navigation. If the goal is to send the quadruped robot from one coordinate location to another, there is enough accumulated expertise to design a mobile robot that could achieve this within a typical building. However, if the goal is to carry out function-driven navigation, we are still eons away from achieving this.

With the development of machine vision technology and AI technology, how to understand the environment like a human being is the new direction for the vision-based autonomous robot.

ACKNOWLEDGMENTS

This paper is supported by the Natural Science Fund for Colleges and Universities in Jiangsu Province, China (Grant No. 18 KJD510008).

REFERENCES

Adhyapak, S., Kehtarnavaz, N. & Nadin, M. (2007). Stereo matching via selective multiple windows. *Journal of Electronic Imaging, 16*(1), 013012.

Alard, C. & Lupton, R. (1998). A method for optimal image subtraction. *Astrophysical Journal, 503*(1), 325–331.

Batavia, P.H. & Singh, S. (2001). Obstacle detection using adaptive color segmentation and color stereo homography. In *Proceedings of 2001 IEEE International Conference on Robotics and Automation* (vol. 1, pp. 705–710).

Canny, J.A. (1986). A computational approach to edge detection. *IEEE Transactions on Pattern Analysis and Machine Intelligence, PAMI-8*(6), 679–698.

Chan, W.L., Choi, H. & Baraniuk, R.G. (2006). Multiscale image disparity estimation using the quaternion wavelet transform. In *2006 IEEE International Conference on Image Processing* (pp. 1229–1232).

Chu, H., Ye, S., Guo, Q. & Liu, X. (2007). Object tracking algorithm based on Camshift algorithm combinating with difference in frame. In *2007 IEEE International Conference on Automation and Logistics* (pp. 51–55).

Dhankhar, P. & Sahu, N. (2013). A review and research of edge detection techniques for image segmentation. *International Journal of Computer Science & Mobile Computing, 2*(7), 86–92.

Dima, C. & Hebert, M. (2005). Active learning for outdoor obstacle detection. In *Proceedings of Robotics: Science & Systems, 8–11 June 2005, Massachusetts Institute of Technology, Cambridge, MA* (pp. 9–16). doi:10.15607/RSS.2005.I.002.

Etemad, K. & Chelappa, R. (1993). A neural network based edge detector. In *IEEE International Conference on Neural Networks* (vol. 1, pp. 132–137). doi:10.1109/ICNN.1993.298518.

Faugeras, O. (1993). *Three-dimensional computer vision: A geometric viewpoint.* Cambridge, MA: MIT Press.

Gandhi, T., et al. (2000). Detection of obstacles on runways using ego-motion compensation and tracking of significant features. *Image & Vision Computing, 18*(10), 805–815.

Hong L. & Chen, G. (2004). Segment-based stereo matching using graph cuts. In *Proceedings of the 2004 IEEE Computer Society Conference on Computer Vision and Pattern Recognition (CVPR 2004)* (vol. 1, pp. I-74–I-81).

Horn, B.K.P. & Schunck, B.G. (1981). Determining optical flow. *Artificial Intelligence, 17*(1–3), 185–203.

Janani, B., Harini, R., Bhattacharjee, J.B. & Thilakavathi, B. (2012). Edge detection algorithm for machine vision system. *Annals Computer Science Series, 10*(1), 106–111.

Kanade, T. (1994). Stereo matching algorithm with an adaptive window: Theory and experiment. *IEEE Transactions on Pattern Analysis & Machine Intelligence, 16*(9), 920–932.

Kremen, R. (February 10, 2015). Boston Dynamic's new quadruped is quiet, robust, and generally cool. *Robot.* Retrieved from http://www.botmag.com/boston-dynamics-new-quadruped-is-quiet-robust-and-generally-cool/

Kunt, M. (1982). Edge detection: A tutorial review. In *Proceedings of IEEE International Conference on Acoustics, Speech, and Signal Processing, 1982* (pp. 1172–1175).

Lowe, D.G. (2004). Distinctive image features from scale-invariant keypoints. *International Journal of Computer Vision, 60*(2), 91–110.

Miller, I. & Campbell, M. (2007). Rao-Blackwellized particle filtering for mapping dynamic environments. In *Proceedings of 2007 IEEE International Conference on Robotics and Automation* (pp. 3862–3869).

Miro, J.V., Dissanayake, G. & Zhou, W. (2005). Vision-based SLAM using natural features in indoor environments. In *2005 International Conference on Intelligent Sensors, Sensor Networks and Information Processing* (pp. 151–156). doi:10.1109/ISSNIP.2005.1595571.

Mur-Artal, R., Montiel, J.M.M. & Tardós, J.D. (2015). ORB-SLAM: A versatile and accurate monocular SLAM system. *IEEE Transactions on Robotics, 31*(5), 1147–1163.

Pan, B. (2011). Recent progress in digital image correlation. *Experimental Mechanics, 51*(7), 1223–1235.

Paz, L.M., Piniés, P., Tardós, J.D. & Neira, J. (2008). Large-scale 6-DOF SLAM with stereo-in-hand. *IEEE Transactions on Robotics, 24*(5), 946–957.

She, A.C. & Huang, T.S. (1994). Segmentation of road scenes using color and fractal-based texture classification. *Image Processing, 1994. Proceedings. ICIP–94. IEEE International Conference IEEE, 1994* (pp. 1026–1030 vol. 3).

Smith, R., Self, M. & Cheeseman, P. (1988). Estimating uncertain spatial relationships in robotics. *Machine Intelligence & Pattern Recognition, 5*(5), 435–461.

Sonka, M., Hlavac, V. & Boyle, R. (1993). 3D vision. In M. Sonka, V. Hlavac & R. Boyle, *Image processing analysis & machine vision* (pp. 373–421). New York, NY: Springer.

Sun, J., Li, Y., Kang, S.B. & Shum, H.Y. (2005). Symmetric stereo matching for occlusion handling. In *Proceedings of 2005 IEEE Conference on Computer Vision and Pattern Recognition (CVPR 2005)* (vol. 2, pp. 399–406).

Sun, J., Zheng, N.N. & Shum, H.Y. (2003). Stereo matching using belief propagation. *IEEE Transactions on Pattern Analysis & Machine Intelligence, 25*(7), 787–800.

Tombari, F., Mattoccia, S., Di Stefano, L. & Addimanda, E. (2008). Classification and evaluation of cost aggregation methods for stereo correspondence. In *Proceedings of 2008 IEEE Conference on Computer Vision and Pattern Recognition (CVPR 2008)* (pp. 1–8). doi:10.1109/CVPR.2008.4587677.

Tougaard, S. (1989). Practical algorithm for background subtraction. *Surface Science, 216*(3), 343–360.

Veksler, O. (2003). Fast variable window for stereo correspondence using integral images. In *Proceedings of 2003 IEEE Computer Society Conference on Computer Vision and Pattern Recognition* (vol. I, pp. I-556-I-561). doi:10.1109/CVPR.2003.1211403.

Venkateswar V. & Chellappa, R. (1995). Hierarchical stereo and motion correspondence using feature groupings. *International Journal of Computer Vision, 15*(3), 245–269.

Wen, J., Fang, X., Cui, J., Fei, L., Yan, K., Chen, Y. & Xu, Y. (2018). Robust sparse linear discriminant analysis. *IEEE Transactions on Circuits & Systems for Video Technology, 1*(1), 99.

Xiao, D.G., Yu, S.S. & Zhou, J.L. (2003). Motion tracking with fast adaptive background subtraction. *Wuhan University Journal of Natural Sciences, 8*(1 A), 35–40. doi:10.1007/BF02902061.

Zhan, C., Duan, X., Xu, S., Song, Z. & Luo, M. (2007). An improved moving object detection algorithm based on frame difference and edge detection. In *IEEE 2007 International Conference on Image and Graphics* (pp. 519–523).

Control systems

Automatic Control, Mechatronics and Industrial Engineering – He & Qing (Eds)
© 2019 Taylor & Francis Group, London, ISBN 978-1-138-60427-8

Research on innovative control strategies of bridgeless PFC based on one-cycle control

Y. Jin, S.C. Liu, C.Y. Wu, Y. Li, Y. Dong & K. Gao
Shanghai Institute of Space Power Sources, Minhang, Shanghai, China

X.J. Li
Shanghai Shangdian Power Engineering Co. Ltd., Shanghai, China

ABSTRACT: Conventional power factor correction topologies like boost PFC, which have a rectifier on the input side, may cause conduction loss and low converting efficiency. Much more bridgeless PFC topologies have been developed and researched to improve efficiency and minimize conduction loss of overall systems. Input voltage inspection, current detection and common mode noise are significant issues in bridgeless PFC topologies. In this paper, an improved OCC bridgeless PFC control strategy, based on a peak current sampling approach, is proposed. In addition, simulations have been carried out through implementation of MATLAB and Simulink to verify a proposed control system. A bridgeless PFC with a low EMI circuit is demonstrated by applying IR1155S as a proof of the design.

Keywords: One-cycle control, bridgeless PFC, efficiency, electromagnetic interference, total harmonic distortion

1 INTRODUCTION

Conventional boost Active Power Factor Correction (APFC) topology with a rectifier always has at least three semiconductor devices operating during switch on or off status. Large conduction loss would exist in low input and high power applications (Figuerido et al., 2010). To improve converter efficiency, a bridgeless boost PFC topology has been proposed in recent years (shown in Figure 1). In this design, there is no bridge rectifier and the inductor is located at the AC side, the converter is considered as a double boost converter when operating in negative half cycle (Darly et al., 2010), and under each operating status, current only flows through two semiconductor devices which leads to low conduction loss. From Figure 1, input and output sides of bridgeless PFC converters are not connected directly, and the traditional detection methods of voltage and current are unavailable, which leads to a challengeable control system compared with conventional APFC circuits (Li et al., 2009; Ribeiro et al., 2010).

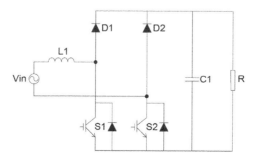

Figure 1. Typical bridgeless boost PFC topology.

A small resistor connected in series at the main circuit, is considered as one of the most common methods to detect inductor current with no need of isolation (Pengju et al., 2009). For bridgeless PFC converters, isolated detection has to be carried out since inductor current and output side was not in common ground. Furthermore, the inductor current in bridgeless PFCs is bidirectional, which is different from the single direction inductor current in traditional PFCs (Choi et al., 2008). A proper control strategy and reliable control system has become one of the significant issues in the research of bridgeless PFC converters (Huber et al., 2008).

2 OCC CONTROL STRATEGIES

2.1 Typical OCC operating principles

Based on boost PFC operating principles of One-Cycle Control (OCC): U_{in} is the Input Voltage after Rectifier; U_o is Output Voltage; T_s is the Switching Period; and DT_s is the Switch Conducting Time. To achieve a power factor correction, U_{in} is given by

$$U_{in} = R_e I_{in} \tag{1}$$

where, R_e is considered as Input Equivalent Resistance of the converter.

As the inductor operates under continuous current mode, input and output voltages can be determined by

$$U_o = \frac{U_{in}}{1-D} \tag{2}$$

where R_s is the Current Detecting Resistor

$$U_m = \frac{U_o R_S}{R_e} \tag{3}$$

The combined equation of Equation 1 and 3 will give

$$R_S I_{in} = \frac{U_m U_{in}}{U_o} \tag{4}$$

The control model will be expressed as Equation 5, where I_{in} is represented as Input Average Current

$$U_m - R_S I_{in} = D U_m \tag{5}$$

2.2 Improved OCC control strategy

The IR1155S controller was designed for AC/DC PFC applications, which may cause large harmonic distortion when the circuit has a small inductance or operates under DCM or CRM modes. To prevent harmonic distortion resulting from peak current sampling, an improved OCC approach has been proposed based on the IR1155 controller (Wang et al., 2008). From Equation 5

$$U_m - R_S I_{Lpeak} = D \frac{U_m - kU_o}{T_s} T_{on} \tag{6}$$

where, k is a coefficient of the output voltage divider resistor.

The relationship of an average and peak current is demonstrated by

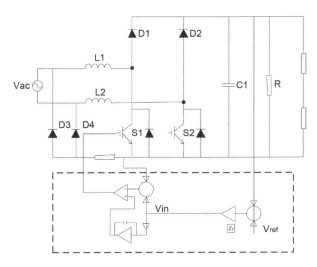

Figure 2.   Improved control strategy of a bridgeless PFC topology.

$$R_S I_{in} = \frac{U_m U_{in}}{U_o} - R_S \frac{1}{2} \Delta I_L$$
$$= \frac{U_m U_{in}}{U_o} - \frac{1}{2} R_S \frac{U_{in}}{L} T_{on}$$
$$= U_m (1-D) - \frac{1}{2} R_S \frac{(1-D)U_o}{L} DT_s \qquad (7)$$
$$I_{LAV} = I_{in} = \frac{U_m}{R_S}(1-D) + (\frac{k}{R_S} - \frac{T_s}{2L})(1-D)U_o D$$

Average current can be derived by taking k as $\dfrac{R_S T_s}{2L}$

$$I_{LAV} = I_{in} = \frac{U_m}{R_S}(1-D) \qquad (8)$$

In order to decrease influence caused by electromagnetic interference, a two-diode bridgeless boost PFC topology was researched. A new PFC control strategy based on an OCC IR1155 is proposed and the overall system is presented in Figure 2. The boost inductor is split and located at the AC side to construct the boost structure (Masumoto et al., 2013). In positive half line cycle, MOSFET S1 and boost diode D1, together with the boost inductor, constructs a boost DC/DC converter. Meanwhile, MOSFET S2 is operating as a simple diode. The input current is controlled by the boost converter and follows the input voltage. During the other half line cycle, the circuit operation operates the same way. Thus, in each half line cycle, one of the MOSFET operates as an active switch and the other one operates as a diode; both the MOSFETs can be driven by the same signal (Su et al., 2011).

3   SIMULATION AND ANALYSIS

The simulation of an improved bridgeless PFC converter is carried out through the implementation of a proposed control strategy in MATLAB and Simulink (see Figure 3). As shown in Figure 4, the input voltage scales down 10 times and the input current is perfectly sinusoidal with input voltage. In addition, Figure 5 demonstrates the output voltage of the overall system with a ripple of 10 V (De Gusseme et al; 2005; El Aroudi et al., 2013).

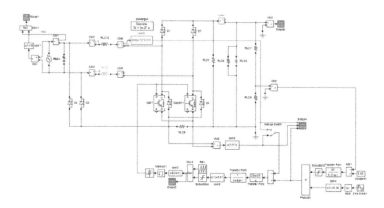

Figure 3.   Simulation model of overall system.

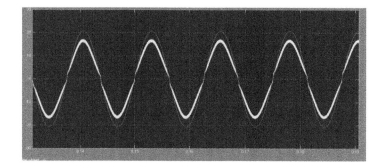

Figure 4.   220 V AC input voltage (blue?) and inductor current (yellow) at 2.5 kW load.

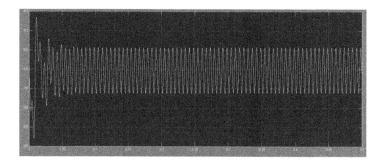

Figure 5.   400 V output voltage at 2.5 kW load.

4   EXPERIMENTAL RESULTS

Based on the analysis above, the bridgeless PFC circuit can simplify the circuit topology and improve the efficiency; the OCC is the most attractive control method for the bridgeless PFC circuit. Figure 6 shows the prototype of the 2.5 kW improved bridgeless PFC. All voltage waveforms were obtained through differential probes. The input power, PF and thermal harmonic distortion were measured by a Tektronix PA1000 single phase power analyzer. The output power was measured by two Keysight 34465A digital multimeters.

Figure 7 presents the experimental curve of 220 V AC input voltage (blue) and inductor current (pink) at 2.5 kW load. The input current perfectly follows the input voltage. Thus, the power factor correction function is achieved by using an OCC.

Figure 6.   2.5 kW improved bridgeless PFC prototype.

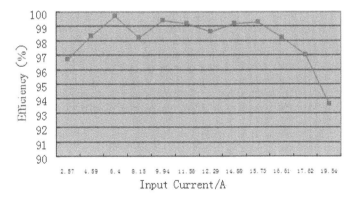

Figure 7.   Experimental curve of 220 V AC input voltage (blue) and inductor current (pink) at 2.5 kW load.

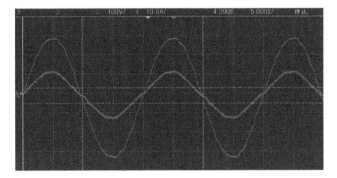

Figure 8.   Efficiency curve of the prototype under the conditions of V_{in} = 220 V AC, V_{out} = 400 V AC and L = 880 uH.

Figure 8 shows the efficiency curves measured at both 220 V AC input voltage with a 0 Ω turn-on gate resistor and a 10 Ω turn-off gate resistor. It can be seen that the overall efficiency of a bridgeless PFC converter is above 96% at nominal load.

Figure 9 shows a power factor curve of the prototype under the conditions of V_{in} = 220 V AC, V_{out} = 400 V AC and L = 880 uH. The power factor reaches around 0.99 from half load to full load variations.

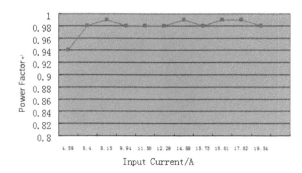

Figure 9. Power factor curve of the prototype under the conditions of V_{in} = 220 V AC, V_{out} = 400 V AC and L = 880 uH.

5 CONCLUSION

Conventional PFC operating principles and control strategies were analyzed. In order to reduce conducting loss and improve converting efficiency, bridgeless PFC topology was proposed gradually, though input voltage detection, current detection and EMI have not been solved effectively. This paper investigated the control logic of a traditional bridgeless PFC converter as well as proposing a new control strategy based on the OCC approach. Through carrying out the simulation and experimental works, the proposed improved bridgeless PFC converter is able to make a low EMI noise. A 2.5 kW bridgeless PFC prototype was developed and the experimental results demonstrate the design of a new control strategy which is able to achieve power factor correction.

REFERENCES

Choi, W.Y., Kwon, J.M. & Kwon, B.H. (2008). Bridgeless dual-boost rectifier with reduced diode reverse-recovery problems for power-factor correction. *IET Power Electronics*, *1*(2), 194–202.

Darly, S.S., Ranjan, P.V., Bindu, K.V. & Rabi, B.J. (2011). A novel dual boost rectifier for power factor improvement. *International Conference on Electrical Energy Systems*, 121–127.

De Gusseme, K., Van de Sype, D.M., Van den Bossche, A.P.M. & Melkebeek, J.A. (2005). Digitally controlled boost power-factor-correction converters operating in both continuous and discontinuous conduction mode. *IEEE Transactions on Industrial Electronics*, *52*(1), 88–97.

El Aroudi, E., Haroun, R., Cid-Pastor, A. & Martínez-Salamero, L. (2013). Suppression of line frequency instabilities in PFC AC-DC power supplies by feedback notch filtering the pre-regulator output voltage. *IEEE Transactions on Circuits and Systems I: Regular Papers, 60*(3), 796–809.

Figuerido, J.P.M., Tofili, F.L. & Silva, B.L.A. (2010). A review of single-phase PFC topologies based on the boost converter. *Proceedings of the IEEE International Conference on Industrial Applications*, Sao Paulo, Brazil, 1–6.

Huber, L., Jang, Y. & Jovanovic, M.M. (2008). Performance evaluation of bridgeless PFC boost rectifiers. *IEEE Transactions on Power Electronics*, *23*(3), 1381–1390.

Li, Q., Andersen, M.A E. & Thomsen, O.C. (2009). Conduction losses and common mode EMI analysis on bridgeless power factor correction. *International Conference on Power Electronics and Drive Systems*, Taipei, 1255–1260.

Masumoto, K., Shi, K., Shoyama, M. & Tomioka, S. (2013). Comparative study on efficiency and switching noise of bridgeless PFC circuits. *Proceedings of the IEEE PEDS*, 613–618.

Pengju, K., Shuo, W. & Lee, F.C. (2008). Common mode EMI noise suppression for bridgeless PFC converters. *IEEE Transactions on Power Electronics*, *23*(1), 291–297.

Ribeiro, H., Borges, B., Pinto, S. & Silva, F. (2010). Bridgeless single-stage full-bridge converter with one cycle control in the output voltage. *Energy Conversion Congress and Exposition*, 2899–2906.

Su, B., Zhang, J.M. & Lu, Z.Y. (2011). Totem-pole boost bridgeless PFC rectifier with simple zero-current detection and full-range ZVS operating at the boundary of DCM/CCM. *IEEE Transactions on Power Electronics, 26*(2), 427–435.

Wang, W., Liu, H., Jiang, S. & Xu, D. (2008). A novel bridgeless buck-boost PFC converter. *Proceedings of the IEEE PESC*, 1304–1308.

Automatic Control, Mechatronics and Industrial Engineering – He & Qing (Eds)
© 2019 Taylor & Francis Group, London, ISBN 978-1-138-60427-8

Design and implementation of a state machine controlled 3D model car based on FPGA + gyroscope

K.R. Ren, L.X. Zhang & H. Tan
College of Electronics and Information Engineering, Tongji University, Shanghai, China

ABSTRACT: With the massive increase of drivers in developing countries, training time in driving school has become limited and better utilization of resources is an emerging problem. This paper introduces the method of using Verilog HDL to develop I²C and UART transmitter state machine based on FPGA and gyroscope. The modeling in Unity3D software on PC is discussed and implemented by controlling a 3D car, which lays the foundation of practical driving for users with Virtual Reality (VR) technology. Simulation results of system functions verify the validity and stability of the system.

Keywords: FPGA, State Machine, VR

1 INTRODUCTION

With the increase of automotive application scenarios, driving has become an indispensable skill. Especially in developing countries such as China and India, the number of driving school students has risen significantly. However, the learning resources in driving school are limited, resulting in emerging problems such as waiting time is longer than actual training time, and the time utilization efficiency is low. This paper therefore proposes an approach to improve time utilization at driving school: to simulate car driving using a new human-computer interaction method.

Taking advantages of virtual reality (VR) technology, drive training can be enhanced with an alternative and more economic approach before getting into actual vehicles. VR forms a new human-computer interaction mode between the virtual world and the real world. By creating dynamic driving scenes in VR, the tool can be used to simulate different real driving conditions while providing trainees practical experience without driving the actual vehicle. Many VR programming tools have been created, in which Unity3D is an application development tool that makes it easy to create 3D video games.

On the other hand, FPGA can flexibly modify logic control according to design requirements and make full use of hardware resources (Niu, 2006). By implementing FPGA in Unity3D, this paper proposes using an FPGA-based gyroscope to control a 3D model car. The design and implementation are presented in the following sections, in which Section 2 illustrates the detailed model architecture design, and Section 3 discusses the functional analysis and state machine design. Section 4 then presents the test validation results, and finally Section 5 concludes the finding of this paper and lays down the foundation for future work.

2 ARCHITECTURE DESIGN

The model architecture development is shown in the following subsections. In order to realize the X-axis position data acquisition from the gyroscope to FPGA, transfer the position information to Unity 3D on the PC-side, and then control the 3D model car using game development software, the design needs to complete the following steps:

2.1 Communication between FPGA and PC

It is established that in order to send data via the serial port, the design needs to follow the prescribed protocol (Yang, 2011). Communication algorithm uses the serial port module to send the position data to the PC end, and uses the PC serial port debugging assistant tool to test whether the received data satisfies the required programming format, and hence realizes the effective data transmission between the FPGA and the PC.

2.2 Communication between FPGA and gyroscope

The second step is the communication between the FPGA and the gyroscope. To achieve this, the design uses I²C synchronous communication protocol (He, 2004), which is shown in Figure 1.

The communication protocol is briefly stated as below:

Data transmission: When SCL is high and the SDA line is stable, the SDA is transmitting bit data; if the SDA changes, it indicates the start or end of the session. Acknowledge (ACK) occurs at SDA low.

Data change: SDA can change only when SCL is low.

The design requires the programming of an I²C communication module to communicate with the gyroscope. In order to obtain the desired angle information from the gyroscope, the internal register of the gyroscope should be configured after power-on, the main function register is initialized, and then the angle information is obtained from the gyroscope through the I²C module.

2.3 Serial data transfer protocol

Since FPGA and PC only use serial communication, in order for the PC to distinguish different received data such as the angle X and the button status, the main control module in the FPGA needs to be programmed to encode before using the serial port to send data. The data processing rules are defined as follows:

Every time the serial port sends a data packet, use a newline character '/n' as the end marker, and use '|' to distinguish different data. The specifications of the data sent by each package are as follows:

|Gyroscope x-axis angle | Acceleration button status | Deceleration button status Newline ('/n').

The gyroscope data is represented by 16 bits. The design maps the 16-bit binary data to a 4-digit quaternary number, and then converts each bit into an ASCII code. The conversion is shown in Figure 2.

The acceleration and deceleration button states are converted to ASCII codes '0' and '1' and sent to the serial port, with '0' for release, and '1' for pressing.

For example, if the sending gyroscope is placed horizontally (the x-axis angle is 0), the acceleration button is pressed, and the deceleration button is released. The data received by the serial port should be as follows:

| 0000 | 1 | 0 ('/n')

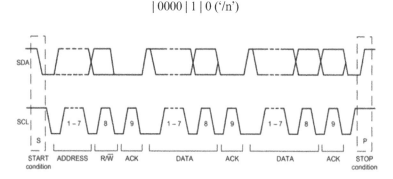

Figure 1. I²C synchronous communication.

2.4 Function realization

In order to achieve a good driving experience, the design uses Unity 3D software for scene development. Based on the existing car and track models in the software, additional data received from the serial port can be decoded and vehicle control codes can be programmed and integrated. At the same time, continuous improvements are made to enhance the game experience.

2.5 Circuit diagram/hardware connection diagram

The circuit has a USB to TTL module, gyroscope MPU6050, and button switch. Each device is connected as shown in Figure 3.

2.6 Overall module layout diagram

The overall module layout diagram is illustrated in Figure 4. There are four main modules including the clock, main control module, communication protocol module, and the top-level module for the serial port send.

clk is the global clock.

uart_tx is the top-level module to send by the serial port, in which the module *uart_send* sends data according to the protocol, and the baud rate module *clock_set_tx* generates the serial port clock signal to control the serial port transmission rate. The two modules use the *uart_tx* as the top-level module for external call.

BIT	0-7	8-23	24-31	32-39	40-47	48-55	56-63	64-71
NOTATION	'|'	X	'|'	A	'|'	D	'\r'	'\n'
FUNCTION	Seperator	Angle of X	Seperator	Acceleration	Seperator	Deceleration	Carriage Return	Line Feed

1 map to a 4-digit quaternary number

2 convert each bit into an ASCII

BYTE	0	1-4	5	6	7	8	9	10
NOTATION	'|'	X	'|'	A	'|'	D	'\r'	'\n'
FUNCTION	Seperator	Angle of X	Seperator	Acceleration	Seperator	Deceleration	Carriage Return	Line Feed

Figure 2.   Serial data transmission data frame format.

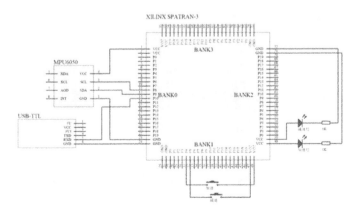

Figure 3.   Hardware connection diagram.

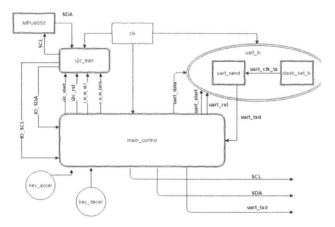

Figure 4.   Overall module layout diagram.

i2c_tran is an I²C communication protocol module with an external interface, which can perform external I/O information control to realize read and write operations in I²C communication.

main_control is the main control module, which is used to control the I²C module to initialize and read data from the gyroscope. It processes and packages the received data and then uses the serial port module to send.

3   FUNCTIONAL ANALYSIS AND STATE MACHINE DESIGN

After the model is established, this section discusses the detailed functional analysis of the model and presents the state machine design to achieve these functions.

3.1   *Serial port transmitter module*

The serial port transmitter module design setting baud rate is 38400, with no parity, and 1stop bit.

To generate a 38400 baud rate, the design uses a clock division to divide the 8 MHz main clock *clk* to produce a 38400 Hz clock signal. At the same time, the enable signal *start* is set, so that the baud rate clock *uart_clk_tx* is generated only when the enable signal appears. When the 8-bit data to be transmitted is sent, the end signal of the data transmission *finish* is received, and the baud rate clock is terminated.

FPGA programming is suitable for parallel operation. In order to enable the module to send data according to the communication protocol, the state machine is designed to realize the switching of different states, that is, the state of the start bit, the data bit and the stop bit. The serial port transmission state machine is shown in Figure 5. Where *uart_txd* is the *txd* data line, *start* is the serial port send enable signal, and *finish* is the serial port send stop signal.

3.2   *I²C communication module*

I²C communication module is used in the design to enable intra-board communication. The I²C communication module contains 9 input and output pins:

$$CLK, SDA, SCL, RST, ENABLE, DATA, W\_R, IO\_ERROR, REGS,$$

where *CLK* is the system clock; *SDA* is the data transmission line; *SCL* is the communication clock; *RST* is the reset signal; *ENABLE* is the chip enable signal; *DATA* is the received data register; *W_R* is the read/write enable signal; *IO_ERROR* is the error signal (gyroscope not responding); *REGS16* is the bit register, in which the high eight bits store the gyroscope register address, and the low eight bits store the initialization register input data.

In order for the module to receive and send data correctly according to the I²C protocol, the program sets 14 state machines. The read and write operations are defined by the state machines to perform their roles, respectively. The communication state machines are shown in Figure 6.

3.3 *Main control module*

The main control module includes the gyroscope control sub-module and the serial port control sub-module.

3.3.1 *Gyroscope control*

The main control module *main_control* uses a frequency divider to generate a series of ordered clock signals to control the I²C and UART modules. The I²C and UART modules work on two timelines respectively, where *i2c_change* is used to switch the working state of the *i2c* module, and *uart_rst* is used to switch the send data in the serial port.

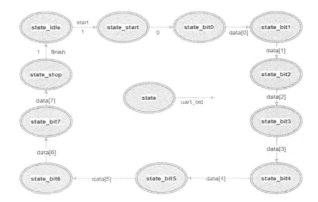

Figure 5. Serial port transmission state machine.

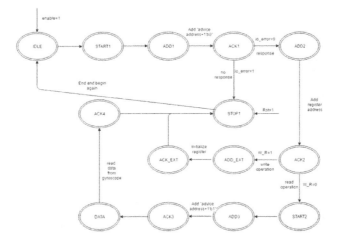

Figure 6. I²C communication state machines.

49

The main control module builds state machines when extracting data from the gyroscope and sending data to the serial port. The gyroscope control state machine is shown in Figure 7. The system is in the *state_idle* state after initial or reset. When the *i2c_change* pulse occurs, the state machine switches to the next state *state_init_pwr*. When state machine is in the following states, *state_idle*, *state_init_pwr*, *state_init_smplrt*, *state_init_config1*, *state_init_gyro*, and *state_init_accel*, the I²C module will be called to write data to the gyroscope register and initialize its power management register. It then loops in the last four state machines, *state_read_gyroH*, *state_read_gyroL*, *state_read_accelH*, and *state_read_accelL*, constantly calling the I²C module to read data from the gyroscope.

3.3.2 *Serial port control*

The serial port control state machine is shown in Figure 8, which is used to send data to the serial port in a certain frame format. The state machine is a loop structure, and the initial

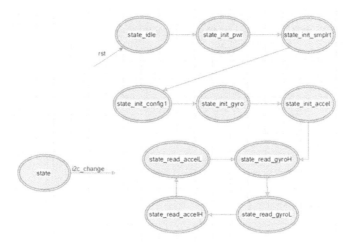

Figure 7.　Gyroscope control state machine.

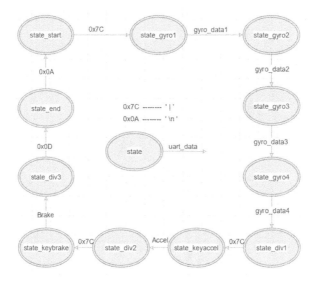

Figure 8.　Serial port control state machine.

50

transmission value is $0 \times 7C$-8 bit separator, which is '|'. Then the data is converted to quaternary data and then converted from bit to bit to ASCII code and sent to the serial port. It is followed by the 8-bit separator $0 \times 7C$, the converted accelerator button value, another 8-bit separator $0 \times 7C$, the converted brake button value, the carriage return $0 \times 0D$, and last the newline character $0 \times 0A$. This sends a set of packed data frames from the serial port to the PC. The serial write port of the PC takes a line of the data each time, and the data is sorted to the corresponding place according to the separator so that parsing and transformation can be realized, hence controlling the car.

3.4 *Unity-based vehicle model control module*

Finally, the Unity-based vehicle model control module is integrated to achieve the model car control.

3.4.1 *Module description*
The design uses the Unity3D physical model to design part of the control code, enhancing the driving experience and maximizing the advantages of the gyroscope.

3.4.2 *Control strategy design*
The program reads the serial data, including the data sent from the FPGA to the gyroscope's X-axis and the two button switch data.

The gyroscope X-axis will be used to control the steering of the car model. The serial port can receive gyroscope X-axis data from $-80°$ to $+80°$, which is sufficient for automotive steering requirements. In addition, the design ignores the $-5°$ to $+5°$ deflection to prevent the small-scale jitter that can cause unexpected steering of the car model. To provide a friendly operating experience, the design reduces steering sensitivity so that it reduces the risk that the vehicle is out of control caused by oversteering.

4   TEST EVALUATION

Test validation results are provided in this section to verify the effectiveness of the model and control strategy. Simulation of system functions and stability of the system are studied in details.

4.1 *Serial port transmitter module*

The serial port starts to send an 8-bit separator $0 \times 7C$, which is '|'. The test result is analyzed by the logic analyzer. The result is shown in Figure 9:

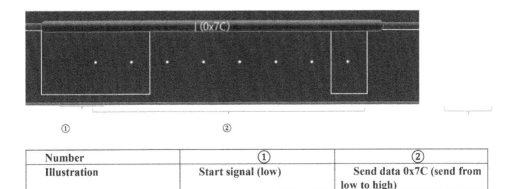

Number	①	②
Illustration	Start signal (low)	Send data 0x7C (send from low to high)

Figure 9.   Serial port transmission test result.

51

4.2 *I²C communication module*

As the gyroscope is connected, the I²C write protocol waveform obtained from the logic analyzer is shown in Figure 10. Write 0×00 into the gyroscope register $0 \times 6B$, the gyroscope sends the response signal, and the data is successfully written.

As the gyroscope is connected, the I²C read protocol waveform obtained from the logic analyzer is shown in Figure 11. Obtained from the 0×44 register, it can be seen that the gyroscope sends a response signal and transmits the data. After receiving the data, a non-response signal NAK is generated, indicating obtaining the data from the gyroscope is successful.

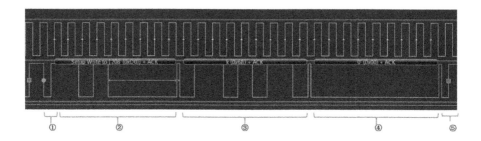

Number	①	②	③	④	⑤
Illustration	Start	Master sends I2Caddr(7bit)+write operation(0), and receives an acknowledge signal (ACK)	Master sends REGaddr(8bit), and receives an acknowledge signal（ACK）	Master sends data（8bit）, and receives an acknowledge signal（ACK）	Stop

Figure 10.　I²C test result of write.

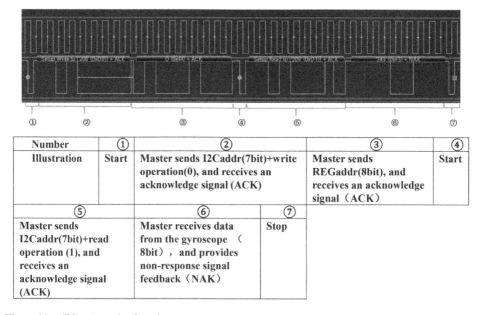

Number	①	②	③	④
Illustration	Start	Master sends I2Caddr(7bit)+write operation(0), and receives an acknowledge signal (ACK)	Master sends REGaddr(8bit), and receives an acknowledge signal（ACK）	Start

⑤	⑥	⑦
Master sends I2Caddr(7bit)+read operation (1), and receives an acknowledge signal (ACK)	Master receives data from the gyroscope （8bit）, and provides non-response signal feedback（NAK）	Stop

Figure 11.　I²C test result of read.

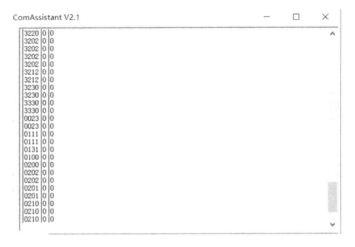

Figure 12.　Test result at the PC serial port.

4.3　*Main control module—serial receiver*

By using debugging assistant, the test results at the PC serial port are shown in Figure 12.

It can be seen that the PC-side serial port debugging assistant test analysis results meet the code requirements.

5　CONCLUSION

This paper proposes a state machine design based on FPGA + gyroscope to control a 3D model car, and discusses its potential to implement for VR technology integrated with life scenes. The model has been verified with high stability and practical real-time performance. However, since VR technology is still in its infancy stage, which results in expensive equipment cost, technical difficulty and insufficient database, this article does not implement VR technology in testing, and related work needs further research in the future.

REFERENCES

He, L. 2004. I2C IntellCBUS system design and applications, Beijing: Beihang University Press.
Niu, T, Wu, B., Jiao, F., and Liu, J. 2006. Design of a kind of UART circuit based on FPGA. Electronic Measurement Technology 29(3): 73–74.
Yang, Y., Ye, P., and Li, L. 2011. Design and implementation of UART based on FPGA. Electronic Measurement Technology 34(7): 80–82.

Automatic Control, Mechatronics and Industrial Engineering – He & Qing (Eds)
© 2019 Taylor & Francis Group, London, ISBN 978-1-138-60427-8

Ultra-low-altitude airdrop fuzzy sliding-mode control for a cargo-stuck fault

W. Wei, X.X. Sun & X.F. Deng

Equipment Management and Unmanned Aerial Vehicle Engineering College, Air Force Engineering University, Xi'an, China

ABSTRACT: To solve the problem of cargo suddenly becoming stuck on the guide rail during the traction stage of ultra-low-altitude reloading of air cargo, a fuzzy sliding-mode variable structure control method with input and output feedback linearization is proposed. Firstly, according to the aircraft dynamics equation, a longitudinal mathematical model of the aircraft under the condition of a cargo-stuck fault is derived, and then the mathematical model is decoupled and linearized using input–output feedback linearization theory. Furthermore, a fuzzy sliding-mode controller with fuzzy control theory and sliding-control method is designed to track the system speed and attitude angle. Finally, the simulation results show that the designed controller has a good control effect.

Keywords: ultra-low-altitude airdrop, cargo stuck, feedback linearization, fuzzy control, sliding-mode control

1 INTRODUCTION

Ultra-low-altitude airdrop (Gurfil et al., 2010; Zang et al., 2013; Li, 2005) refers to large-scale transport aircraft descending smoothly to a height of three to ten meters above the ground, and equipment then being delivered to designated areas. Ultra-low-altitude airdrops have increasingly been valued by experts and scholars because of their long operating range, strong maneuverability, fast response, good concealment characteristics, and timely and efficient support. In the case of an ultra-low-altitude airdrop, the pilot's reaction time must be short because of the close proximity of the aircraft to the ground. Therefore, it is necessary to consider the problem of the safety of the aircraft when a sudden fault occurs, such as a cargo-stuck fault.

There is much research on reloading airdrop technology, both in China and further afield. Sun et al. (2016), Ouyang and Ding (1992) and Liu et al. (2013) used a method of separation modeling to study deeply the relationship between cargo and transporting aircraft, and comprehensively considered various influencing factors to establish a higher-accuracy model, closer to the actual airdrop. Liu et al. (2015) considered the airdrop model, which is decoupled and linearized by the use of feedback linearization theory. The designed second-order sliding-mode controller can track the attitude angle and speed of the carrier. Furthermore, Zhang et al. (2014) and Yang and Lu (2012a) proposed a control method with a sliding-mode disturbance observer control and backstepping control technology for reloading the airdrop system. In addition, on the basis of the equation of motion of the aircraft, Dai et al. (2013) analyzed the force between the cargo and the aircraft when the cargo is stationary, sliding or stuck, and established a dynamic equation for the conveyor under the corresponding conditions, designing a fuzzy control to control the airdrop process.

On the basis of the above, this paper designs a fuzzy sliding-mode controller based on feedback linearization for the ultra-low-altitude airdrop model of a cargo-stuck fault. Firstly, the cargo-stuck model is linearized by utilizing feedback linearization theory. Then, two fuzzy logic controllers are designed to adaptively adjust the control parameters of the sliding-mode controller. The result has strong robustness and anti-interference ability while eliminating the chatter of the sliding-mode controller. Finally, the simulation results show that the designed method has good robustness and can meet the task performance indicators.

2 AIRDROP KINETIC MODEL WITH A CARGO-STUCK FAULT CONDITION

2.1 *Cargo dynamics model*

Combined with the basic principles of theoretical mechanics, the force of the cargo-stuck fault is analyzed. The interaction force between the cargo and the aircraft can be obtained from the dynamic equation of the cargo (Sun et al., 2016). The acceleration of the cargo is analyzed and obtained by the acceleration synthesis theorem (Yang & Lu, 2012b). The absolute acceleration a_a of the cargo is the vector sum of the implied acceleration a_i, the relative acceleration a_r and the coriolis acceleration a_c, that is:

$$a_a = \underbrace{\frac{dV}{dt} + \frac{d\Omega}{dt} \times r_{oc} + \Omega \times (\Omega \times r_{oc})}_{a_i} + \underbrace{\frac{\tilde{d}^2 r_{oc}}{dt^2}}_{a_c} + \underbrace{2\Omega \times \frac{dr_{oc}}{dt}}_{a_r} \tag{1}$$

where V is the carrier flight speed vector, Ω is the aircraft's rotational angular velocity vector and r_{oc} is the displacement of the cargo from the origin of the cargo coordinate system.

Regardless of lateral movement, the absolute acceleration a_a is decomposed along the velocity axis to obtain:

$$\begin{cases} a_c|_{x_b} = \dot{V} + q^2 r_{oc} \cos \alpha - \ddot{r}_{oc} \cos \alpha + \dot{q} r_{oc} \sin \alpha + 2q\dot{r}_{oc} \sin \alpha \\ a_c|_{z_b} = -V\dot{\gamma} - q^2 r_{oc} \sin \alpha + \ddot{r}_{oc} \sin \alpha + \dot{q} r_{oc} \cos \alpha + 2q\dot{r}_{oc} \cos \alpha \end{cases} \tag{2}$$

In addition, the following result is obtained by the acceleration theorem:

$$m_c a_c|_{x_b} = F_{cx} - F_p - m_c g \sin \gamma \tag{3}$$

$$m_c a_c|_{z_b} = -F_{cz} + m_c g \cos \gamma \tag{4}$$

where $F_{cx} = \mu F_{cz}$, $F_p = m_c g \lambda$, μ is the friction coefficient and λ is the traction ratio.
Substituting Equation 2 into Equations 3 and 4, we obtain:

$$F_{cx} = F_p + m_c r_{oc} \sin \alpha \dot{q} + m_c \dot{V} + (m_c g \sin \theta + m_c q^2 r_{oc} - m_c \ddot{r}_{oc}) \cos \alpha \\ -(m_c g \cos \theta - 2m_c q \dot{r}_{oc}) \sin \alpha \tag{5}$$

$$F_{cz} = -m_c r_{oc} \cos \alpha \dot{q} + m_c V \dot{\gamma} + (m_c g \cos \theta - 2m_c q \dot{r}_{oc}) \cos \alpha \\ +(m_c g \sin \theta + m_c q^2 r_{oc} - m_c \ddot{r}_{oc}) \sin \alpha \tag{6}$$

where:

$$\ddot{r}_{oc} = a_c = \dot{V} \cos \alpha + V \sin \alpha \dot{\gamma} + g \sin \theta - \mu g \cos \theta + \mu F_p \sin \alpha / m_c + r_{oc} q^2 \\ +F_p \cos \alpha / m_c + \mu(\dot{V} \sin \alpha - V \cos \alpha \dot{\gamma} + \dot{q} r_{oc} + 2q\dot{r}_{oc}) \tag{7}$$

The disturbance torque of the cargo relative to the aircraft is:

$$M_c = \left(F_{cz} \cos \alpha - F_{cz} \sin \alpha \right) \cdot r_{oc} = m_c r_{oc} (g \cos \theta - \dot{V} \sin \alpha + V \dot{\gamma} \cos \alpha - \dot{q} r_{oc}) \tag{8}$$

In terms of the cargo-stuck fault, the kinematic equation of the cargo sliding on the guide rail is:

$$\begin{cases} r_{oc} = \begin{cases} \int A_c t dt & r_{oc} < R_z \\ R_z & r_{oc} = R_z \end{cases} \\ A_c = \begin{cases} \ddot{r}_{oc}, & r_{oc} < R_z \\ 0, & r_{oc} = R_z \end{cases} \end{cases} \tag{9}$$

2.2 *Aircraft kinematics equation*

Consider the cargo as a mass point and only in longitudinal motion. The force relationship of the aircraft is shown in Figure 1.

In Figure 1, o is the center of mass of the aircraft, m_b is the mass of the aircraft, g is the acceleration of gravity, V is the flight speed of the aircraft, θ, γ and α represent the pitch angle, the climb angle and the angle of attack, respectively. The load on the aircraft mainly involves gravity G, lift L, air resistance D, thrust T, cargo pressure on the aircraft F_{cz} and friction F_{cx}, and the torque experienced mainly consists of pitching aerodynamic moment M_y, generated by aerodynamic force and cargo movement. Disturbance torque is M_c.

Because the height and speed of the airdrop changes little, the formula for calculating the engine thrust can be simplified as:

$$T = T_0 + T_m \delta_p \tag{10}$$

where T_m is the maximum thrust of the engine, δ_p is the throttle-opening coefficient.

Because the altitude of the aircraft during the airdrop process is not much changed, the aerodynamic forces and aerodynamic moments can be linearly approximated as in Equation 11:

$$\begin{cases} D = \bar{q}S[C_{D0} + C_{D\alpha}(\alpha - \alpha_0) + C_{D\delta_e}\delta_e] \\ L = \bar{q}S[C_{L0} + C_{L\alpha}(\alpha - \alpha_0) + C_{L\delta_e}\delta_e] \\ M_y = \bar{q}Sc_A[C_{m0} + C_{m\alpha}(\alpha - \alpha_0) + C_{mq}qc_A/2V + C_{m\delta_e}\delta_e] \end{cases} \tag{11}$$

where $\bar{q} = \rho V^2/2$ is the pull start pressure, S is the carrier wing area, δ_e is the elevator skewness, and C_D, C_L and C_m are the resistance, lift and pitching moment coefficients, respectively.

Considering the uncertainty factors in the actual situation, the airdrop model, Equation 10 can be transformed into the following form:

$$x = f(x) + \Delta f(x) + B(x)u \tag{12}$$

where:

$$x = [V, \alpha, q, \theta], \; u = [\delta_e, \delta_p], \; B(x) = [b_1, b_2] = \begin{bmatrix} b_{11} & b_{12} & b_{13} & b_{14} \\ b_{21} & b_{22} & b_{23} & b_{24} \end{bmatrix}^T, \; f(x) = [f_1, f_2, f_3, f_4]^T.$$

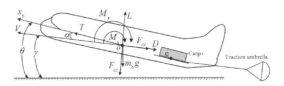

Figure 1.   Aircraft force analysis.

$$
\begin{cases}
f_1 = (m_b + m_c)^{-1}[-\bar{q}S(C_{D0} + C_{D\alpha}(\alpha - \alpha_0)) + T_0 \cos\alpha - m_b g \sin\gamma \\
\quad - m_c r_{oc} \sin\alpha f_3 - (\Lambda_1 \cos\alpha - \Lambda_2 \sin\alpha + F_p)] \\
f_2 = [(m_b + m_c)V]^{-1}[-T_0 \sin\alpha - \bar{q}S(C_{L0} + C_{L\alpha}(\alpha - \alpha_0)) \\
\quad + m_b g \cos\gamma - m_c r_c \cos\alpha f_3 + \Lambda_1 \sin\alpha + \Lambda_2 \cos\alpha)] \\
f_3 = \bar{q}S c_A \Lambda_3^{-1}[C_{m0} + C_{m\alpha}(\alpha - \alpha_0) + C_{mq} q c_A / 2V] \\
\quad + m_c r_{oc}[(m_b + m_c)\Lambda_3]^{-1}[F_p \sin\alpha - \Lambda_2 - m_b g \cos\gamma \cos\alpha \\
\quad + m_b g \sin\gamma \sin\alpha + \bar{q}S \sin\alpha(C_{D0} + C_{D\alpha}(\alpha - \alpha_0)) \\
\quad + \bar{q}S \cos\alpha(C_{L0} + C_{L\alpha}(\alpha - \alpha_0))] + r_c \Lambda_3^{-1}(\Lambda_2 - F_p \sin\alpha) \\
f_4 = q
\end{cases} \tag{13}
$$

$$
\begin{cases}
b_{11} = -(m_b + m_c)^{-1}(m_c r_{oc} \sin\alpha b_{31} + \bar{q}S C_{D\delta_e}) \\
b_{12} = -(m_b + m_c)^{-1} T_m \cos\alpha \\
b_{21} = -(m_b + m_c)^{-1}(m_c r_{oc} \cos\alpha b_{31} + \bar{q}S C_{L\delta_e}) \\
b_{22} = -[(m_b + m_c)V]^{-1} T_m \sin\alpha \\
b_{31} = \bar{q}S c_A C_{m\delta_e} \Lambda^{-1} + \bar{q}S m_c r_{oc}(m_b + m_c)^{-1}\Lambda_3^{-1}(C_{D\delta_e} \sin\alpha + C_{L\delta_e} \cos\alpha) \\
b_{32} = b_{41} = b_{42} = 0
\end{cases} \tag{14}
$$

$$
\begin{cases}
\Lambda_1 = m_c g \sin\theta + m_c q^2 r_{oc} - m_c \ddot{r}_{oc} \\
\Lambda_2 = m_c g \cos\theta - 2m_c q \dot{r}_{oc} \\
\Lambda_3 = I_y + m_c r_{oc}^2 - m_c^2 r_{oc}^2 / (m_b + m_c)
\end{cases} \tag{15}
$$

Because the aircraft is very close to the ground during an ultra-low-altitude airdrop, the outer-ring Proportional–Integral–Derivative (PID) control link is designed to adjust the flight altitude of the aircraft through the altitude change relationship $\dot{H} = V \sin(\theta - \alpha)$ to ensure the flight safety.

3   LINEARIZATION AND DECOUPLING

In Equation 12, the state quantities are coupled to each other. Therefore, Equation 12 needs to be decoupled and linearized by using the input and output precise linearization method. The system output is selected as $H(x) = [V \quad \theta]^T$ and the relative order is obtained separately.

For the output status V, we have:

$$
\begin{aligned}
L_{g_1} V &= [1 \quad 0 \quad 0 \quad 0]g_1 = g_{11} \neq 0 \\
L_{g_2} V &= [1 \quad 0 \quad 0 \quad 0]g_2 = g_{12} \neq 0
\end{aligned}
$$

Then, the relative order of state V is $\gamma_1 = 1$.

For the output status θ, one has:

$$
\begin{aligned}
L_{g_1}\theta &= [0 \quad 0 \quad 0 \quad 1]g_1 = g_{41} = 0; L_{g_1} L_f \theta = [0 \quad 0 \quad 1 \quad 0]g_1 = g_{31} \neq 0 \\
L_{g_2}\theta &= [0 \quad 0 \quad 0 \quad 1]g_2 = g_{42} = 0; L_{g_2} L_f \theta = [0 \quad 0 \quad 1 \quad 0]g_2 = g_{31} \neq 0
\end{aligned}
$$

Then, the relative order of state θ is $\gamma_2 = 2$.

Due to the total relative order $\gamma_1 + \gamma_2 = 3 < 4$, the system is controllable. Through feedback linearization theory, we have:

$$
[y_1^{(\gamma_1)} \quad y_2^{(\gamma_2)}]^T = [\dot{V} \quad \ddot{\theta}]^T = B + E(x)U \tag{16}
$$

where:

$$
B = [L_f^{\gamma_1} y_1(x) \quad L_f^{\gamma_2} y_2(x)]^T = [f_1 \quad f_3]^T \tag{17}
$$

58

$$E(x) = \begin{bmatrix} L_{g_1} L_f^{\gamma_1-1} y_1 & L_{g_2} L_f^{\gamma_1-1} y_1 \\ L_{g_1} L_f^{\gamma_2-1} h_m & L_{g_2} L_f^{\gamma_2-1} h_m \end{bmatrix} = \begin{bmatrix} g_{11} & g_{12} \\ g_{31} & g_{32} \end{bmatrix} \tag{18}$$

According to Equation 16, the following state feedback control law is designed as:

$$U = -E^{-1}(x)B(x) + E^{-1}(x)v \tag{19}$$

Substituting Equation 19 into Equation 16, the following dynamic equation is obtained:

$$[y_1^{(\gamma_1)} \quad y_2^{(\gamma_2)}]^{\mathrm{T}} = [v_1 \quad v_2]^{\mathrm{T}} \tag{20}$$

Then, the decoupling and linearization is completed. Hence, the system can be controlled to design $[v_1 \quad v_2]^{\mathrm{T}}$.

4 SLIDING-MODE CONTROLLER DESIGN

Let the expected outputs of V and θ be V_d and θ_d, respectively; then the tracking error of the system can be expressed as:

$$\begin{cases} e_1 = V - V_d \\ e_2 = \theta - \theta_d \end{cases} \tag{21}$$

According to Equation 21, the following sliding-surface function is designed as:

$$\begin{cases} S = [s_1 \ s_2 \ \cdots \ s_m]^{\mathrm{T}} \\ s_i = e_i^{(\gamma_i-1)} + c_{i(\gamma_i-1)} e_i^{(\gamma_i-2)} + \cdots + c_{i1} e_i \end{cases} \tag{22}$$

The parameters are adaptively selected by the fuzzy controller.
Substituting γ_1 and γ_2 into Equation 22 and taking the derivative, we get:

$$\begin{cases} s_1 = e_1 = V - V_d \\ \dot{s}_1 = \dot{e}_1 = \dot{V} = v_1 \end{cases} \tag{23}$$

$$\begin{cases} s_2 = \dot{e}_2 + ce_2 = \dot{\theta} - \dot{\theta}_d + c(\theta - \theta_d) \\ \dot{s}_2 = \ddot{e}_2 + c\dot{e}_2 = \ddot{\theta} - \ddot{\theta}_d + c(\dot{\theta} - \dot{\theta}_d) \end{cases} \tag{24}$$

Selecting the following exponential approach law:

$$\dot{S} = -\varepsilon \, \mathrm{sign}(S) - KS \tag{25}$$

and combining Equations 19–25, we derive the following sliding-mode control law:

$$\begin{cases} v_1 = -k_1 s_1 - \varepsilon_1 sign(s_1) \\ v_2 = -k_2 s_2 - \varepsilon_2 sign(s_2) + \ddot{\theta}_d - c(\theta - \theta_d) \end{cases} \tag{26}$$

where ε_1, ε_2 is the adaptive selection by the fuzzy controller.
Returning Equation 26 back to Equation 19 yields:

$$U = \begin{bmatrix} g_{11} & g_{12} \\ g_{12} & g_{32} \end{bmatrix} \begin{bmatrix} -f_1 - k_1 s_1 - \varepsilon_1 sign(s_1) \\ -f_3 - k_2 s_2 - \varepsilon_2 sign(s_2) + \ddot{\theta}_d - c(\theta - \theta_d) \end{bmatrix} \tag{27}$$

5 FUZZY CONTROLLER DESIGN

This section introduces the design of the fuzzy sliding-mode controller with adaptive sliding-mode surface slope, adaptive control-gain fuzzy controller and the double fuzzy sliding-mode variable structure controller.

5.1 Sliding-surface slope fuzzy controller design

When designing the fuzzy logic sliding-mode controller with adaptive sliding-mode surface slope, the slope of the sliding surface is calculated dynamically based on the state of the system and its derivative. The structure of the controller is shown in Figure 2.

The fuzzy controller input is the error and the derivative of the error. The membership function of the fuzzy input is shown in Figure 3a and 3b, using NB, NM, NS, Z, PS, PM and PB to represent Negative Big, Negative Medium, Negative Small, Zero, Positive Small, Positive Medium and Positive Big, respectively. Output is the negative value of the slope of the sliding surface, and the membership function of the output precision is shown in Figure 3c. Take the value of slope of the traditional sliding surface as VVS, VS, S, M, B, VB and VVB from left to right, representing Very Very Small, Very Small, Small, Medium, Big, Very Big and Very Very Big. To ensure the stability of the control system, the negative value of the slope of the sliding surface must be a positive number.

The fuzzy rule settings are given in Table 1.

The fuzzy rule in Table 1 is based on the sliding-mode function $s_i = c_i e_i + \dot{e}_i = 0$, the input error and the derivative of the error. The slope of the sliding-mode surface, which is closest to the system state in real time, ensures that the system can reach the sliding surface faster. To explain the fuzzy rules, three typical rules (of the red-marked values in Table 1) are expressed as follows:

1, if $e_i = NB$ and $\dot{e}_i = Z$ THEN $c_i = VVS$
2, if $e_i = NB$ and $\dot{e}_i = PB$ THEN $c_i = M$
3, if $e_i = Z$ and $\dot{e}_i = PB$ THEN $c_i = VVB$

In rule 1, when the error is negative and the derivative of the error is approximately zero, the negative value of the slope of the sliding surface is chosen to be a very small value to bring the sliding surface closer to the system. In rule 2, when the error is negative and the derivative of

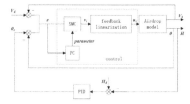

Figure 2. Structure diagram of the fuzzy sliding-mode controller.

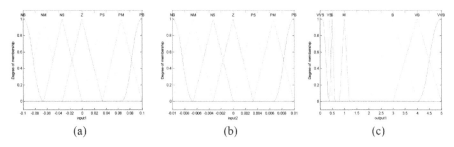

| (a) | (b) | (c) |

Figure 3. Membership function: (a) error; (b) derivative of error; (c) sliding-surface slope.

the error is positive, the negative value of the slope of the sliding surface is close to one, so that the sliding surface is close to the system. In rule 3, when the error is approximately zero and the derivative of the error is positive, the slope of the sliding surface is negative. Take a very large value to bring the sliding surface closer to the system. For the other rules, the appropriate slope of the sliding surface can also be selected to make the sliding surface follow the system.

5.2 Control-gain fuzzy controller design

This section introduces the design of the fuzzy sliding-mode controller with control-gain adaptation. The structure is the same as that shown in Figure 2. The input of the fuzzy controller mainly takes the system state error and the derivative of the error, and the time-varying control gain is the output through fuzzy logic reasoning.

The fuzzy input functions of membership are also shown in Figure 3a and 3b. The output membership function is shown in Figure 3c, where M is defined as a trigonometric function centered on the traditional control gain.

Table 1. Fuzzy rules of the fuzzy controller.

	de_i						
e_i	NB	NM	NS	Z	PS	PM	PB
NB	M	S	VS	VVS	VS	S	M
NM	B	M	S	VS	S	M	B
NS	VB	B	M	S	M	B	VB
Z	VVB	VB	B	M	B	VB	VVB
PS	VB	B	M	S	M	B	VB
PM	B	M	S	VS	S	M	B
PB	M	S	VS	VVS	VS	S	M

Table 2. Fuzzy rules for controlling gain.

	de_i						
e_i	NB	NM	NS	Z	PS	PM	PB
NB	M	B	VB	VVB	VB	B	M
NM	S	M	B	VB	B	M	S
NS	VS	S	M	B	M	S	VS
Z	VVS	VS	S	M	S	VS	VVS
PS	VS	S	M	B	M	S	VS
PM	S	M	B	VB	B	M	S
PB	M	B	VB	VVB	VB	B	M

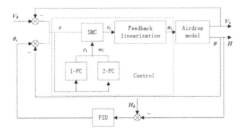

Figure 4. Structure diagram of the double fuzzy sliding-mode controller.

The fuzzy controller increases the control gain when the error is large, so that the system can reach the sliding surface faster. When the error is very small, the adjustment gain is especially small to avoid overshoot. In this way, fuzzy inference rules are established. The fuzzy rule settings are shown in Table 2.

When the system error is largely positive and the error derivative is approximately zero, the control gain takes a large value to accelerate the system state to the sliding surface. When the error is approximately zero and the error derivative is largely positive, the control gain should be taken to a very small value to avoid overshoot.

5.3 *Design of a double fuzzy sliding-mode variable structure controller*

Combining the two fuzzy controllers introduced in Sections 5.1 and 5.2, a dual fuzzy sliding-mode controller is proposed. The controller structure is shown in Figure 4. The fuzzy controller adjusts the sliding-mode surface slope and control gain in real time based on the system state error and its derivative.

6 SIMULATION

The following three kinds of fuzzy controllers are combined with the sliding-mode controller to carry out comparative simulation to verify the control effect of the designed controller. The initial conditions are selected as follows:

$$H_0 = 10m, V_0 = 80m/s, \alpha_0 = 3.9°, \theta_0 = 3.9°, m_b = 25t, m_c = 4t, l = 10m,$$
$$\mu = 0.02, T_0 = 32kN, T_M = 145kN.$$

The simulation results are shown in Figure 5.

Using a comparison analysis of the results in Figure 5a, 5b, 5c and 5d, it can be seen in the carrier response curves of the three fuzzy sliding-mode controllers—the overshoot, convergence speed and steady state error—that the control effect of the double fuzzy sliding-mode controller represented by the blue curve is obviously better than the adaptive slope fuzzy

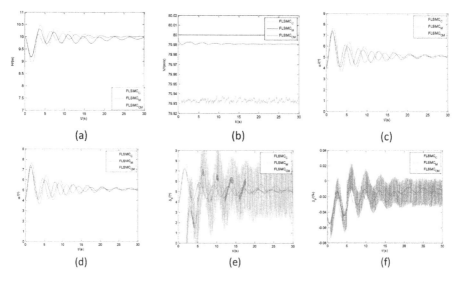

Figure 5. Simulation effect of three fuzzy sliding-mode controllers: (a) carrier height variation response curve; (b) carrier speed change response curve; (c) aircraft angle of attack change response curve; (d) carrier pitch angle change response curve; (e) carrier elevator rudder variation curve; (f) carrier throttle-opening curve.

sliding-mode controller represented by the red curve, and the adaptive control-gain fuzzy sliding-mode controller represented by the black curve. In Figure 5e and 5f, it can be seen that the control signal of the double fuzzy sliding-mode controller represented by the blue curve is smoother.

In summary, the double fuzzy sliding-mode controller can better solve the shortcomings of the sliding-mode controller with large chatter. In addition, it has the advantages of faster convergence speed and better convergence effect. The comparison shows that the designed controller has better dynamic response characteristics and achieves the expected control effect.

7 CONCLUSION

In this paper, a fuzzy controller with sliding-mode slope adaptive characteristics and a fuzzy controller with control-gain adaptability are designed to overcome the inherent shortcomings of the sliding-mode controller and improve its control effect. The designed sliding-surface slope adaptive fuzzy controller not only ensures the global robustness of the system but also has the advantage of being insensitive to unknown uncertainty. The designed control-gain adaptive fuzzy controller can eliminate the chatter of the control signal and accelerate the stability of the system. Combining the advantages of these two fuzzy controllers, a dual fuzzy sliding-mode controller is designed. Finally, a simulation analysis shows that the resulting double fuzzy sliding-mode controller has a better control effect on the cargo-stuck fault of an airdrop. In addition, it has good tracking performance for system speed and attitude angle and can eliminate the shortcomings inherent in the sliding-mode controller.

REFERENCES

Abdelhameed, M.M. (2005). Enhancement of sliding mode controller by fuzzy logic with application to robotic manipulators. *Mechatronics*, *15*, 439–458.

Dai, Z.S., Lu, Y.P. & Yang, Y. (2013). Fuzzy flight control for ultra-low altitude airdrop in fault state. *Ordnance Industry Automation*, *32*, 16–20.

Gurfil, P., Feldman, S. & Feldman, M. (2010). Coordination and communication of cooperative parafoils for humanitarian aid. *IEEE Transactions on Aerospace and Electronic Systems*, *46*, 1747–1761.

Hu, S.B. & Lu, M.X. (2013). Feedback linearization double fuzzy sliding mode control for multi-links robot. *Mechanical Science and Technology for Aerospace Engineering*, *32*, 105–115.

Li, G.Y. (2005). Status and trends of the large foreign military cargo planes. *Aeronautic Manufacturing Technology*, *12*, 36–43.

Liu, R., Sun, X.X. & Dong, W.H. (2013). Modeling of flight dynamics and analyzing of simulation for airdrop in atmospheric disturbance. *Flight Dynamics*, *31*, 24–28.

Liu, R., Sun, X.X. & Dong, W.H. (2015). Dynamics modeling and control of a transport aircraft for ultra-low altitude airdrop. *Chinese Journal of Aeronautics*, *28*, 478–487.

Ouyang, S.X. & Ding, C.S. (1992). The study on the dynamic characteristics of aircraft with cargos moving in its cargo cabin. *Flight Dynamics*, *10*, 77–86.

Sun, X.X., Xu, G.Z., Liu, R., Dong, W.H. & Qi, P.C. (2016). Dynamics model of airdrop process for air transportation cargo. *Journal of Traffic and Transportation Engineering*, *16*, 125–131.

Yagiz, N. & Hacioglu, Y. (2009). Robust control of a spatial robot using fuzzy sliding modes. *Mathematical and Computer Modelling*, *49*, 114–127.

Yang, Y. & Lu, Y.P. (2012a). Backstepping sliding mode control for super-low altitude heavy cargo airdrop from transport plane. *Acta Aeronautica et Astronautica Sinica*, *33*, 2301–2312.

Yang, Y. & Lu, Y.P. (2012b). Dynamics modelling for super low attitude parachute extraction system on transport airplane. *Journal of Nanjing University of Aeronautics and Astronautics*, *44*, 294–300.

Zang, J., Liu, H., Liu, T. & Ni, X. (2013). Object-oriented mission modeling for multiple transport aircraft. *International Journal of Aeronautical and Space Sciences*, *14*, 264–271.

Zhang, C., Chen, Z. & Wei, C. (2014). Sliding mode disturbance observer-based backstepping control for a transport aircraft. *Science China Information Sciences*, *57*, 1–16.

Automatic Control, Mechatronics and Industrial Engineering – He & Qing (Eds)
© 2019 Taylor & Francis Group, London, ISBN 978-1-138-60427-8

Guaranteed cost control for uncertain discrete switched systems by using sojourn-probability-dependent method

L.N. Wei, E.G. Tian & K.Y. Wang
School of Electrical Engineering and Automation, Nanjing Normal University, Nanjing, China

ABSTRACT: In this paper, the guarantee cost control problem is studied for a class of discrete-time switched system with state delays and parameters uncertainties. Different from the existing methods of analysis of switching systems, we construct a new type of switching system model by using sojourn probability information. Considering the uncertainties of the system parameters, sufficient conditions of both quadratic stability and the existence of guaranteed cost control law of the system are derived in terms of the linear matrix inequality method. Finally, a numerical example is given to illustrate the effectiveness of the proposed method.

Keywords: Discrete switched systems, Sojourn probability, Guarantee cost control

1 INTRODUCTION

For a practice system, stability is a basic requirement to ensure the system running. It cannot reflect the essential characteristics of the system, thus other cost indicators need to be considered too. Generally, the control system is required to reach closed-loop asymptotic stability and achieve a certain performance level. One way to deal with this requirement is to establish an integral quadratic cost function, which means the guarantee cost control (Chang SS L. 1972). In the actual system, the cost function constrains the control variables. And this requires the input variables not to exceed a certain value; otherwise it will waste energy and make it difficult to control. The guaranteed cost control can handle this class of problem, making the control system design consistent with current production process. Thus, it is necessary to study the guaranteed cost control of the systems.

A lot of achievements have been obtained in the study of the guaranteed cost control. The guaranteed cost control of a class of pulse switching systems with normal boundary uncertainty was considered (Xu H. 2008). It dealt with the reliable guaranteed cost control of delta operator-converted linear switching systems, which employed the moving average time method (Hu H. 2016). While the switching rules which have a significant impact on system performance levels are still commonly used, such as dwell time switching, Markov process switching, etc. For example, the researchers studied the problem of nonfragile guaranteed cost control for discrete time Takagi-Sugeno fuzzy Markovian jump system (Wu Z.G. 2017). The problem of robust nonfragile guaranteed cost control was proposed for a class of uncertain Markovian jump systems with time-varying delays (Liu G. 2017). But in some specific systems, the probability of process transitions in Markovian jump systems is very difficult or impossible to measure (H. Chen 2015). However, some researchers only studied the guarantee cost control system regardless of switched system (Chang SS L. 1972, Yu L. YL 1997). Some researchers studied the guarantee cost control of switched system but ignored the parameters uncertainties (Xu H. 2008, Hu H. 2016, and Wu Z.G. 2017). The guarantee cost control based on sojourn probability of uncertain discrete switched system has not been studied. This motivates the current study.

This paper studies the guaranteed cost control problem for a class of discrete switched systems with uncertainty parameter and state delay. Different from most existing methods, a new type of switching system model is constructed. Then a corresponding cost function is defined by using the sojourn probability information. Considering the system parameters uncertainties, sufficient conditions for the system to achieve quadratic stability and the existence of the guaranteed cost control law are given by using the linear matrix inequality method. Finally, a numerical example is used to verify the effectiveness of the proposed method.

2   PROBLEM FORMULATION

We consider a discrete time-delay switching system with parameters uncertainties

$$\begin{cases} x(k+1) = A_{\sigma(k)}(k)x(k) + A_{d\sigma(k)}(k)*x(k-d_{\sigma(k)}) + B_{\sigma(k)}u(k) \\ x(k) = \phi(k), k = -d, -d+1, \cdots, 0 \end{cases}, \tag{1}$$

where $x(k) \in R^n, u(k) \in R^m$, denotes the state, the input, respectively. $\phi(k)$ denotes system initial value. $\sigma(k): Z^+ = \{0,1,2,\cdots\} \to \Omega = \{1,2,\cdots,N\}$ denotes the system switching rules. $\sigma(k) = i \in \Omega$ denotes that the ith subsystem is the active subsystem. $d = \max\{d_i\}$ denotes the switching system delay, assuming it a time constant. Suppose that the delay of each subsystem is constant in this paper. $A_i(k), A_{di}(k), B_i, C_i, D_i$ denotes the parameter of the ith subsystem and satisfies $A_i(k) = A_i + \Delta A_i(k), A_{di}(k) = A_{di} + \Delta A_{di}(k)$, where $\Delta A_i(k)$ and $\Delta A_{di}(k)$ denotes the unknown uncertainty parameter and satisfies $[\Delta A_i(k)\ \Delta A_{di}(k)] = HF(k)[E_{1i}\ E_{2i}]$.

H, E_{1i}, E_{2i} denotes the known constant matrix with proper dimensions of the ith subsystem, and satisfies: $F^T(k)F(k) \le I$, where $F(k)$ is a unknown parameters matrix, i.e.

$$[A_i(k)\quad A_{di}(k)] = [A_i\quad A_{di}] + HF(k)[E_{1i}\quad E_{2i}]. \tag{2}$$

Consider a state feedback controller of the form

$$u(k) = K_{\sigma(k)}x(k), \tag{3}$$

where $K_{\sigma(k)}$ denotes the gain of the feedback controller to be designed, $\sigma(k)$ denotes switching rules. Substitute (3) into (1) and the following equations are derived

$$\begin{cases} x(k+1) = (A_{\sigma(k)} + B_{\sigma(k)}K_{\sigma(k)})x(k) + A_{d\sigma(k)}(k)x(k-d_{\sigma(k)}) \\ x(k) = \phi(k), k = -d, -d+1, \ldots, 0 \end{cases}. \tag{4}$$

Assumption 2.1: Assuming that the system sojourn probability in each subsystem is known, i.e.

$$P_r\{\sigma(k) = i\} = \alpha_i, i \in \Omega, \sum_{i=1}^{N} \alpha_i = 1,$$

where $\alpha_i \in (0,1)$ denotes the sojourn probability of the ith subsystem.

Remark 2.1 (Tian E. 2014): The switching rules of the Markovian jump systems depend on the current state and the information before jumping. But the switching rule based on the sojourn probability does not need that. The sojourn probability can be obtained by statistical method: $\alpha_i = \lim_{k \to \infty} \frac{k_i}{k}$, where k_i denotes the number of times of $\sigma(k) = i$ in the interval $[1, k]$ ($k \in \Omega$).

Definition 2.1: Define Bernoulli discrete variables $\alpha_i(k): Z^+ \to \{0,1\}$:

$$\alpha_i(k) = \begin{cases} 1, \sigma(k) = i \\ 0, \sigma(k) \ne i \end{cases}, i \in \Omega, k \in Z^+.$$

According to the relationship between $\sigma(k)$ and $\alpha_i(k)$, it can be obtained that for any $k \in Z^+$, $\mathbb{E}\{\alpha_i(k)\} = P_r\{\sigma(k) = i\} = \alpha_i$, where $\sum_{i=1}^N \alpha_i = 1, \sum_{i=1}^N \alpha_i(k) = 1$.

According to Assumption 2.1 and Definition 2.1, the system (4) can be converted into

$$\begin{cases} x(k+1) = \sum_{i=1}^N \alpha_i(k)\{(A_i(k) + B_i K_i)x(k) + A_{di}(k)x(k-d_i)\} \\ x(k) = \phi(k), k = -d, -d+1, \ldots, 0 \end{cases}. \tag{5}$$

Define that $\xi^T(k) = [x^T(k) \quad \psi^T(k,d_i)], y(k) = x(k+1) - x(k)$, where $\psi^T(k,d_i)$ can be expressed as $\psi^T(k,d_i) = [x(k = d_1) \quad \cdots \quad x(k-d_N)]$. Then the system (5) can be simplified as

$$\begin{cases} x(k+1) = \sum_{i=1}^N \alpha_i(k)\Gamma_{1i}\xi^T(k) \\ x(k) = \phi(k), k = -d, -d+1, \ldots, 0, \\ y(k) = \sum_{i=1}^N \alpha_i(k)\Gamma_{2i}\xi^T(k) \end{cases} \tag{6}$$

where $\Gamma_{1i} = [A_i(k) + B_i K_i \quad v_i], (v_i = [\ldots, A_{di}(k), \ldots]), \Gamma_{2i} = [A_i(k) - I + B_i K_i \quad v_i]$.

Supposing that the cost function for system (5) is

$$J = \sum_{k=0}^{\infty}[x^T(k)Qx(k) + u^T(k)Ru(k)], \tag{7}$$

where $Q \in R^{n \times n}, R \in R^{n \times m}$ denotes the given positive definite weighting matrix respectively.

Definition 2.2 (Hu S. 2012): For the system (5), if there is a feedback control law $u(k) = K_i x(k)$, the cost value of the cost function satisfies $J \leq J^*$, where J^* is a certain constant. J^* is called the cost upper bound of the system (5), and $u(k)$ is called the cost control laws of the system (5).

Definition 2.3: For a given constant $\gamma > 0$, symmetric positive definite matrices Q and R, if all of the uncertainty parameters satisfy equation (2), the following conditions are true:

1. When $\omega(k) = 0$, the closed-loop system (5) is asymptotically stable;
2. When $\omega(k) = 0$, the function (7) satisfies $J \leq J^*$, where J^* is a certain positive constant;

Then control law $u(k) = K_i x(k)$ is called the guaranteed cost control law of the system (5).

Lemma 2.1 (I. Petersen 1987): Given a matrix $W = W^T$, and matrices of appropriate dimensions H, E, then the inequality is true: $W + HF(k) + E^T F^T(k)H^T < 0$. Where a necessary and sufficient condition for all $F(k)$ satisfying $F^T(k)F(k) \leq I$ is that there is a positive number $\rho > 0$ such that the following formula holds: $W + \rho HH^T + \rho^{-1}E^T E < 0$.

3 MAIN RESULTS

3.1 *System guarantee performance analysis*

First, regardless of the parameters uncertainties, that is, having A_i, A_{di} instead of $A_i(k)$, $A_{di}(k)$, the following results can be obtained.

Theorem 3.1: For the given cost function (7) of the system (5), if there is positive definite matrix P and matrices $S_i, T_i (i \in \Omega)$ satisfying the following matrix inequalities

$$\begin{bmatrix} \Sigma_{11} & * & * \\ \Sigma_{21} & \Sigma_{22} & * \\ \Sigma_{31} & 0 & \Sigma_{33} \end{bmatrix} < 0, \tag{8}$$

where

$$\Sigma_{11} = \begin{bmatrix} \Pi_{11} & * & * & * \\ \Pi_{21} & \Pi_{22} & * & * \\ 0 & \Pi_{32} & -Q^{-1} & * \\ 0 & \Pi_{42} & 0 & \Pi_{44} \end{bmatrix}, \Pi_{11} = -P + \sum_{i=1}^{N} S_i - \sum_{i=1}^{N} T_i, \ \Pi_{21} = [T_1, T_2, \dots, T_N]^T,$$

$$\Pi_{22} = diag\{-S_1 - T_1, -S_2 - T_2, \dots, -S_N - T_N\}, \Pi_{32} = [I \quad 0], \Pi_{42} = [K_i \quad 0],$$

$$\Pi_{44} = diag\{-R^{-1}, \dots, -R^{-1}\}, \Sigma_{21} = \begin{bmatrix} \sqrt{\alpha_1}\tilde{\Gamma}_{11} \\ \sqrt{\alpha_2}\tilde{\Gamma}_{12} \\ \vdots \\ \sqrt{\alpha_N}\tilde{\Gamma}_{1N} \end{bmatrix}, \Sigma_{31} = \begin{bmatrix} d_1\mathbb{N} \\ d_2\mathbb{N} \\ \vdots \\ d_N\mathbb{N} \end{bmatrix}, \ \mathbb{N} = \begin{bmatrix} \sqrt{\alpha_1}\tilde{\Gamma}_{21} \\ \sqrt{\alpha_2}\tilde{\Gamma}_{22} \\ \vdots \\ \sqrt{\alpha_N}\tilde{\Gamma}_{2N} \end{bmatrix},$$

$$\tilde{\Gamma}_{1i} = [A_i + B_iK_i \quad \tilde{v}_i \quad 0 \quad 0], \ \tilde{v}_i = [\dots, A_{di}, \dots], \ \tilde{\Gamma}_{2i} = [A_i - I + B_iK_i \quad \tilde{v}_i \quad 0 \quad 0],$$

$$\Sigma_{22} = diag\{-P^{-1}, -P^{-1}, \dots, -P^{-1}\}, \ \Sigma_{33} = diag\{-T_1^{-1} \dots -T_1^{-1}, \dots, -T_N^{-1} \dots -T_N^{-1}\}$$

Then the system (5) is asymptotically stable, and the cost function (7) has an upper bound, and

$$J \le J* = x^T(0)Px(0) + \sum_{i=1}^{N} x^T(-d_i)S_ix(-d_i) + d_i \sum_{i=1}^{N} \sum_{S=-d_i}^{-1} y^T(s)T_iy(s). \tag{9}$$

And the proof process refers to references (Yuan Yue-hua 2014).

3.2 *System guaranteed cost control*

The following step gives the condition that the system (5) has guaranteed cost control. Unlike the Theorem 3.1, we consider the parameters uncertainties in the constructed switching system model in this part.

Theorem 3.2: For system (5), if there exists a feedback gain matrix K_i, symmetric matrices $P>0, S_i>0, T>0$ and a set of positive numbers $\mu_i(i \in \Omega)$, satisfying the following inequality

$$\begin{bmatrix} \Omega_{11} & * \\ \Omega_{21} & \Omega_{22} \end{bmatrix} < 0, \tag{10}$$

where

$$\Omega_{11} = \begin{bmatrix} \Sigma_{11} & * & * \\ \Sigma_{21} & \Sigma_{22} & * \\ \Sigma_{31} & 0 & \Sigma_{33} \end{bmatrix}, \Omega_{21} = [\sqrt{\alpha_1}\gamma_1 \quad \sqrt{\alpha_2}\gamma_2 \quad \cdots \quad \sqrt{\alpha_N}\gamma_N]^T,$$

$$\gamma_i = \begin{bmatrix} 0 & H^i & d_1H^i & \cdots & d_NH^i \\ \Delta^i & 0 & 0 & \cdots & 0 \end{bmatrix}, H^i = [0 \dots \mu_iH \dots 0], \Delta_i = [E_{1i} 0 \dots E_{2i} \dots 0],$$

$$\Omega_{22} = diag\{-\mu_1I, -\mu_1I, -\mu_2I, -\mu_2I, \dots, -\mu_NI, -\mu_NI\},$$

then it can be seen that the system (5) is robust and mean square stable, and $u(k) = K_i\chi(k)$ is the guaranteed cost control law of the system (5). The function (7) has an upper bound and satisfies

$$J \le J^* = x^T(0)Px(0) + \sum_{i=1}^{N} x^T(-d_i)S_ix(-d_i) + d_i \sum_{i=1}^{N} \sum_{S=-d_i}^{-1} y^T(s)T_iy(s).$$

68

And the proof process refers to references (Yuan Yue-hua 2014).

Remark 3.1: For the existence of the inverse of the variable in the inequality, equations (8) and (10) are not strictly linear matrix inequalities. Therefore, we will use a cone complementarity linearization method in Laurent EI (Chaoui L.E. 1997) to find the gain K_i of the system.

The following algorithm can be derived by using the cone complementarity linearization method.

Algorithm 3.1:

Step 1: Give a set of initial values $\alpha_i, d_i (i \in \Omega)$;

Step 2: Having $\bar{P}, \bar{T}_i (i \in \Omega)$ instead of P^{-1}, T_i^{-1} in equation (8), (10), respectively, we obtain a new inequalities (8)', (10)'. Find a set of feasible solutions $(P, \bar{P}, T_i, \bar{T}_i)$ that satisfying (8)' and (10)',

$$\begin{bmatrix} P & I \\ I & \bar{P} \end{bmatrix} > 0, \begin{bmatrix} T_i & I \\ I & \bar{T}_i \end{bmatrix} > 0 (i \in \Omega). \tag{11}$$

Let $m = 1$;

Step 3: Solve the minimum problem: $\min tr(P_m \bar{P} + \bar{P}_m P + \sum_{i=1}^{N} (T_{im} \bar{T}_i + \bar{T}_{im} T_i))$, and solve the variables $(P, \bar{P}, T_i, \bar{T}_i (i \in \Omega))$ of inequalities (8)', (11) or (10)', (11);

Step 4: If (12) holds for any sufficiently small variable $\varepsilon > 0$, the calculation is completed. Otherwise, let $m = m+1$, and then judge: if $m < c$ (the number of iteration steps given in advance), return to step 2; if $m = c$, end.

$$\{tr(P\bar{P} + \sum_{i=1}^{N} (T_i \bar{T}_i)) - (1 + N)n\} < \varepsilon, \tag{12}$$

where n is the dimension of the square matrix P;

Step 5: Adjust $\alpha_i, d_i (i \in \Omega)$, and repeat steps 1 to 4;

Step 6: Output the control gain K_i.

4 ILLUSTRATIVE EXAMPLE

Considering the switching system (5) with two sub-systems and the system cost function (7), the parameters are as follows,

$$A_1 = \begin{bmatrix} -0.52 & -0.1 \\ 0 & 1.21 \end{bmatrix}, A_2 = \begin{bmatrix} 0.57 & 0.29 \\ 0.21 & 0.40 \end{bmatrix}, A_{d1} = \begin{bmatrix} 0.1 & -0.05 \\ 0 & 0.023 \end{bmatrix}, A_{d2} = \begin{bmatrix} -0.0125 & -0.023 \\ 0 & 0.014 \end{bmatrix}$$
$$I = \begin{bmatrix} 1 & 0 \\ 0 & 1 \end{bmatrix}, B_1 = \begin{bmatrix} 0.1 & 0 \\ 0 & 0.2 \end{bmatrix}, B_2 = \begin{bmatrix} 0.2 & 0.2 \\ 0 & 0.1 \end{bmatrix}, Q = \begin{bmatrix} 1.16 & -0.25 \\ -0.25 & 0.83 \end{bmatrix}, R = \begin{bmatrix} 1.47 & 0 \\ 0 & 1.47 \end{bmatrix}.$$

Taking $d_1 = d_2 = 4, \alpha_1 = 0.3, \alpha_2 = 0.7$, the controller gain can be obtained by Theorem 3.1,

$$K_1 = \begin{bmatrix} -0.0371 & -0.0050 \\ -0.0269 & -0.0100 \end{bmatrix}, K_2 = \begin{bmatrix} -0.8013 & -0.6552 \\ -0.2531 & -0.0274 \end{bmatrix}.$$

When define the initial condition $\phi(k) = [0.3 \quad 0.1]^T$, the system state response curve obtained by using the controller gain is shown in Figure 1. The cost upper boundary is $J^* = 0.1652$.

Now considering the parameters uncertainties, the system parameters are as follows

$$H = \begin{bmatrix} 0.8 & 0 \\ 0.4 & 0.6 \end{bmatrix}, E_{11} = \begin{bmatrix} 0.02 & 0.03 \\ 0.01 & 0.02 \end{bmatrix}, E_{12} = \begin{bmatrix} 0.02 & 0.03 \\ 0.05 & 0.02 \end{bmatrix}, E_{21} = \begin{bmatrix} 0.01 & 0.04 \\ 0.02 & 0.02 \end{bmatrix}, E_{22} = \begin{bmatrix} 0.02 & 0.02 \\ 0.03 & 0.04 \end{bmatrix}.$$

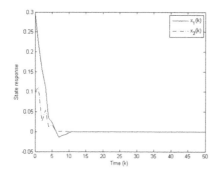 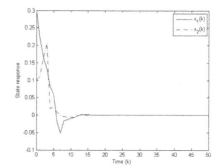

Figure 1. Switching system state response curve without parameters uncertainties.

Figure 2. Switching system state response curve with parameters uncertainties.

Taking $d_1 = d_2 = 4$, and $\alpha_1 = 0.3, \alpha_2 = 0.7$, the controller gain can be found by Theorem 3.1,

$$K_1 = \begin{bmatrix} -0.0126 & -0.0121 \\ -0.0134 & -0.0248 \end{bmatrix}, K_2 = \begin{bmatrix} -0.0134 & -0.0068 \\ -0.0635 & -0.0076 \end{bmatrix}.$$

When taking the initial condition $\phi(k) = \begin{bmatrix} 0.3 & 0.1 \end{bmatrix}^T$, the system state response curve obtained by using the controller gain is shown in Figures 2. The cost upper boundary is $J^* = 0.1704$.

From Figure 1 and Figure 2, one can see clearly is that the switching system with parameters uncertainties can achieve almost the same stability with the system regardless of that. That is, the method used in this paper is effective. The designed guaranteed cost control law can ensure the favorable control performance of the whole system in the case of subsystem delay.

5 CONCLUSION

This paper focuses on a class of discrete time-delay switching systems with parameters uncertainties, and studies the problem of guaranteed cost control. By using the sojourn probability information method, we have constructed a new switching system model, and defined the corresponding cost function. By using the linear matrix inequality method, we can obtain the sufficient conditions for the system quadratic stability and the existence of the guaranteed cost control law when considering the system parameters uncertainties. The simulation results have been used to verify that the method can meet the expected performance of the system. For future work, we will build upon the presented results to adaptive event-trigger control system.

REFERENCES

Chang SS L., Peng TK C. 1972. Adaptive guaranteed cost control of systems with uncertain parameters. IEEE Transaction on Automatic Control, 17(4): 474–483.

Chen, H., M. Liu and S. Zhang. 2015. Robust Finite-time control for discrete markovian jump systems with disturbances of probabilistic distributions. Entropy, 17: 346–367.

Ghaoui L.E., Oustry F., Aitrami M. 1997. A cone complementarity linearization algorithm for static output-feedback and related problems. IEEE Transactions on Automatic Control, 42(8): 1171–1176.

Hu H., Jiang B., Yang H. 2016. Reliable guaranteed-cost control of delta operator switched systems with actuator faults: mode-dependent average dwell-time approach. Control Theory & Applications IET, 10(1):17–23.

Hu S., Yin X., Zhang Y., et al. 2012. Event-triggered guaranteed cost control for uncertain discrete-time networked control systems with time-varying transmission delays. Control Theory & Applications, 6(18):2793–2804.

Liu G., Wei Y., Ma Q., et al. 2017. Robust non-fragile guaranteed cost control for singular Markovian jump time-delay systems. Transactions of the Institute of Measurement & Control, 2017: 014233121769614.

Petersen, I. 1987. A stabilization algorithm for a class of uncertain linear systems. System & Control Letters, 8(4): 351–357.

Tian E., Yue D., Yang T.C. 2014. Analysis and synthesis of randomly switched systems with known sojourn probabilities. Information Sciences, 277(2):481–491.

Wu Z.G, Dong S., Shi P., et al. 2017. Fuzzy-Model-Based Nonfragile Guaranteed Cost Control of Nonlinear Markov Jump Systems. IEEE Transactions on Systems Man & Cybernetics Systems, PP(99):1–10.

Xu H., Teo K.L., Liu X. 2008. Robust stability analysis of guaranteed cost control for impulsive switched systems. IEEE Press.

Yu L. YL., Wang J. WJ, Chu J. CJ 1997. Guaranteed cost control of uncertain linear discrete-time systems. 5:3181–3184.

Yuan Yue-hua, Yue Dong, Tian En-gang. 2014. Robust H-infinity control for discrete switched systems by using sojourn-probability-dependent method. Control Theory & Applications, 31(2), 175–180.

Automatic Control, Mechatronics and Industrial Engineering – He & Qing (Eds)
© 2019 Taylor & Francis Group, London, ISBN 978-1-138-60427-8

DNN-HMM-based automatic speech recognition system for intelligent LED lighting control

J.L. Xian, W.X. Cai, H.X. Pan, N.Z. Chen & X.Y. Chen
School of Physics and Optoelectronic Engineering, Foshan University, Guangdong, China

Y.W. Sun
Midea Kitchen Appliances Division, Guangdong, China

D. Yan
School of Physics and Optoelectronic Engineering, Foshan University, Guangdong, China

ABSTRACT: This paper presents an Automatic Speech Recognition (ASR) system that can be efficiently used for personalized lighting control. Using a Deep Neural Network-Hidden Markov Model (DNN-HMM) algorithm, the system implements the dimming, color temperature and color adjustment of Light-Emitting Diodes (LEDs) according to 31 sets of human command words. Tests were conducted at distances of 20 cm, 50 cm and 100 cm in quiet and noisy environments. Results show that the mean accuracy of speech recognition is 94% and the response time is 0.7 s, which indicates good potential for commercialization.

Keywords: Automatic speech recognition, DNN-HMM, LED, lighting control

1 INTRODUCTION

Light-Emitting Diodes (LEDs) have been widely used for general illumination purposes such as home lighting, office lighting, traffic lighting, and so on (Schubert, 2006; Krames et al., 2007; Chen et al., 2012; Pimputkar, 2009). Controlling lights typically requires human interfaces that incorporate the use of standard devices such as hand- or finger-actuated switches or knobs and touch plates. For colored lights, additional dimensions of control complicate this. With the popularization of smart illumination (Schubert & Kim, 2005), Automatic Speech Recognition (ASR) can be a replacement for traditional control (Wang et al., 2016; Lei et al., 2014; Wang et al., 2011; Rabiner & Juang, 1993; Yu & Li, 2016). This paper deals with a new type of intelligent speech recognition system for modification of LED switching and control of hue, dimming and color temperature. It is based on an embedded offline recognition engine and uses a Deep Neural Network-Hidden Markov Model (DNN-HMM) algorithm for deep learning. The ASR system shows nearly real-time response and stable localized voice recognition.

2 EXPERIMENT

As shown in Figure 1, the ASR system for lighting control includes an ASR module, an LED driver module, LED lamps and a power supply module. When a user gives particular command words, the ASR module responds immediately, delivering the processing signal to the LED drive module to control the switch, color, dimming and color temperature.

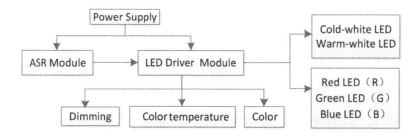

Figure 1.   ASR system for intelligent LED lighting control.

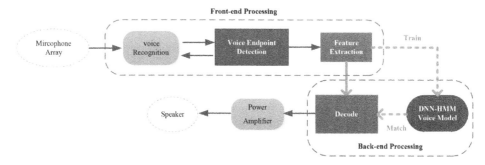

Figure 2.   Principle of automatic speech recognition module.

2.1 *Principle of the ASR system*

As shown in Figure 2, the ASR module includes a microphone array, a front-end processing module, a back-end processing module and a loudspeaker. The front-end processing module and the back-end processing module are the core of the speech recognition system. A Chip-intelli speech recognition chip is integrated inside. The noise-reduction module receives the phase difference from the microphone array, and then filters the sound signals to minimize background noise. The noise-reduced signal is used for speech endpoint detection in the ASR module. Feature extraction is conducted through the DNN-HMM model. The extracted features are matched with the speech database after decoding. Finally, the output is converted to analog signals through the power amplifier circuit and the speaker.

Feature extraction is a key part of the speech recognition system. The feature is displayed in the envelope of the short-term power spectrum of the speech. Mel-Frequency Cepstral Coefficients (MFCCs) are based on the human auditory model. The frequency axis is converted from linear scale to mel frequency in the frequency domain, and is then switched to the cepstrum domain to obtain cepstral coefficients, which will accurately describe the characteristics of this envelope. The specific relationship between the mel frequency and the actual frequency is:

$$Mel(f) = 2595 \times \log_{10}\left(1 + \frac{f}{700}\right) \qquad (1)$$

Pattern matching is the core of the speech recognition system. The feature vector of the input speech is compared with the template library. The highest similarity is output as the recognition result. DNN models the posterior probability of the input signal. Then, it combines the posterior probability with the prior probability to obtain the observed probability of the state. The input sample is x and the output state is s; $P(s|x)$ represents the output of the DNN posterior probability, namely:

$$P(x|s) = \frac{P(s|x)P(x)}{P(s)} \qquad (2)$$

where $P(s)$ is the prior probability of the modeling unit, and $P(x)$ is the prior probability of the observed sample. The observation probability can be decoded in combination with HMM and language models.

2.2 LED driver module

As shown in Figure 3, the LED drive module takes the Microcontroller Unit (MCU) as the core. When it receives command signals from the ASR module, the MCU will control the LED lamps according to the set program. It then uses the Pulse-Width Modulation (PWM) method to mix colors to realize the desired brightness, color temperature, and color. In addition, the LED driver circuit provides a suitable current for the LED array.

2.2.1 Principle of color temperature, dimming and color adjustment

Due to the visual persistence of the human eye, when the LED blinking frequency is higher than 85 Hz, it will be considered to be constantly bright. Thus, by turning LEDs on and off, we can change the duty cycle to adjust the brightness. The output voltage changes accordingly:

$$U_{O\_AVG} = U_p \cdot D \tag{3}$$

$$D = \frac{t_{on}}{t_{on} + t_{off}} \tag{4}$$

where U_{O_AVG} is the average output voltage, D is the duty cycle, U_p is the voltage amplitude of PWM voltage, t_{on} is the on-time, and t_{off} is the off-time. Based on Equations 3 and 4, it can be concluded that if on-time is increased, the duty ratio will be increased accordingly. When the amplitude of the voltage stays unchanged, the output voltage will increase and the lamp will become brighter.

The system uses common cathode RGB (Red/Green/Blue) LED lamps. The illuminating color can be converted into chromaticity coordinates or tristimulus values, corresponding to the standard chromaticity diagram of the CIE (International Commission on Illumination). The PWM method is used to adjust the duty ratio of RGB, so that different colors are mixed. Similarly, the color temperature can be adjusted according to the ratio of the warm and cold LED lamps. The warm-color temperature LED lamp used is 4000 K, and the cool-color temperature is 6500 K.

2.2.2 Principle of the LED drive circuit

Figure 4 shows a schematic diagram of part of the drive circuit. The PWM control signal is output by the MCU. The load terminal is connected by five LEDs; the middle part is the LED driver circuit. Among them, Q1 is a switch tube, used to protect the MCU.

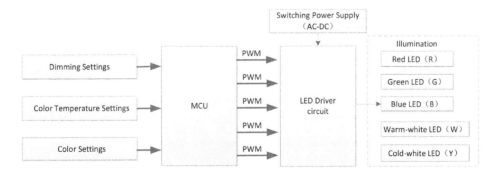

Figure 3.   Principle of the LED driver module.

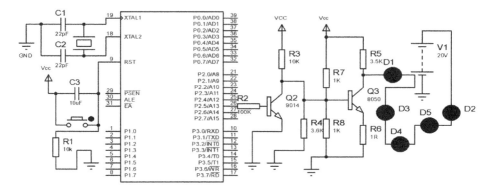

Figure 4. Schematic diagram of part of the drive circuit.

A common-emitter amplifier circuit with Q2 is used to amplify the current, achieving the working current of the LED.

3 RESULTS AND ANALYSIS

3.1 *Adjustment of lighting brightness, color temperature and color*

The system realizes the communication between the ASR module and the MCU. The LEDs are controlled by command signals from the MCU, there by realizing the control of dimming, color temperature and color of illumination.

As shown in Figure 5, a–f are the six dimming levels captured in a dark environment, which darken from left to right. When the speech recognition module received a voice command such as 'Brighter', 'Darker', 'Minimum brightness' or 'Maximum brightness', the PWM circuit was adjusted to change the brightness of the LED lamps.

As shown in Figure 6, a–f are the six color temperature levels captured in a dark environment, which warm from left to right. When the speech recognition module received command terms such as 'Natural white' or 'Warm white', the duty cycle of the warm and cold LED lamps was adjusted to reach the corresponding color temperature level.

Figure 7 shows pictures of the 16 colors from the system in a dark environment. When the ASR module recognized a corresponding color command, such as 'Correctred' or 'Lightgreen', the system adjusted the RGB LED duty cycles to mix the color, thus obtaining 16 different lighting colors.

3.2 *The response and speech recognition rate test*

Ten participants were invited into the experiment. There were 31 sets of command words, and each was spoken three times at a distance of 20 cm, 50 cm, and 100 cm. The experimental environment was not absolutely quiet and simulated everyday use.

Response time is the main factor affecting the speech recognition system experience. The response time was measured from the end of the human speech to the beginning of the loudspeaker response. Currently, the response time of network speech recognition in the industry is about 1.5 s. In this test, the response time was 0.6 seconds, and 0.7 seconds in the noisy environment. The result shows that the local scheme used in our system involves less delay than network speech recognition.

The statistical results of the recognition rate test are shown in Table 1. The 31 sets of command words were divided into four types: wake-up, dimming, color temperature and color. The mean accuracy of speech recognition was 94%. The first three types of recognition rate test results were higher than this, but the color recognition was relatively poor.

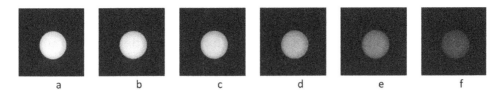

Figure 5.   Picture of LED brightness adjustment.

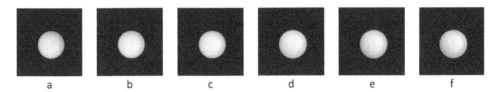

Figure 6.   Picture of LED color temperature adjustment.

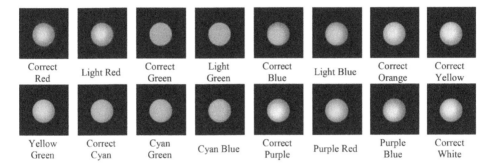

Figure 7.   Picture of LED color adjustment.

Table 1.   Statistical recognition rate of command words (%).

Command word	Distance (cm)		
	20	50	100
Wake-up	100	94	90
Dimming	99	100	90
Color temperature	100	100	93
Color	92	85	82

It was found that at a distance of 20 cm, the recognition rate of partially adjusted colors was less than 80%, and the recognition rates of 'Light blue' and 'Correct yellow' were 78%. This is because the Chinese sounds of 'light' and 'green', 'yellow' and 'red' are similar to a certain extent, and the system was prone to errors when they were spoken.

At a distance of 50 cm, the recognition rates of 'Correct red', 'Correct green', 'Light blue', 'Correct yellow' and 'Correct Purple' were below 80%. As the distance increased, the noise effect was more obvious and recognition errors increased. In the case of a distance of 100 cm, the recognition rate dropped slightly more: the recognition rate was below 80% for command words such as 'Maximum brightness' in dimming, and most ('Correct red', 'Light green', 'Light blue', 'Correct orange', 'Light green', 'Light blue', 'Correct yellow', 'Yellow green', 'Correct cyan', 'Correct purple' and 'Fruit purple') of the color adjustment type.

When analyzing recognition errors, we found that they were concentrated among words such as 'correct', 'light', 'green', 'yellow' and 'red'. The distance factor was a possible cause for the decrease in the recognition rate. With an increase of working distance, background noise plays a more important part in all the signals acquired by the microphone array and the recognition rate deteriorates. Further, because most of the testers had a Cantonese accent in their spoken Mandarin, this may be another reason for the error in the recognition of these pronunciations.

These minimization of these noise and dialect effects will be optimized in our sub sequent research. We will add dialect recognition to improve the recognition rate, which will also be improved by incorporating more samples into our speech library.

4 CONCLUSION

This paper presented an intelligent speech recognition system for LED lighting control. It realized human–machine voice interaction, thus controlling the switching, dimming, color temperature and color of LED lamps. It has great significance in solving the intelligent illumination control problem and promotes the commercialization of intelligent lighting based on voice control.

ACKNOWLEDGMENTS

This study was conducted under Special Funds for the Cultivation of Scientific and Technological Innovation for College Students (GJ41301), the National Natural Science Foundation of China (61704027), the Cooperative Foundation for Production, Teaching and Research from the Ministry of Education (201701027003) and the Scientific Research Starting Foundation of Foshan University (GG040928).

REFERENCES

Chen, K.J., Chen, H., Tsai, K., Lin, C., Tsai, H., Chien, S., … Kuo, H. (2012). Light-emitting devices: Resonant-enhanced full-color emission of quantum-dot-based display technology using a pulsed spray method. *Advanced Functional Materials*, *22*(24), 5138–5143.

Krames, M.R., Shchekin, O.B., Mueller-Mach, R., Mueller, G.O., Zhou, L., Harbers, G. & Craford, M.G. (2007). Status and future of high-power light-emitting diodes for solid-state lighting. *Journal of Display Technology*, *3*(2), 160–175.

Lei, Y., Scheffer, N., Ferrer, L. & McLaren, M. (2014). A novel scheme for speaker recognition using a phonetically-aware deep neural network. In *IEEE International Conference on Acoustics, Speech and Signal Processing (ICASSP 2014), Florence, 4–9 May 2014*. New York: IEEE. doi:10.1109/ICASSP.2014.6853887.

Pimputkar, S. (2009). Prospects for LED lighting. *Nature Photonics*, *3*(4), 180–182.

Rabiner, L. & Juang, B.H. (1993). *Fundamentals of speech recognition*. London: Prentice-Hall International.

Schubert, E.F. (2006). *Light-emitting diodes*. London: Cambridge University Press.

Schubert, E.F. & Kim, J.K. (2005). Solid-state light sources getting smart. *Science*, *308*(5726), 1274–1278.

Wang, X., Takaki, S. & Yamagishi, J. (2016). Investigating very deep highway networks for parametric speech synthesis. In *Proceedings of 9th ISCA Speech Synthesis Workshop, 13–15 September 2016, Sunnyvale, California* (pp. 166–171). doi:10.21437/SSW.2016-27.

Wang, S.R., Huang, S. & Yuan, F. (2011). Design and implementation of speech recognition system based on SPCE061A. *Advanced Materials Research*, *187*, 389–393.

Yu, D. & Li, D. (2016). *Automatic speech recognition: A deep learning approach*. London: Springer.

Automatic Control, Mechatronics and Industrial Engineering – He & Qing (Eds)
© 2019 Taylor & Francis Group, London, ISBN 978-1-138-60427-8

Rotary inverted pendulum system tracking and stability control based on input-output feedback linearization and PSO-optimized fractional order PID controller

Y. Yang, H.H. Zhang & R.M. Voyles
School of Engineering Technology, Purdue University, West Lafayette, Indiana, USA

ABSTRACT: This paper is aimed to propose a controller which can achieve the tracking control of the rotary arm and the stability control of the inverted pendulum in the Rotary Inverted Pendulum (RIP) system. An Input-Output Feedback Linearized (IOFL) RIP system dynamics can be divided into two separate portions: one is the linearized dynamic portion associated with the pendulum arm, and the other is the un-linearizable portion associated with the rotary arm. For the unlinearizable portion, a fractional order $PI^\lambda D^\mu$ (FOPID) controller is implemented to enhance the tracking control of the rotary arm with controller parameters determined from solving a prescribed optimization problem using Particle Swarm Optimization (PSO) algorithm. For the linearized portion, the state feedback control gains are obtained by conventional pole-placement method. It is revealed that a much more robust and effective tracking and stabilizing performance can be guaranteed with the FOPID involved in the controller design when compared with a previously designed synthesized IOFL controller for the RIP system. The new FOPID controller developed in this paper for the RIP system can be used as a reference to the tracking and stabilizing control of other complicated dynamic systems since the RIP system is a testbed for controller design of the fourth order Single-Input-Multiple-Output (SIMO) dynamic system.

1 INTRODUCTION

Rotary inverted pendulum (RIP) is an electro-mechanical system which is composed of two linked arms placed on a rigid stand. RIP system has two degree of freedoms and it has unique value to serve as a test platform for evaluating nonlinear controllers.

Recent researchers have focused their attention to the modeling of dynamic system using fractional order calculus, mainly in the cases of the system transients, such as switching on/off, as well as the other essential features that conventional methods and transforms fail to capture. For example, a comparison study of integer and fractional order ultra-capacitor models in Dzieliński et al. (2011) showed that the fractional order description of the ultra-capacitor dynamics validates a good match with the experimental data points. In the same way, a fractional calculus model was presented for the complex dynamics in biological tissues, and demonstrated its high precision and conciseness of fractional order model in bioengineering research areas (Magin 2010). Meanwhile, Benson et al. (2013) applied fractional calculus in hydrologic modeling and its simulation, whose results implied the natural existence of fractional order dynamic systems. Similarly, the fractional order models of basic mechanical and electrical elements, such as the spring, damper, capacitor, resistor and inductor, are developed by utilizing fractional order calculus theories (Magin 2006). The fractional order model can provide much comprehensive information of the original system since it takes account of the transitional or interim states between two integer order states. However, due to the ambiguity and the deficient understanding of the physical meanings behind the fractional order differentiation and integration, it is challenging to formulate fractional order mathematical principles towards the natural phenomena explicitly. Instead of proposing a list of assumptions to force the system dynamics to be fractional, and designing a fractional order controller based on the fractional order system, a novel approach is to directly apply a fractional order controller to an integer order system dynamics. The fractional order controller involves more parameters to be adjusted, for example, for the $PI^\lambda D^\mu$ controller, not only the proportional gain, derivative gain and integral gain need to be tuned, the fractional order index λ and μ also require to be properly set. The flexibility of adjusting more parameters of the fractional order controller implies more chance of making the overall system to deliver a better performance.

Over the past decades, several common dynamic systems which are hard to be expressed in the form of fractional order differential equations are explored by researchers. Podlubny (1998) studied the step response of the $PI^\lambda D^\mu$ (FOPID) controller in a LTI system plant, the results give the evidence that FOPID has higher robustness than the traditional PID controller. Correspondingly, the cases when the FOPID controller is applied to nonlinear integer order systems are also investigated. For instance, the fractional order parameters λ and μ are adjusted by trial and error to control a single-link RIP system (Liao et al. 2015). Wan et al. (2015) proposed the adjustable fractional order adaptive control (AFrAC) with which the fractional order β can be dynamically determined through optimization. Dwivedi et al. (2017) incorporated the frequency domain analysis to fractional order controller design processes. All these methods prevail only in some specific dynamic systems, but unfortunately, these fractional controller design methods lose their effectiveness when they are applied to the RIP system.

Previously, a synthesized input-output feedback linearization (IOFL) controller is presented to achieve the stabilization of the RIP system, but its tracking control performance is not investigated (Yang & Zhang 2018). As a continuous work, this paper will focus on RIP system tracking and stability control. For the nonlinear IOFL transformed subsystem, a $PI^\lambda D^\mu$ is presented to ensure the tracking performance of the rotary arm. The fractional order parameters of the $PI^\lambda D^\mu$ are optimized with the Particle Swarm Optimization (PSO) algorithm (Chong & Zak 2012). Meanwhile, The Ourstaloup filter is introduced to approximate the FOPID operator.

2 MODELING OF THE ROTARY INVERTED PENDULUM SYSTEM

2.1 *Dynamics and the affine form of the rotary inverted pendulum system*

The rotary inverted pendulum (RIP) system is a single-input-multi-output (SIMO) dynamic system. The structure of the RIP system is quite simple: a servomotor is mounted on a stand stage, the rotary arm is positioned horizontally with one end attached on the motor and the other end of the rotary arm is hinged with the end of the pendulum arm. The motor will drive the rotary arm to circularly move and the rotational movement of the rotary arm will force the swaying of the pendulum arm. The structure of the RIP system is simple with physical model and the coordinate system shown in Figure 1. The detailed description of the coordinate system has been given in Yang & Zhang (2018).

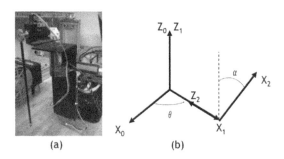

(a)　　　　　　　　　(b)

Figure 1.　(a) The physical model of the RIP system. (b) The coordinate system of the RIP system.

With the help of Lagrange dynamics, the nonlinear system dynamics is given in Equations (1) and (2), where the coefficients are summarized in Yang & Zhang (2018).

$$\left(J_r + m_1 l_{1c}^2 + C_2 \sin^2 \alpha + I_{2xx} \cos^2 \alpha + m_2 l_1^2 \right) \ddot{\theta} + 2\left(C_2 - I_{2xx} \right)$$
$$\dot{\theta}\dot{\alpha} \sin\alpha \cos\alpha + C_3 \ddot{\alpha} \cos\alpha - C_3 \dot{\alpha}^2 \sin\alpha = -b_\theta \dot{\theta} + \tau_m + \tau_{C\theta} \tag{1}$$

$$C_1 \ddot{\alpha} + C_3 \ddot{\theta} \cos\alpha - \left(C_2 - I_{2xx} \right) \dot{\theta}^2 \sin\alpha\cos\alpha - C_4 \sin\alpha = -b_\alpha \dot{\alpha} + \tau_{C\alpha} \tag{2}$$

2.2 *Fractional calculus definitions and its application rules*

In this paper, Riemann-Liouville (RL) fractional derivative is taken as a tool to develop the FOPID. For an arbitrary order RL fractional derivative of real order μ, where $0 \leq n-1 \leq \mu < n$, $n \in \mathbb{N}$. The RL derivative is defined in Equation (3) (Podlubny 1998).

$$_a D_t^\mu f(t) = \frac{d^n}{dt^n} \, _a D_t^{\mu-n} f(t) \tag{3}$$

In this definition, the case for $-1 \leq \mu < 0$ needs to be noted since $-1 \leq \mu - n < 0$ in the right side of Equation (3). Then, the Cauchy integral can be introduced to generalize the derivative order to any real order. The Cauchy integral is shown in Equation (4), where $-\lambda < 0$ and $\lambda = m - \mu, m \geq n$.

$$_a D_t^{-\lambda} f(t) = \frac{1}{\Gamma(\lambda)} \int_a^t (t-\tau)^{\lambda-1} f(\tau) d\tau \tag{4}$$

The Laplace transform of RL fractional derivative has a form of Equation (5).

$$\mathcal{L}\{_0 D_t^\mu f(t)\} = s^\mu F(s) \tag{5}$$

To approximate the fractional order operators using integer order transfer functions in frequency domain, Oustaloup filter (Oustaloup 1993, 1995) was designed. A modified version of Oustaloup filter (Xue 2017) is employed in this paper to estimate the effects of the fractional order operators as shown in Equation (6).

$$s^\gamma \approx \left(\frac{dw_h}{b} \right)^\gamma \left(\frac{ds^2 + bw_h s}{d(1-\gamma)s^2 + bw_h s + d\gamma} \right) \prod_{k=1}^N \frac{s + w_k'}{s + w_k} \tag{6}$$

where $w_k' = w_b w_u^{(2k-1-\gamma)/N}$, $w_k = w_b w_u^{(2k-1+\gamma)/N}$, $K = w_h^\gamma$ and the weighting factors are b = 10 and d = 9, by default.

3 CONTROLLER DESIGN

The affine form state of the nonlinear dynamics (Yang & Zhang 2018) implies the following control law can transform the nonlinear dynamics into partially linear.

$$u = \tau_m = \frac{\det(M)v - c + d}{-C_3 \cos\alpha} \tag{7}$$

Applying pole placement method to the linearized subsystem gives gains $K_3 = -800$ and $K_4 = -60$ which are associated with α and $\dot{\alpha}$. (Yang & Zhang 2018).

The $PI^\lambda D^\mu$ controller is proposed with the control law as Equation (8).

$$v_1(t) = K_p e_1(t) + K_i D^{-\lambda} e_1(t) + K_d D^\mu e_1(t) \tag{8}$$

where $e_1(t)$ is the state error defined by $e_1(t) = \theta_r(t) - \theta(t)$, θ_r is the reference rotary arm angle and θ is the actual rotary arm angle. Laplace transform of Equation (8) gives the transfer function of the $PI^\lambda D^\mu$ controller.

$$C_1(s) = K_p + \frac{K_i}{s^\lambda} + K_d s^\mu \tag{9}$$

The embedded control law in Equation (7) should have the form as below:

$$v(t) = v_1(t) - K_3\alpha - K_4\dot{\alpha} \tag{10}$$

Based on the control law given in Equation (10), five parameters associated with the $PI^\lambda D^\mu$ need to be adjusted. An optimization problem can be formulated and the cost function of it is defined as follows:

$$J(K) = w_1 M_p + w_2 E_{ss} + w_3(t_s - t_r) + w_4 \int_0^T e_1^2(t)\,dt + w_5 \int_0^T t e_1^2(t)\,dt \tag{11}$$

where overshoot M_p, steady state error E_{ss}, settling time t_s, rising time t_r are related to the pendulum arm angle (θ) response, the state error $e_1(t)$ is associated to the rotary arm angle (α) response. w_1, w_2, w_3, w_4 and w_5 are weighted factors selected by trial and error to balance the objective terms. In this paper, the weighted factors are chosen as follows: $w_1 = 10$, $w_2 = 1$, $w_3 = 0.5$, $w_4 = 1000$, $w_5 = 200$. Therefore, the optimization problem is established with $K = (K_p\ K_i\ K_d\ \lambda\ \mu)$ forming the parameter space. To ensure the high efficiency of searching process in solving this minimization problem, a roughly approximated ranges of values for optimal parameters are empirically specified, i.e., $K_p \in [5,15]$, $K_i \in [0,5]$, $K_d \in [5,15]$, $\lambda \in [0.5,0.999]$, $\mu \in [0.5,0.999]$.

PSO is a soft computing algorithm which is based on randomized evolutionary population redistribution in the parameter space. (Chong & Zak 2012) Note that in the algorithm, the cognitive coefficient $c_1 = 2.05$, the social coefficient $c_2 = 2.05$, the constriction coefficient $\kappa = 0.729$. In step 2, the particle velocity need to be clamped, i.e., $v = \min\{v_{max},\ \max\{-v_{max},\ v\}\}$ with $v_{max} = 0.500$. Consider swarm size to be 50 and the number of iterations is 200. The optimal result is finally obtained as follows:

$$K_p = 10.017 \quad K_i = 0.104 \quad K_d = 12.277 \quad \lambda = 0.864 \quad \mu = 0.852$$

Based on Equation (6), the fractional order operators in the $PI^\lambda D^\mu$ controller are approximated as Equation (12).

$$s^{-0.864} \approx \frac{0.0015034 s(s+1111)(s+877.7)(s+128.8)(s+18.91)(s+2.775)(s+0.4074)(s+0.0598)}{(s+596.1)(s+167.2)(s+24.55)(s+3.603)(s+0.5289)(s+0.07762)(s+0.01139)(s-0.0007776)}$$

$$s^{0.852} \approx \frac{2222 s(s+1111)(s+169.2)(s+24.83)(s+3.645)(s+0.535)(s+0.07852)(s+0.01153)}{(s+7508)(s+867.6)(s+127.4)(s+18.69)(s+2.744)(s+0.4027)(s+0.05911)(s+0.0007668)}$$

$$\tag{12}$$

Substitute the optimized parameters and Equation (12) into Equation (9), the optimized $PI^\lambda D^\mu$ can be obtained.

4 SIMULATION RESULTS

To test the performance of the optimized $PI^\lambda D^\mu$ controller, a simulation experiment is conducted with Simulink. The simulation of the RIP system involves two parts: one is the system dynamics module to emulate the nonlinear dynamics of the RIP system after IOFL transformation, the other one is the fractional order PID controller module consisting of a $PI^\lambda D^\mu$ controller with its parameters adjusted by solving an optimization problem using PSO algorithm. In this Simulink model, the feedback sensor block and the actuator block are neglected given that the sensor and actuators exhibit much higher frequency bandwidth than other components of the system. To get the optimal parameters of the FOPID controller block, the overall Simulink block is coupled with an optimization loop program. A step input signal (amplitude = 1, step time = 0) and a sinusoidal signal (amplitude = 1, frequency = 1.5 π rad/s, phase = 0 rad) are considered as the input signals to test the rotary arm's tracking performance.

The system responses are divided into four parts: θ response and $\dot{\theta}$ response corresponding to the rotary arm, α response and $\dot{\alpha}$ response corresponding to the pendulum arm. In this paper, the simulation results with original synthesized IOFL controller constructed in Yang & Zhang (2018) are employed to compare with the simulation results of the optimized $PI^\lambda D^\mu$ controller proposed in section 3. Figure 2(a) and Figure 2(b) present the simulation responses when the input signal is a step function. In Figure 2(b), the system is configured with the original synthesized IOFL controller. As it is shown in Figure 2(b), the rotary arm (θ response) eventually return to the zero position ($\theta(\infty) = 0$). However, the rotary arm finally moves to the step signal's steady state position ($\theta(\infty) = 1$) after overcoming an overshoot which is shown in Figure 2(a). Similarly, when the system input is a sinusoidal signal, the system responses are shown in Figure 3. It turns out that the θ has a sinusoidal mimicking behavior with respect to

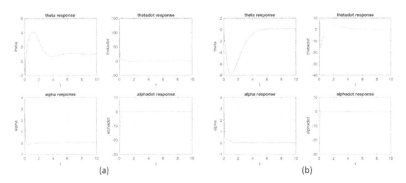

Figure 2.   (a) System responses with FOPID controller and unit step input. (b) System responses with original synthesized IOFL controller and unit step input.

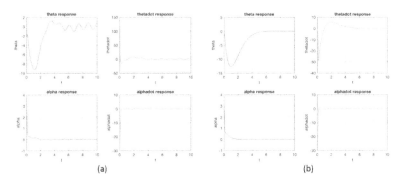

Figure 3.   (a) System responses with FOPID controller and sinusoidal input. (b) System responses with original synthesized IOFL controller and sinusoidal input.

the input signal as shown in Figure 3(a), but the tracking capability is lost when the system is equipped with the original synthesized IOFL controller as shown in Figure 3(b).

5 CONCLUSION

Beyond the random disturbances on the rotary arm, with unit step as well as sinusoidal signal imposed to the rotary arm, the simultaneous control of the tracking and stabilization of the RIP system is achieved. The RIP system dynamics is first treated with input-output feedback linearization, because one portion of the transformed system dynamics can be seen as a second-order linear time invariant system characterizing α responses (the pendulum arm). Since the other portion is a second-order nonlinear dynamics representing θ responses (the rotary arm), the transformed RIP dynamics is equipped with a fractional order PID controller, out of which the five parameters are optimally determined with PSO algorithm. Compared with the system performance resulted from the original synthesized IOFL controller, it is displayed that not only the global asymptotical stability is guaranteed by the new optimized $PI^\lambda D^\mu$ controller, but also the tracking control of the rotary arm is engaged. As a trade-off, the new approach needs to identify the optimal values of five parameters in the $PI^\lambda D^\mu$ controller by utilizing PSO algorithm to minimize the specified loss function, while for the original synthesized IOFL controller, only two parameters' optimal values need to be estimated using genetic algorithm. The controller developed in this paper can be generalized to applications that require the system dynamics to be stabilized and the indicated components of the system need to have capability to track a specified input signals.

ACKNOWLEDGMENTS

This work was supported in part by the National Science Foundation through grant CNS-1726865.

REFERENCES

Benson, D.A., Meerschaert, M.M. & Revielle, J., 2013. Fractional calculus in hydrologic modeling: A numerical perspective. Advances in Water Resources, Volume 51, pp. 479–497.

Chong, E.K.P. & Zak, S.H., 2012. An Introduction to Optimization. 4 ed. Hoboken, New Jersey: John Wiley & Sons.

Dwivedi, P., Pandey S. & Junghare, A., 2017. Performance Analysis and Experimental Validation of 2-DOF Fractional-Order Controller for Underactuated Rotary Inverted Pendulum. Arabian Journal for Science and Engineering, 42(12), pp. 5121–5145.

Dzieliński, A., Sarwas, G. & Sierociuk, D., 2011. Comparison and validation of integer and fractional order ultracapacitor models. Advances in Difference Equations.

Liao, W., Liu, Z., Wen, S., Bi, S. & Wang, D., 2015. Fractional PID based stability control for a single link rotary inverted pendulum. Beijing, China, IEEE.

Magin, R., 2006. Fractional calculus in bioengineering. 1 ed. s.l.:Begell House.

Magin, R., 2010. Fractional calculus models of complex dynamics in biological tissues. Computers & Mathematics with Applications, 59(5), pp. 1585–1593.

Oustaloup, A. & Melchior, P., 1993. The great principles of the CRONE control. Systems Engineering in the Service of Humans, Volume 2, pp. 118–129.

Oustaloup, A., 1995. La dérivation non entière: Théorie, synthèse et applications. Paris: Hermes Science Publications.

Podlubny, I., 1998. Fractional Differential Equations, Volume 198. 1 ed. s.l.: Academic Press.

Wan, Y., Zhang, H.H. & French, M., 2015. Adjustable Fractional Order Adaptive Control on Single-delay Regenerative Machining Chatter. Journal of Fractional Calculus and Applications, 6(1), pp. 185–207.

Xue, D., 2017. Fractional-Order Control Systems: Fundamentals and Numerical Implementations. 1 ed. Berlin, Boston: De Gruyter.

Yang, Y. & Zhang, H.H., 2018. Stability Study of LQR and Pole-Placement Genetic Algorithm Synthesized Input-Output Feedback Linearization Controllers for a Rotary Inverted Pendulum System. International Journal of Engineering Innovations and Research, 7(1), pp. 62–68.

Automatic Control, Mechatronics and Industrial Engineering – He & Qing (Eds)
© 2019 Taylor & Francis Group, London, ISBN 978-1-138-60427-8

Research on an HCI cognitive performance test method of a distributed control system

K. Yu
China Ship Development and Design Center, Wuhan, China

ABSTRACT: The Human-Computer Interface (HCI) of a Distributed Control System (DCS) is the most important interaction medium between human and industrial systems. The cognitive performance of DCS HCI can directly affect the operational safety and efficiency of industrial systems. The cognitive behavior of DCS HCI is studied to clarify the interaction process and influencing factors during DCS operation and monitoring tasks. The test indexes and methods based on eye movement data and subjective fatigue data are proposed. The DCS HCI cognitive performance test procedure is designed by combining eye tracking tests and subjective fatigue tests. The DCS HCI cognitive performance of a black background and a gray background is tested and compared. The results show that the gray background DCS HCI has an advantage in cognitive performance, and the black background DCS HCI brings higher cognitive fatigue and lower efficiency. This method is significant in reducing the cognitive load and improving the interaction efficiency for DCS HCI.

Keywords: distributed control system, human-computer interface, cognitive performance, eye movement data

1 INTRODUCTION

A Distributed Control System (DCS) is a distributed computer control system based on a microprocessor. The remarkable characteristics of a DCS are centralized operation and monitoring of industrial processes, while control tasks are completed by different computer control devices. The DCS is currently the most widely used in the industrial control field, and has been widely used in aviation, transportation, the chemical industry, manufacturing, energy, nuclear power and other industrial control fields.

Usually, a DCS concentrates information and human-computer interaction equipment in a centralized control room for information display and operation. In the current advanced DCS, the human-computer interface (HCI) based on computer technology, has gradually become the core interaction medium between the operators and the industrial system for information display, control input, management and so on.

The operators' monitoring and operation is based on the video display units, interactive graphical HCI, special keyboard, mouse or track ball. The form of the DCS HCI elements include general appearance, groups, points, trends, warnings, operation guidance and flow-charts, to make the monitoring and operation centralized and convenient. DCS HCI task failure may lead to industrial system operation safety problems. DCS HCI is responsible for the normal, abnormal and post-accident display and control of industrial systems. At the same time, huge potential risks concentrated on a few operators for the centralized monitoring and operation, and it plays a direct and vital role in the efficiency and safety of the industrial control system (Guofang et al., 2016). For example, the Three Mile Island and the Chernobyl nuclear power plant accidents are closely related to the human-machine interaction system. Therefore, it is necessary to research the cognitive performance of DCS HCI in industrial systems, so as to ensure that the cognitive performance of DCS HCI matches well with the demand of manipulation and the ability of operators.

The research of HCI cognitive performance has been paid more and more attention to improve operational performance and reduce cognitive pressure (Wu et al., 2017). Traditional HCI cognitive performance research methods include user interviews, experimental observation, questionnaire surveys and so on. For example, Park and Yung (2006) and Jou et al., (2009) used the Task Load Index (TLX) scale to study the emergency tasks of industrial systems, the cognitive performance and load of specific tasks in the main control room. Jing et al., (2018) studied the influence of HCI shape coding on information search and target cognition performance. Its limitations lie in relying on subjective feeling and experience to evaluate the cognitive performance. The result is a subjective feeling value, and it is difficult to accurately describe the state of cognitive performance. The test results are subjective and lack the defects of users' objective data support.

In recent years, human physiological data has been used in HCI cognitive research. For example, lo Storto (2013) used eye movement data in website interface cognitive performance and cognitive pressure research. Ha et al., (2006) used eye movement data to study the cognitive load of accident diagnosis tasks in industrial systems. Ha et al., (2016) used eye tracking data for operator thinking in a nuclear power plant. Eye movement data was used to measure visual attention and interest points (Balcombea et al., 2017). Based on the direct correlation between physiological data and HCI monitoring and the operation task process, the test data is objective and real-time, which belongs to the accurate cognitive research method. HCI researches based on physiological parameters have become a trend.

This paper mainly studies the DCS HCI cognitive behavior model and the cognitive performance test method. The effective DCS HCI cognitive performance test method is proposed.

2   DCS HCI COGNITIVE BEHAVIOR MODEL

The DCS HCI of industrial systems provides information display in the way of simulating flow chart, bar chart, numerical meter, trend curve, table, button, dialog box and so on. It can complete dynamic data display, operation and alarm and so on. The information display of DCS HCI is closely connected with the principle of the industrial process system. The dynamic and static information of DCS HCI should be monitored and the coupling relationship is complex.

The operator's cognition process for the DCS HCI includes the perception, memory, thinking processing, selection, decision making and other activities of HCI information. DCS HCI information interaction involves information stimulation, information transmission feedback, and output and result correlation (Jingling, 2014). The cognitive behavior model of DCS HCI is as show in Figure 1.

a. First, the DCS HCI stimulus information is presented in the visual vision and audition mode, such as device icons, process simulation diagrams and other changes, such as flickering output. The operators register the sensation of HCI stimulation through the vision and audio perception organs and store the input stimulus information.

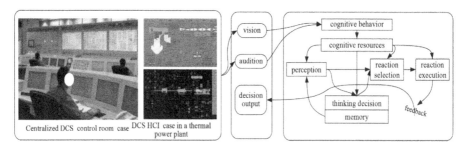

Figure 1.   Cognitive behavior model of DCS HCI.

b. Operators' perception is the processing of sensory information formed by stimulation of DCS HCI, so as to form a complete information reflection.
c. After the sensory and perceptual integration of stimulus information, the operators began to think and make decisions based on the information. This stage involves the working memory and long term memory of the operators.
d. The operators are able to respond to the stimulus by choosing the form of response choice. In some cases, the operators will choose not to respond to the HCI stimulus, but the information is stored in their long-term memory, and then their short-term memory is retrieved from the long-term memory to enter the execution of the short-term memory.
e. After the process of information output has been processed, the operators decide the operation reaction activity on DCS HCI stimulation.

The information process requirements of DCS HCI are severe and require more cognitive resources. Operators' cognitive resources are limited. The DCS HCI exceeds the limits of their cognitive resources and will inevitably lead to low monitoring efficiency, high psychological pressure and mis-operation rate (Li et al., 2016).

3 DCS HCI COGNITIVE PERFORMANCE TEST INDEXES

The DCS HCI cognitive performance test is based on the combination of eye movement measurement and subjective evaluation. The subjective scale and objective physiological data of the target object in the interaction process on DCS HCI experience are obtained through the study of the operator's eye movement and subjective evaluation in the simulation task experiment.

a. The eye movement data refers to duration data and count data of eye movements. Test indexes: φ. The fixation count: the count of eye fixation points to be paid attention to on DCS HCI during the task process. κ. Duration data includes first fixation duration, total fixation duration and total visit duration of Regions of Interest (ROI). Here, fixation duration refers to the time used for eye movement to stay at a specific ROI and to search for the fixations. λ. ROI visit count represents the number of eye visits to a region. The value is less than or equal to the number of fixation counts. And the value is more than 1, representing the seeing back behavior of operators.
b. Fatigue evaluation scale: psychological evaluation can be measured, analyzed and evaluated by scientific, objective and standard measurement methods. It is the most simple, portable and subjective evaluation method. The fatigue evaluation scale is designed by the equidistant table to evaluate the psychological pressure, fatigue, eye swelling and comfort of the operators.

4 DCS HCI COGNITIVE PERFORMANCE TEST EXPERIMENT

4.1 Test experiment method

In the test experiment, the tester simulates the task process of DCS HCI. Condensate water HCI of a thermal energy industry is selected as the test object. The screen size displaying the condensate water HCI is 22 inches and the resolution is 1280*800. The DCS HCI task process is performed by simulation testers, at the same time, eye movement data and subjective fatigue tests are measured in real time.

Objective: In the field of DCS, the most commonly used HCI background colors include black and gray. The cognitive performance of a black background (RGB:0,0,0) and a gray background (RGB:222,222,222) DCS HCI are tested and compared.

Test equipment: eye movement test equipment is the Tobii Pro X3–120 portable screen eye tracker, the eye movement data acquisition software is Tobii Studio.

Fatigue tests used the fatigue evaluation scale.

The experimental task flow is as follows:

a. The testers rest for 2 minutes before the start of the experiment to prepare for the best working condition.

b. Eye movement calibration at the end of 2 minutes.
c. Start the task of simulating the condensate water HCI. The time is 15 minutes. During this duration, the tester's eye movement data is recorded in real time.
d. After the implementation of the task, fatigue is evaluated. The subjective psychological evaluation index from low to high with 0~10 as the score and orally reporting fatigue.
e. Complete the data record and analyze it.

4.2 *Test experiment method case*

The test scene is shown in Figure 2. The visualized fixation and ROI statistics and analysis chart of the DCS HCI for a black background and a gray background of the thermal power industry system as shown in Figure 3.

The fixation statistics chart shows the tester's sight position, sequence and observation time in a specific region on the DCS HCI. Therefore, the main characteristics of fixation trajectories are to reveal the fixation time sequence or location and observation time. The numbers in the circle represent the order of the tester to observe the interface, and the time of observation is usually expressed as the duration of the fixation point, which is represented by the dots of different diameters on the fixation path chart. The longer the observation time is, the greater the diameter of the dots.

On the DCS HCI, the region that needs the tester to focus attention is divided into ROIs to test whether the tester's performance meets the design intention and goal. This experiment divides 14 ROIs into the DCS HCI. (1) time to first fixation of ROI: the shorter the time of an ROI is found, the sooner the representative is searched, or more easy it is to search; (2) fixation count before first fixation of ROI is similar to the first fixation time, describing the count of fixations before the discovery of a specific region, and the smaller the number, the faster the target is searched. A comparison chart of time to first fixation of, and fixation

Figure 2.   DCS HCI test scene.

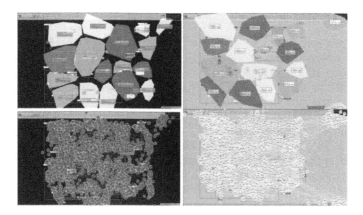

Figure 3.   Visualized fixation and ROI statistics and analysis chart of the DCS HCI (black background and gray background).

count before fixation duration of ROI on DCS HCI between the black background and the gray background is shown in Figure 4 and Table 1.

According to the chart of time to first fixation of ROI on a DCS HCI with a black background, the tester's visual search order is from the ROI 1 to the ROI 14 smooth, basically conforming to the visual observation logic order in the HCI design. The fixation sequence of DCS HCI of the gray background is approximately the same as the black background, but there is a difference: the first fixation of ROI 1, 4, 5, 6, 7, 10, 14 of the gray background is more rapid and direct, as shown in Table 1. And the time to first fixation of is shorter, and the fixation count before first fixation of ROI is less. For ROI 3, 8, 9, 12 and 13, the time to first fixation of is shorter and the fixation count before first fixation of ROI is less for the black background HCI, but for the ROI 2 and 11, although the time to first fixation of is shorter in the black background HCI, but the fixation count before first fixation of ROI is less.

The first fixation duration represents the first time the tester's attention time to a specific region on DCS HCI is long, the first observation time of the HCI is long, indicating that the correlation of the elements of HCI is strong, as shown in Figure 5.

According to the first fixation duration, there is a great difference between the black background and the gray background. The length distribution of each ROI in the black background is shorter than the gray background, which indicates that the correlation between the elements in the gray background is stronger.

Total fixation duration represents the test user' total attention time to a specific region on DCS HCI. The comparison chart of total fixation duration of ROI on DCS HCI between the black background and the gray background is shown in Figure 6.

According to the total fixation duration, the total fixation duration is longer on the gray background HCI, indicating that the continuous attention of the tester is more stable on the gray background HCI. For ROI 1, 2, 3, 7, 10, 11 and 12, the tester's attention is significantly higher than the remaining ROI.

The visits count represents the count of visits to a specific region. The count is less than equal to the fixations count, and the count is more than 1 representing the occurrence of the return.

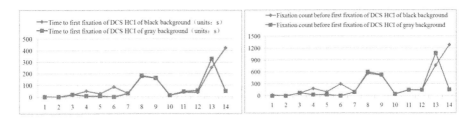

Figure 4. Comparison chart of time to first fixation and fixation count before first fixation of ROI on DCS HCI between the black background and the gray background.

Table 1. Comparison of first fixation duration and fixation count before first fixation of ROI on DCS HCI between the black background and the gray background.

	ROI 1	ROI 2	ROI 3	ROI 4	ROI 5	ROI 6	ROI 7
First fixation duration (s)	5.32	−0.03	−3.1	43.8	20.4	86.9	0.56
Fixation count before first fixation	17	0	−7	153	68	297	6
	ROI 8	ROI 9	ROI 10	ROI 11	ROI 12	ROI 13	ROI 14
First fixation duration (s)	−3.06	−3.06	0.03	−18.6	−14.2	−75.9	373.7
Fixation count before first fixation	−40	−12	2	1	−32	−307	1124

The more difficult it is to recognize and extract information of DCS HCI; the longer the fixation duration indicates that the tester needs to spend more time to extract the information. The fixations count usually indicates that the tester has more meaningless fixations and lower cognitive performance. As shown in Figure 7, according to the visits counts, the visits counts are more uniformly relative to the fixations count to each ROI. But still there is a higher degree of attention to ROI 1, 2, 3, 7, 10, 11, and 12. Compared with the black background, visits counts to each ROI of the gray background is more and more uniform, indicating that the gray background DCS HCI is more comfortable for the tester, allowing lower cognitive load and higher attention.

After the DCS HCI eye movement test, subjective fatigue evaluation was carried out. In this experiment, the tester is required to report the mental pressure, fatigue, eye swelling and comfort, in the form of an oral report after the execution of the simulation task, with the equal distance between 0–10. The lower the score, the lower the pressure, the easier on the eye it is, and the more comfortable the eyes are. On the other hand, the fatigue degree is higher. From Figure 8, it can be seen very intuitively that the black background will cause the maximum fatigue for the tester. For the gray background, the fixations assignment is more uniform and the fixations time is longer.

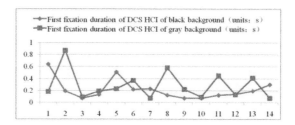

Figure 5.   Comparison chart of first fixation duration of ROI on DCS HCI between the black background and the gray background.

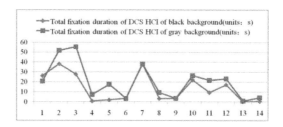

Figure 6.   Comparison chart of total fixation duration of ROI on DCS HCI between the black background and the gray background.

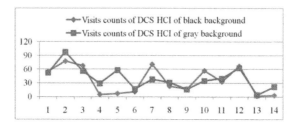

Figure 7.   Comparison chart of visits counts of ROI on DCS HCI between the black background and the gray background.

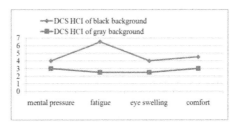

Figure 8. Comparison chart of subjective fatigue evaluation of DCS HCI between the black background and the gray background.

5 CONCLUSIONS

The cognitive performance of DCS HCI is researched. The HCI cognitive behavior model and its influencing factors are defined. A cognitive performance test method based on eye movement data and subjective fatigue evaluation is proposed for DCS HCI. Aiming at the comparison of the cognitive performance test between a black background and a gray background HCI, the test method and process is designed. HCI eye movement data and subjective fatigue evaluation data are collected. The test results show that a gray background HCI has obvious advantages in efficient cognition, low pressure load and attention concentration. The feasibility of this method for a DCS HCI cognitive performance test is verified. In the further research, eye movement and cerebral blood oxygen measurements will be used to verify the validity of synchronous measurement mode data in cognitive research.

REFERENCES

Balcombea, K., Fraserb, I., Williamsa, L. & McSorleyaa, E. (2017). Examining the relationship between visual attention and stated preferences: a discrete choice experiment using eye-tracking. *Journal of Economic Behavior & Organization, 144*, 238–257.
Ha, C.H., Kim, J.H., Lee, S J. & Seong, P.H. (2006). Investigation on relationship between information flow rate and mental workload of accident diagnosis tasks in NPPs. *IEEE Transactions on Nuclear Science, 53*(3), 1450–1459.
Ha, J.S., Byon, Y.J., Baek, J & Seong, P.H. (2016). Method for inference of operators' thoughts from eye movement data in nuclear power plants. *Nuclear Engineering and Technology, 48*, 129–143.
Jing, L., Yu, S. & Wu X. (2018). Effects of shape character encodings in the human-computer interface on visual cognitive performance. *Journal of Computer-aided Design & Computer Graphics, 30*(1), 163–179.
Jingling, Z. (2014). The research of digital human-computer interface evaluation method of operation and monitoring system for nuclear power plants. *Ph.D Thesis, Harbin Engineering University*, 22–24.
Jou, Y.-T., Yenn, T.-C., Lin, C.J., Yang, C.-W. & Chiang, C.-C. (2009). Evaluation of operators' mental workload of human-system interface automation in the advanced nuclear power plants. *Nuclear Engineering and Design, 239*, 2537–2542.
Li, G., Shi, P., Zhang, K. (2016). Study on drivers' cognitive performance of urban road guide sign layout. *China Safety Science Journal, 26*(08), 100–104.
Li, Y., Li, Z. & Huiqiao, J.I.A. (2016). The study of the type of operation task and time pressure on cognitive load of digitized nuclear power plant operator's monitor behavior. *Applied Energy Technology*, 52–58.
Lo Storto, C. (2013). Evaluating ecommerce websites cognitive efficiency: an integrative framework based on data envelopment analysis. *Applied Ergonomics, 44*(6), 1004–1014.
Park, J. & Jung, W. (2006.) A study on the validity of a task complexity measure for emergency operating procedures of nuclear power plants-comparing with a subjective workload. *IEEE Transactions on Nuclear Science, 53*(5), 2962–2970.
Wu, X. & Shuzhang, Z. (2017). How to reduce the cognitive load in interaction design. *Journal of Hubei Correspondence University 30*(8), 110–111.

Electrical engineering

Automatic Control, Mechatronics and Industrial Engineering – He & Qing (Eds)
© 2019 Taylor & Francis Group, London, ISBN 978-1-138-60427-8

Analysis of electrical power stability in a distribution grid undergoing live maintenance

T. Jin, W. Zhang, X.F. Shi & S.K. Chen
State Grid Zhejiang Electric Power Co. Ltd., Hangzhou, Zhejiang, China

L.Y. Wang & J.C. Gao
School of Electrical Engineering, Wuhan University, Wuhan, Hubei, China

ABSTRACT: Bypass operations were first proposed by the Japanese power sector in the 1970s. Through half a century of engineering exploration and development, bypass operations have achieved relatively sound systems. The bypass operation takes three main forms: bypass cable, transformer vehicle, and generating car. The overhaul of distribution network transformers is the major objective of transformer vehicle work. At present, the study of transformer vehicles has been limited to improving the continuity of power supply. There are fewer analyses of user-side power quality during operations to ensure power supply reliability. In order to provide users with higher-quality power, we use MATLAB software to simulate and analyze the change of power waveform in a distribution network during replacement operations involving a transformer vehicle. On the basis of a controlling variable method, we identify the variables that produce least impact on the power quality of the user side, thereby ensuring high quality and reliability of power supply.

1 INTRODUCTION

Electrical energy has become an indispensable component of production and people's lives. Power quality is therefore an important factor affecting people's quality of life. In order to improve the continuity of electrical energy, power supply companies carry out many innovations. Bypass operations are the main way to improve the continuity of power quality (Shi & Chen, 2017; Hu et al., 2016). The working principle of a bypass operation is load transfer. Load transfer involves the transfer of the distribution transformer load to the transformer vehicle. The transformer vehicle acts like a bridge between the power network and the user. A bypass operation can reduce the impact on users of power outages caused by maintenance, thereby improving the quality of the power supply (Hong et al., 2016; Wei et al., 2012). At the same time, the transformer vehicle can also be used for temporary construction and power supply and can prevent inconvenience and safety hazards caused by long take-up lines during construction. The transformer vehicle has good worth: it can share load during peak periods of electricity consumption and guarantee power supply quality.

Research on transformer vehicles mainly started with field construction reports and, at present, is based on the experience of repeated operations in engineering practice through live working. Analysis and research into their influence on customer power quality is still insufficient. In order to provide users with more stable and high-quality electricity, this paper analyzes the problem of cutting off the impulse current at different time nodes, the integrity of the voltage waveform on the user side, and identifying the optimal access point (Xu et al., 2016; Wang & Tang, 2003; He et al., 2015; Yang et al., 2013). In this paper, MATLAB simulation software (MathWorks, Inc., Natick, MA, USA) is used to build a simple and functional power grid model, and the influence of the transformer vehicle on different power grid stages is analyzed. The simulation starts from the regular maintenance of substations in the distribution network. It realizes the complete operational process for the transformer vehicle in the distribution network without interruption of operation. During the process, simulation data for the user-side voltage waveform and the current change in the transformer vehicle substation are obtained. The influence of load transfer on users and the distribution network itself is analyzed.

2 MODEL FRAMEWORK CONSTRUCTION

In this paper, the basic grid model is used as the simulation framework. This basic framework has the advantages of simple structure, complete function and obvious simulation effect. In the network, we can clearly see the change of power quality on the user side during the whole process of load transfer, and can search for the optimal access point through variable adjustment of the control variables. The construction of a complex and large-scale representative power network would involve more variables and a slower simulation process, making it harder to obtain a large number of comparative simulation results.

Figure 1 shows a basic power system simulation model, which contains a power generation module. This simulated module can generate a basic three-phase power with a frequency of 50 Hz and a voltage of 30 kV. The initial voltage is converted into 500 kV ultra-high voltage by a booster substation. The power is transmitted over a standard length of high-voltage cable, which is protected by three-section current. A distribution substation is installed at the end of the voltage transmission line. The transformation ratio of substations is 50:1. Converting the ultra-high voltage of 500 kV on the high-voltage transmission line to the 10 kV ring network cabinet voltage used in the distribution network, the voltage of the 10 kV distribution network is converted to low-voltage output through a 25:1step-down transformer.

In the construction of the power grid system, the distribution network step-down transformer is divided into separate sub-modules, which is conducive to observation and research. The simulated transformer sub-module is shown in Figure 2. The core part of the sub-module is a 25:1 three-phase step-down transformer. It includes a three-phase circuit breaker which simulates the operation of the transformer and a three-phase grounding point. The construction of the transformer vehicle is basically the same as that of this step-down transformer module.

In order to simulate the influence of the load shifting of the transformer vehicle to the user side and the power grid more accurately, the parameter configuration of the whole power grid should be close to that of the real power grid. In the configuration of power plants, a three-phase power supply module is used to simulate the power generation directly. The module provides three-phase electrical stability and waveform integrity. The capacity of the

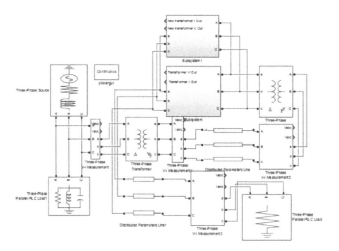

Figure 1.   Basic power system simulation model.

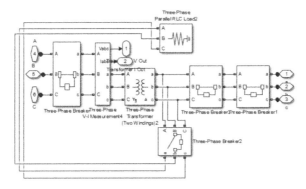

Figure 2. Simulation model of transformer sub-module in a distribution network.

three-phase power supply is 60 MVA, the line voltage is 30 kV, and the rated frequency is 50 Hz. The step-up transformer capacity is 60 MVA and the rated frequency is 50 Hz. The primary-side rated voltage is 30 kV, and the secondary-side rated voltage is 500 kV. The length of the high-voltage transmission line is 300 km. The value of resistance, capacitance, and reactance per unit length is uniform. The national standard high-voltage transmission line has the same specified value. The capacity of the transformer at the end of the high-voltage transmission line is 60 MVA. Electrical energy is wasted during transmission over high-voltage transmission lines. Therefore, the rated voltage of the primary side of the step-down transformer is 450 kV, and the rated voltage of the secondary side is 10 kV. In the transformer on the distribution network, the simulated user-side consumption is small and the user side is not complicated. Therefore, the rated capacity of the distribution network is small, with a capacity of 50 kVA.

3 POWER QUALITY ANALYSIS

3.1 *Transformer maintenance simulation*

The aim is to comprehensively study voltage and current changes in the power network during load shifting. The simulation is for different directions and different categories of simulation, including two types of transformer periodic maintenance and simulation of a transformer aging and being updated. The setting of breaker can be set up in simulation. In order to obtain the experimental data more quickly, the use of a three-phase power supply instead of power plants can effectively reduce the system stability preparation time, ensuring that the whole system can be stabilized in 0.2 seconds.

There are different ways to repair and replace, but the result of the simulation is the same as that of the real world: the working principle is load shifting. In the setting of circuit breaker parameters, the time of the original state can be changed by the initial breaking state of the circuit breaker. In order to obtain a large amount of simulation data conveniently, the whole simulation process is set to last 5 seconds. After 0.2 seconds, the power system voltage is basically stable and the simulated process can start. At 0.21 seconds, the transformer vehicle will be connected to the system to carry out voltage synchronization, so as to prepare for the removal of the transformer. At 0.3 seconds, the right-hand side of the transformer close to the user-side circuit breaker is cut off, and at 0.33 seconds the left-hand side circuit breaker of the transformer is cut off. The transformer is detached from the power grid and the transformer discharge is completed as soon as possible, preventing internal residual power in the transformer from being discharged elsewhere and increasing maintenance risk. Between 0.33 seconds and 4 seconds, the transformer's maintenance, replacement and other operations are conducted. At 4.3 seconds, the transformer is reconnected to the power grid to carry out voltage synchronization and allow the removal of the transformer vehicle. At

4.5 seconds, the transformer vehicle is removed from the system and its transformer is discharged as soon as possible, so as to complete the process.

3.2 *Simulation results and analysis*

This paper develops a simulation of the different effects of load transfer on the power grid and its users under different conditions. The standard unitary value is used for comparison. When the maintenance plan is received, the power grid is working normally for the users. The maintenance unit builds the transformer vehicle and lays the bypass cable. The transformer vehicle is allowed to settle into a stable working state for a period of time in parallel to the transformer in the distribution network. Following stable operation of the power system at 0.2 s, the 0.19 s transformer vehicle has just changed the access system. After 0.3 s, the transformer vehicle changes the waveform stability, with removal of transformer circuit-breakers on both sides of the distribution network at 0.32 s.

The initial grid connection and removal simulation is shown in Figure 3. As can be seen from the figure, when the transformer vehicle's transformer is first connected to the power system, the waveform of the transformer vehicle (the bottom trace) is very disorderly and there is no stable voltage fluctuation following the small grid. However, this has little impact on the main power network and almost no impact on the user side. It proves that the transformer vehicle is merged into the power system when it is in normal operation; changing the vehicle will not cause obvious power interference to the user side, nor will it pollute the main power network. Between 0.19 and 0.3 s on the horizontal axis shown in Figure 3, the waveform of the transformer vehicle (the bottom trace) is unstable. The transformer in the distribution network should not be removed at this time. After 0.3 s, the waveform of the transformer vehicle is basically stable, and the transformer in the distribution network can be removed. When removing transformer in load. It can be seen from Figure 3 that the removal of the distribution network transformer at 0.32 s has a greater impact on the transformer vehicle under the same load conditions. At this time, the three-phase current value of the transformer vehicle varies markedly, resulting in a large current inrush, which lasts 10–15 ms. The user side also has the same effect, and the user-side voltage reflects a transient distortion. However, there is no obvious effect on the stability of the overall user-side voltage or waveform integrity. After 0.34 s, the transformer in the distribution network is completely removed without voltage. After grounding discharge, this transformer is repaired and replaced.

Transformer maintenance and replacement process in distribution network. The distribution load of the small power grid distribution network is transferred to the transformer vehicle. The user side does not experience waveform distortion in the whole process, nor overvoltage or other adverse conditions. The working state of the transformer vehicle is stable, and the working state of the power grid is stable. Once the distribution transformer has been overhauled or replaced, it needs to be reconnected to the distribution network. When it is replaced at 4.3 seconds (0.2 s on the horizontal axis in Figure 4), the transformer of the

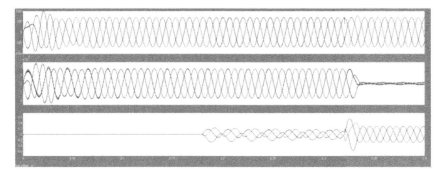

Figure 3.   Simulation of initial load transfer.

distribution network is merged into the distribution network and will cause transient distortion on the user side. The larger the distortion is, the smaller the impact on the transformer vehicle. When the transformer vehicle changes into a single phase, the voltage will change and the remaining two phases will work normally. Once the distribution network transformer can supply stable power to the user side, at 4.32 seconds (0.22 s in Figure 4), the transformer vehicle can be removed from the distribution network. In the process of removing the load from the transformer vehicle, there is no obvious effect on the voltage waveform of the user side, nor on the transformer in the distribution network. The transformer will be discharged from the distribution network to remove the transformer. The transformer maintenance and transformer replacement operation is then complete.

3.3   *Analysis of factors affecting cutting angle*

The aim of this section is to evaluate the influence of different disconnection angles on the user side of the distribution transformer and explore when to remove the distribution transformer and minimize the impact on the user side. Based on a single-phase benchmark, the effect of transformer removal from the distribution network at different single-phase periodic nodes on the user side is evaluated, on the basis of a power frequency voltage $f = 50$ Hz, a voltage period $T = 0.02$ s, and an angular frequency of about 314. The simulation diagrams of the influence of removing the distribution network transformer on the user-side voltage and the transformer vehicle at different angles are shown in Figures 5 through 8. When changing the control angle of the user side, the single-phase angle is variable. The effect of different disconnection angles on the user side and the impact on the transformer vehicle can be seen. From top to bottom, each figure shows the user-side voltage, the high-voltage side of the distribution network transformer, and the impact on the current of the transformer vehicle.

Comparing the degree of influence of different cutting angles on the user side and the distribution network, the results show that the 0°, 90°, 180°, and 270° disconnections all produce instantaneous distortion on the user side. This phenomenon is small and its duration is very

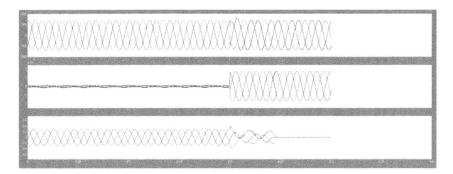

Figure 4.   Simulation of completion of load transfer operation.

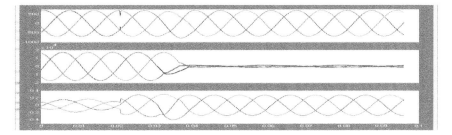

Figure 5.   Simulation of 0.33 s (0°) disconnection of the distribution network transformer.

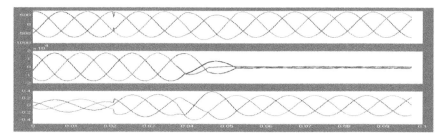

Figure 6.   Simulation of 0.335 s (90°) disconnection of the distribution network transformer.

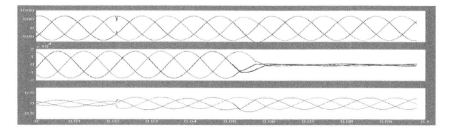

Figure 7.   Simulation of 0.34 s (180°) disconnection of the distribution network transformer.

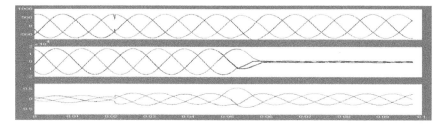

Figure 8.   Simulation of 0.345 s (270°) disconnection of the distribution network transformer.

short. When the power grid is cut off at 90° and 270°, the voltage fluctuation is greater. The larger impact current inside the transformer vehicle can cause greater impact on the power quality of the user side. The influence of the 0° and 180° disconnections on the power distribution substation on the user side is relatively small. When the impulse current produced by the transformer vehicle is small, the substation with the distribution network removed has the least influence on the power quality of the users.

4   CONCLUSION

In order to improve the power quality on the user side, we studied the change of user-side voltage and impulse current of the transformer vehicle in the maintenance operation of a distribution substation. In this paper, the operational process associated with a transformer vehicle was simulated with MATLAB. The following conclusions are drawn from the comparison of power quality changes on the user side:

1. The transformer vehicle has almost no effect on the voltage waveform of the user side and distribution network when it is connected to a stable distribution network. This characteristic enables the transformer vehicle to share part of the load at the peak-load stage for the

distribution transformer, improving the continuity and reliability of power supply. From the simulation charts, the distribution network of different capacity transformer vehicles with no load and no load is obtained. The time from no-load to stable operation varies. Distribution network transformer removal can only be carried out in a phase of stable operation; otherwise, it will lead to power supply disorder on the user side. The shock current of the transformer vehicle may be too large and can even lead to the collapse of the distribution network with the consequence of power failure for users.

2. Once stable operation of the transformer vehicle has been achieved, the distribution transformer can be disconnected. The effect of different disconnection angles on the user side and the transformer vehicle varies, as can be seen from the simulation diagrams. When the disconnection angle is 0°, large voltage fluctuation on the user side will cause certain damage to users' electrical appliances. The transformer vehicle will produce a large impact current, which will reduce the service life of the transformer vehicle. When the transformer is removed at 180°, the effect on the user side will be minimal. No overvoltage is generated, and the impact current of the transformer vehicle is also small. The phase diagram of three-phase power is 180°. In addition, the two phases of electricity are equal to the opposite voltage amplitude. At this point, the removal or introduction of transformers to the entire distribution network and the user side of the transformer vehicle will have less impact.

REFERENCES

He, S. (2015). Analysis of load transfer characteristics of distribution line bypass operation. *Electrical Applications*, *18*, 16–22.

Hong, F., Shi, Q. & Lou, J. (2016). Research on the optimal load transfer path for distribution network maintenance. *Technology and Economic Guide*, *2016*(31), 72.

Hu, D., Wan, S. & Wang, H. (2016). Application of mobile transformer vehicle in rapid repair of distribution network. *China Science and Technology Information*, *2016*(17), 55–56.

Shi, C. & Chen, X. (2017). Application of a bypass operation in the maintenance of 10 kV cable lines. *Science and Technology & Innovation*, *2017*(6), 131–132.

Wang, Z. & Tang, G. (2003). Single load shifting of power distribution systems based on shortest path arborescence. *Electric Power Automation Equipment*, *2003*(6), 25–27.

Wei, L. (2012). Replacement (maintenance) of 10 kV distribution equipment on the column by-pass operation. *Henan Electric Power Technology*, *1*, 4–6.

Xu, X. (2016). Research on load transfer technology for regional power grid. *Application of Electronic Technology*, *z1*, 36–38.

Yang, F. (2013). Operating equipment and its application for overhead-cable hybrid lines. *Hubei Electric Power*, *37*(3), 11–12.

Automatic Control, Mechatronics and Industrial Engineering – He & Qing (Eds)
© 2019 Taylor & Francis Group, London, ISBN 978-1-138-60427-8

FEA based electromagnetic analysis of induction motor rotor bars with improved starting torque for traction applications

M. Sundaram, M. Mohanraj, P. Varunraj & T. Dinesh Kumar
Department of EEE, PSG College of Technology, Coimbatore, Tamil Nadu, India

Shubham Sharma
CSIR-Central Leather Research Institute, Regional Center for Extension and Development, Jalandhar, Punjab, India

ABSTRACT: The requirement of electric vehicle is inexorable to save energy and to protect the environment. Induction motor is very simple to manufacture and it's easy to control, with its improved characteristics are best suited motor for traction applications. This paper focus on design and optimization of induction motor with Less Cost and High Performance (LCHP). The performance of the induction machine depends on its dimensional constraints. The starting torque of the induction motor is an important aspect in traction application which is improved with the double cage rotor is presented in this paper. An analytical modelling is carried out and its validated by using the FEA analysis.

Keywords: Starting torque, double cage, FEA

1   INTRODUCTION

The efficiency of the electric machine is varied depending on the type of machine and its characteristics. When choosing an electric machine for an EV it is not only the efficiency that matters other factors such as cost, reliability etc. are important as well. The various types of electric machines used in EVs can be divided into two group commutator machines and commutator less machines. The commutator machines are primarily the DC machines. The DC machines need commutator and brushes to feed current to the armature. The brushes are worn over time and this makes DC machines more maintenance, heavy than other electric machines. The other group of electric machines is as mentioned the commutator less machines. Two of the most used commutator less machines are the induction machine (IM) and the permanent magnet synchronous machine (PMSM). The benefit of using IMs is low cost, high reliability and maintenance free operation (Zeraoulia et al. 2006). Compared to the DC machine, the IM has additional advantages such as light weight nature, small volume and high efficiency (Casadei et al. 2010). The benefit of using PMSM is its high efficiency and high-power density. However, the PMSM also suffers from some disadvantages such as cost because of the price of rare-earth magnets, safety in case of wreck and possibility of magnet demagnetization because of temperature or control failure (AbdElhafez et al. 2017). Induction motors are extensively used for the traction applications, due to its simple structure and easy control. In terms of efficiency and power density the BLDC motor is the best choice but when reliability and controllability come to the picture Induction motor proves its ability. Tradeoff in Induction motor design is carried out with different frequencies to determine the optimal the performance of the motor (Wang et al. 2005). The thermal and electrical characteristics to be analyzed and validated with different heating temperatures (Ulu, Korman & Kömürgözm 2017). The performance of the motor can be improved with cooling techniques, along with designing the mechanical constraints of the motor (Kim et al. 2013). Optimization has been carried out to improve the machine performance such as starting torque and efficiency (Jeon et al. 2011).

2   STATUS OF ELECTRIC VEHICLES IN INDIA

India's mission is to make all vehicles as electric powered, Fuel Powered automotive vehicles have become an inevitable one for transportation from one place to another. In the developing era, utmost importance has been given to make a better environment with very low emission. In India vehicle increased from 0.3 million in 1951 to 58.3 million in 2001–02. An electric car is powered by an electric motor instead of a gasoline engine. The electric motor gets energy from a controller, which regulates the amount of power-based on the driver's use of an accelerator pedal. The electric car (also known as electric vehicle or EV) uses energy stored in its rechargeable batteries, which are recharged by common household electricity. It doesn't pollute the world; hence the clean and clear energy can be utilized for the traction. India plans to make completely electric vehicles in 2030, to provide pollution free environment, as of now India has 4350 electric cars on the road (Wang et al. 2005, Ulu et al. 2017, Kim et al. 2013 & International Energy Agency IEA). Global automotive sector consumes around 30% in the world, in which this is the 2nd largest producer of CO_2. In 1997 India consumed around 57% of energy and in 2010 it's increased to 85%. It's expected to reach 92% by the year of 2020.Today the automobile sector in India is the upcoming sector which will boom up in the future. This sector has been growing at a CAGR in excess of 15% over the last 5–7 years. It contributed 2.77 to 6% in improving the national GDP rate. India has 19 passenger car manufacturers, 14 commercial vehicles, 16 2/3 wheelers and 12 tractors along with these 5 manufacturers of Engines.

3   DESIGN AND ANALYSIS OF INDUCTION MOTOR

Induction motor is the best choice for traction applications. The characteristics of the different motors are compared and the results are depicted for the conclusion of choosing he motors.

The electric motor is the heart of the traction vehicles, the motor can be designed by considering the major mechanical loading such as yoke thickness, frame size, parameters such as poles, frequency, rated speed, slip, the electrical loading parameters such as number of conductors. The magnetic circuit of the motor is constituted by the lamination of the stator, the rotor, and the air-gap. The energy conversion is assisted by the flux in the air-gap, driven by magneto-motive-force (mmf) produced in the stator winding. The mmf required to drive the flux is influenced by the reluctance of the magnetic circuit. The reluctance of the magnetic circuit is determined by the length and the relative permeability of the material as given in Equation 1.

$$s = \frac{l}{\mu_o \mu_r A} \tag{1}$$

where S = Reluctance, l = Length of the magnetic circuit, μ_o = Permeability of free space, μ_r = Relative permeability and A = Area of the magnetic circuit. The mean length of the

Table 1.   Comparison of different motors.

Characteristics	DC motor	Induction motor	PM motor	SR motor
Power Density	2.5	3.5	5	3.5
Efficiency	2.5	3.5	5	3.5
Controllability	5	5	4	3
Reliability	3	5	4	5
Technological Maturity	5	5	4	4
Cost	4	5	3	4

magnetic circuit is influenced by the shape of the slot and the air-gap. Material composition of stamping influences the relative permeability and the saturation factor coefficient. Reduced reluctance circuit needs less mmf to force the flux, and hence results in less magnetizing current. Hence, the output power, power factor, and efficiency improve significantly. The electromagnetic power (Pm), otherwise called as the air-gap power, is determined by Equation 2.

$$P_m = 3I_2^2 \frac{R_2}{S} \tag{2}$$

The electromagnetic torque (Tm) is calculated using Equation 3.

$$T_m = \frac{P_m}{\omega} \tag{3}$$

where, 'ω' denotes the synchronous speed in rad./s. Equation 4 gives the mechanical shaft output torque.

$$(T_{sh}) \cdot T_{sh} = T_m - T_{fw} \tag{4}$$

where, T_{fw} denotes the frictional and windage torque. The output power (Po) is shown in Equation 5.

$$P_o = T_{sh} - \omega_r \tag{5}$$

where, $\omega_r = \omega(1 - s)$ denotes the rotor speed in rad./s.

Equation (6) is used to calculate the input power to the motor (P_i).

$$P_i = P_o + P_{fw} + P_{rc} + P_{cl} + P_{sc} + P_l \tag{6}$$

where, P_{fw}, P_{rc}, P_{cl}, P_{sc}, and P_l denote the frictional and windage losses, the rotor copper loss, the iron-core loss, the stator copper loss, and the stray loss, respectively.

The power factor is determined by Equation (7).

$$\cos\varphi = \frac{P_1}{3V_1 I_1} \tag{7}$$

Equation 8 is used to determine the efficiency

$$\eta = \frac{P_o}{P_i} \times 100 \tag{8}$$

Induction motor can be modelled by using the equations, further the performance analysis and the starting torque improvement cane be feasible with the improvement in the rotor shapes. The single cage and double cage rotors of the induction motor is constructed and the performance is compared.

4   COMPARSION BETWEEN SINGLE CAGE AND DOUBLE CAGE ROTORS

The 15 kW Induction motor is designed for meeting for the traction applications, the slip is chosen as 4% and no of pole pair is 2. The stack height of the machine is designed as 165 mm. The optimal design of the rotor slot is designed, compared with the double cage bar with better efficiency, better insulated fill factor.

Table 2. General specifications of the motor.		Table 3. Rotor design specifications.	
Supply type	Voltage driven	Inner diameter	33.5 mm
Rated Power	15 kW	Outer Diameter	108 mm
Supply Voltage	38 V	Slot Depth	20.4 mm
Synchronous Speed	1500 rpm	Tang Angle	20
Rated Slip	4%	Slot Opening Width	0.8 mm
Air Gap	0.5 mm	Tooth Tip Thickness	0.5 mm
Stack Height	165 mm	Tooth Width	6.135 mm
No of Poles	4		

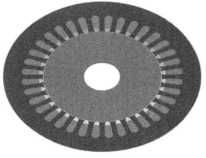

Figure 1.   Induction motor.

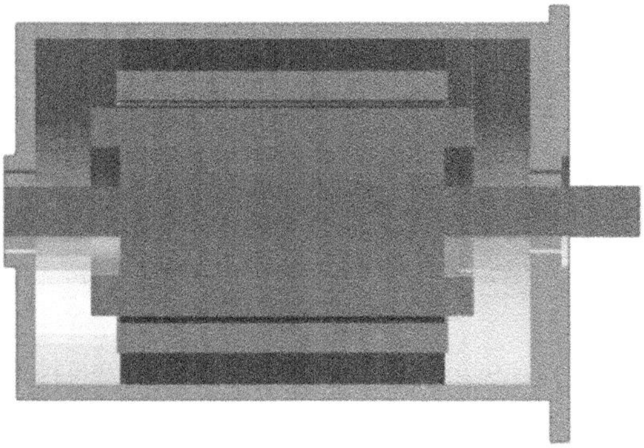

Figure 2.   Axial view if IM.

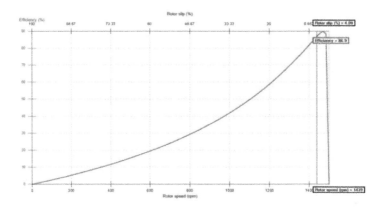

Figure 3.   Efficiency of IM.

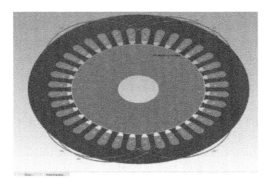

Figure 4.   Winding design.

Table 4. Stator design.		Table 5. Winding design.	
Inner diameter	109 mm	Slot Liner thickness	0.18 mm
Outer Diameter	180 mm	Coil separator thickness	0.2 mm
Slot Depth	19.157	Slot wedge thickness	3 mm
Slot Opening width	3 mm	Strand diameter	2.4 mm
Tooth Tang angle	50 mm	Coil span	9
Tooth Tip Thickness	0.5 mm	Number of layers	1
Tooth Width	5 mm	Number of turns	11

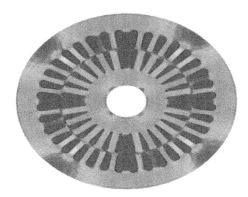

Figure 5. Flux density plot.

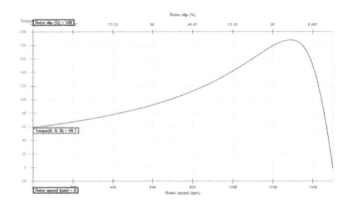

Figure 6. Speed vs Torque.

The Induction motor with single cage is found to be 59.1 N-m and the rated torque of the Induction motor is found to be 108 N-m. The induction motor with single cage is analyzed and performance is compared. In general the double cage induction motor is capable of producing high starting torque and less current and better efficiency during running condition. Further the double cage rotors have two classifications with different material for both the cages and other one with different material for both the cages. In other case a gap is allowed between the two rotor cages

The performance of the machine like torque, starting current, power factor, all these parameters depends on the material which we choose. The properties of the material have the direct effect on the performance on the machine. The BH Curve of the material shows the properties of the material, the performance of the material depends on the material properties and its characteristics. The two materials M19 and M47 have been considered for choosing the

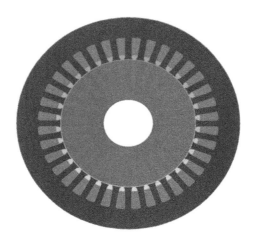

Figure 7.   Double cage rotor bar.

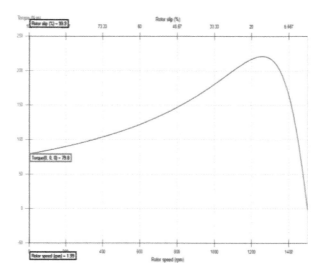

Figure 8.   Starting torque of double cage rotor.

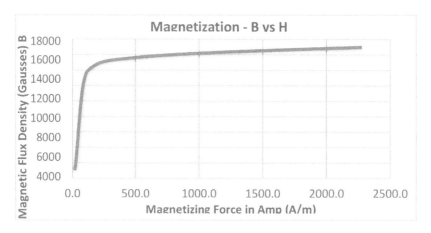

Figure 9.   M-47 fully-processed non-oriented silicon steel [.0185 inch (.47 mm, 26 gauges)].

Table 6. Comparison of aluminium and copper bar.

Rotor material	Aluminium	Copper
Loss—Total (kW)	3.54	2.42
RMS current density (A/mm²)	15.21	11.89
Rotor bar mass	0.8513	2.811
Efficiency (%)	85.56	87.04

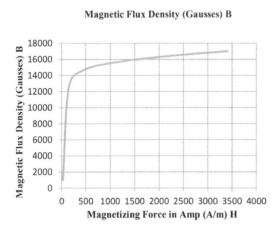

H	B
23.5	1000
34.0	2000
48.8	4000
69.3	7000
97.2	10000
132	12000
168	13000
262	14000
604	15000
995	15500
1577	16000
2382	16500
3405	17000

Figure 10. M-19 fully-processed non-oriented silicon steel (.0185 inch (.47 mm, 26 gauges).

analysis. Two different rotor materials have been chosen for the analysis, and the performance of the machine is presented.

5 CONCLUSIONS

The Induction motor for the traction application is designed and analyzed using FEM analysis. The starting torque for single cage rotor is 59.1 N-m, and the torque for double cage induction motor is found to be 79.8 N-m, the performance of the induction motor is found to be improved with the double cage rotor bar. Along with these the rotor bar material is compared and the performance is shown above, it's clear that the Induction motor produces better efficiency with copper bar. Thus, the induction motor can be best opted for the traction applications with minimizing the cost of the special motors.

REFERENCES

AbdElhafez, A.A., Aldalbehia, M.A., Aldalbehia, N.F., Alotaibi, F.R., Alotaibia, N.A., and Alotaibi, R.S. 2017. Comparative Study for Machine Candidates for High Speed Traction Applications. International Journal of Electrical Engineering 10(1): 71–84.

Casadei, D., Mengoni, M., Serra, G., Tani, A., Zarri, L., & Cabanas, M.F. 2010. Energy-efficient control of induction motors for automotive applications, XIX International Conference on Electrical Machines-ICEM; Proceedings of a meeting, Rome, Italy, 6–8 September 2010. IEEE Explore.

International Energy Agency (IEA), Clean Energy Ministerial, and Electric Vehicles Initiative (EVI) (June 2017). "Global EV Outlook 2017: Two million and counting" (PDF). IEA Publications. Retrieved 2018-01-20. Available at pp. 5–7, 12–22, 27–28, and Statistical annex., pp. 49–51.

Jeon, K.W., & Hahn, S.C. 2011. Optimal Design of Induction Motor Rotor Slot Shape for Electric Vehicle by Response Surface Method. Journal of the Korean Institute of Illuminating and Electrical Installation Engineers 25(11): 58–66.

Kim, B., Lee, J., Jeong, Y., Kangi, B., Kim, K., Kim, Y. & Parki, Y. 2013. Development of 50 kW Traction Induction Motor for Electric Vehicle (EV), IEEE Vehicle Power and Propulsion Conference; Proceedings of a meeting, Seoul, South Korea, 9–12 Oct. 2012. IEEE Explore.

Ulu, C., Korman, O. & Kömürgözm, G. 2017. Electromagnetic and Thermal Analysis/Design of an Induction Motor for Electric Vehicles, IEEE 8th International Conference on Mechanical and Aerospace Engineering (ICMAE); Proceedings of a meeting, Prague, Czech Republic, 22–25 July 2017. IEEE Explore.

Wang, T., Zheng, P., Zhang, Q., & Cheng. S. 2005. Design characteristics of the induction motor used for hybrid electric vehicle. IEEE Transactions on Magnetics 41(1): 505–508.

Zeraoulia, M., Benbouzid, M.E.H. and Diallo, D. 2006. Electric Motor Drive Selection Issues for HEV Propulsion Systems: A Comparative Study. IEEE Trans on Vehicular Technology 55(6):1756–1764.

Automatic Control, Mechatronics and Industrial Engineering – He & Qing (Eds)
© *2019 Taylor & Francis Group, London, ISBN 978-1-138-60427-8*

Fault analysis based on coil currents of a circuit breaker in a converter station

J.B. Li & R. Hu
Hubei Electric Power Research Institute, Wuhan, China

A. Hu
Wuhan University, Wuhan, China

Y.X. Wen, Y.L. Li & Z.Q. Xiong
State Grid Ezhou Electric Power Supply Company, Ezhou, China

L.K. Chen
State Grid Xianning Electric Power Supply Company, Xianning, China

ABSTRACT: The high-voltage circuit breaker is commonly used electrical equipment, and the opening/closing coil is the key component of its operating mechanism. The coil current reflects the motion characteristic of the circuit breaker. This paper explains the structure of the opening/closing coil, and then describes a method to diagnose a circuit breaker fault based on detecting its opening/closing current. The method is applied to a 550 kV circuit breaker in a converter station. It shows that coil current detection of a circuit breaker is an effective method of identifying some hidden mechanical defects in the circuit breaker. Based on this method, some previously undiscoverable mechanical faults could be diagnosed, and the result is useful to the security and stability of the power system.

Keywords: high-voltage circuit breaker, opening/closing coil, coil current detection, fault diagnosis

1 INTRODUCTION

The opening/closing coil is an important component in the operating mechanism of a high-voltage circuit breaker. As a primary control unit in the circuit breaker, the structure of the opening/closing coil is generally a kind of solenoid electromagnet. When a current passes through the opening/closing coil, a magnetic flux is generated, and the iron core inside the coil is magnetically attracted, thereby controlling the circuit breaker in the completion of the opening or closing operation. Therefore, the function of the opening/closing coil is to convert the electrical energy from the power source into magnetic energy, and then into mechanical work output through the action of the iron core (Xu et al., 2000).

During the opening and closing process, the coil current changes with time. The current waveform reflects the status of the opening/closing coil itself and the control mechanism during operation. It contains a large amount of information that can be used for mechanical fault diagnosis. By monitoring the current of the opening/closing coil, the starting time of the operating mechanism, the movement time of the iron core, and the energization time required by the coil can be calculated. This can be utilized to judge the distance of travel of the iron core, the core binding, the coil status (such as whether there is a turn-to-turn short circuit) and the action status of the operating mechanism (Dai & Jiao, 2011; Bian, 2012; Chen, 2009).

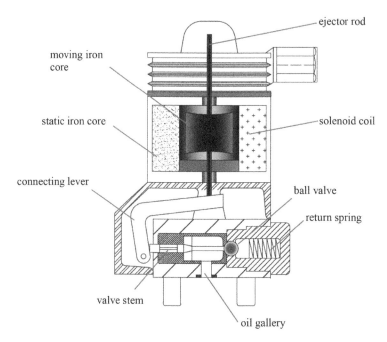

Figure 1. Structure of opening/closing coil.

2 STRUCTURE OF OPENING/CLOSING COIL

The opening coil and closing coil have the same structure, as shown in Figure 1. In principle, the coil consists of three individual parts: the electrical part, the mechanical part and the hydraulic part. The electrical part is mainly composed of a solenoid coil, moving iron core, static iron core and ejector rod. The mechanical part includes a connecting lever and its working chamber. The hydraulic part includes a valve stem, ball valve, return spring and oil gallery.

The ejector rod is fixed in the moving iron core. When the solenoid coil passes current, the iron core is subjected to electromagnetic force and drives the ejector rod downward. The bottom end of the ejector rod presses the connecting lever and opens the ball valve through the valve stem. Then, the high- and low-pressure oil galleries of the circuit breaker hydraulic mechanism are penetrated. When the current in the solenoid coil disappears, under the hydraulic pressure generated by the ball valve and the return spring, the connecting lever and the ejector rod are reset, and the valve is thus closed.

3 CURRENT OF OPENING/CLOSING COIL

Laboratory and on-site inspection experience show that the current waveform of an opening/closing coil has good repeatability (Meng et al., 2006). The normal current waveform is shown in Figure 2.

According to the current waveform of the opening/closing coil, the current has two peak points and one valley point, and the waveform can be divided into four stages according to the movement of the iron core (Song et al., 2011; Yuan et al., 2011; Zhang et al., 2013; Zhong et al., 2011):

- Stage 1: T_0–T_1. T_0 is the starting point of the circuit breaker opening and closing operation. T_1 is the starting moment of the core movement, when the current in the coil rises enough to drive the movement of the core. This stage is characterized by an exponential rise in current and the iron core is at rest. T_1 is related to the controlling power voltage and the coil resistance.

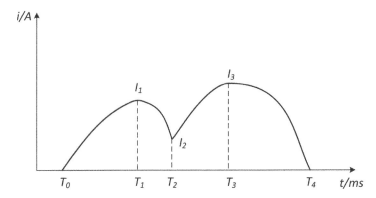

Figure 2.   Current waveform of opening/closing coil.

I_1 is the starting moment current of the iron core. This value is the maximum current during the T_0–T_1 section and reflects the initial position of the iron core and the resistance at idle.

- Stage 2: T_1–T_2. The core moves to produce a counter-electromotive force and the coil current drops. T_2 is the valley point of the control current and represents the actuation by the core of the load of the operating mechanism, thus significantly decelerating or stopping its movement. It is the total movement time of the core.
  As the counter-electromotive force is added to the circuit, the coil current will decrease. I_2 is the core stop moment current, and is the minimum current during the T_1–T_2 section. It reflects the magnitude of resistance during over-travel and the maximum movement speed of the core.
- Stage 3: T_2–T_3. The iron core stops moving and the current rises exponentially. I_3 is the maximum working current of the coil. This value is the maximum current between T_2 and T_3. It is the maximum working current of the coil when the moving iron core stops. It reflects the DC resistance of the coil, and whether there is a short circuit inside the coil or a loose-joint failure.
- Stage 4: T_3–T_4. When the auxiliary switch is turned off, an arc is generated and jumps between the auxiliary switch contacts. The arc voltage rises rapidly, forcing the current to decrease until it is extinguished at time T_4.

Taking T_0 as the zero point of the commanding time, the mechanical characteristics of the circuit breaker can be obtained by extracting the four time-characteristic parameters of T_1, T_2, T_3 and T_4 and the three current-characteristic parameters of I_1, I_2 and I_3 from the current waveform (Mu et al., 2013).

4   DETECTION OF OPENING/CLOSING COIL CURRENT

Since the 550 kV Gas-Insulated Switchgear (GIS) equipment of a ±500 kV converter station was first put into operation, failures during the circuit breakers' reclosing have occurred many times, which has seriously affected the safe and stable operation of the power grid. In the most recent fault, an instantaneous ground fault occurred in line C, and the protection action tripped the 5052 and 5053 circuit breakers. Then, after the protection was issued, the 5052 and 5053 circuit breakers' C-phase reclosing failed. The inconsistent protection action tripped three phases.

In order to analyze more deeply the cause of the circuit breaker failure, the mechanical fault of the circuit breaker is diagnosed on the basis of the current-detection technology of the opening/closing coil. The compensation-loop current sensor is used to detect the coil current signal. The detection object includes the closing coils and the opening coils of the 5052 and 5053 C phases, and the closing and opening coils of the 5053 A phase for comparison.

Table 1. Detection results of opening/closing coil current and resistance.

Coil number	T_1	T_2	T_3	T_4	I_1	I_2	I_3	R
Closing coil								
5053 A	3.8	1.7	4.7	12.4	1.42	0.79	2.13	36.7
5052 C	4.0	1.8	4.5	12.7	1.46	0.79	2.13	36.8
5053 C	3.7	1.7	4.6	12.2	1.44	0.83	2.08	37.7
Opening coil								
5053 A	3.9	1.8	4.6	12.2	1.46	0.83	2.08	37.6
5052 C	3.9	**1.3**	4.7	12.8	**1.21**	0.75	2.17	36.1
5053 C	3.9	**1.3**	4.8	12.5	**1.21**	0.75	2.08	37.6

The time-characteristic and current-characteristic values and the coils' direct resistance are shown in Table 1.

According to the test results, the characteristic value I_3 of each opening and closing coil are basically equal, and the direct resistance of the coils is normal (36 Ω ±10%). Therefore, short circuits inside the coils and loose-joint failures can be discounted.

Comparing the measurement results, the current waveforms of the opening coils of the 5052 and 5053 C phases, in which the reclosing failure occurs, are different from other coils, and the characteristic values T_2 and I_1 (highlighted in bold type in Table 1) are smaller than in other phases.

Through on-site inspection of the circuit breaker operating mechanism, it is found that the operating mechanism is clean, without debris or oil leakage. Therefore, the opening coils of the 5052 and 5053 C phases may have an internal fault.

5 DISASSEMBLY ANALYSIS OF OPENING/CLOSING COIL

The two opening coils were disassembled. Oil traces were found on the ejector rods of the 5052 and 5053 C-phase opening coils, as illustrated in Figure 3. After each ejector rod was inserted downward, the movement of the ejector rod was relatively slow when it rebounded, and it could not automatically be returned normally.

Combined with the current waveform detection result of the opening and closing coils, it is judged that the ejector rods of the 5052 and 5053 C-phase opening coils are not returning to position before the fault. The iron core and the ejector rod are in the intermediate stroke position before the opening coil is energized, and the electromagnetic attraction force is thus larger than in the idle position. Therefore, the required current (I_1) for driving the core and the ejector rod to overcome the resistance is smaller. At the same time, due to the incomplete return, the travel distance of the core is shorter, so the core's movement time (T_2) is shorter.

Under normal working conditions, when the circuit breaker is closing, the closing coil will be energized. Then, the closing control valve is opened, and the high-pressure oil flows from the closing control valve to the lower side of the piston rod, driving the piston rod upward, and then the circuit breaker is closed. At this time, on one side of the opening coil is high-pressure oil, and on the other side is low-pressure oil (Rao et al., 2012; Rao, 2012).

If there exists a fault in the opening coil, causing the ejector rod and the iron core to fail to return completely, the valve in the opening coil cannot be closed, and the high- and low-pressure hydraulic circuits cannot be completely separated. During the closing process, some of the high-pressure oil flows back to the low-pressure fuel tank through the opening coil, so that the driving force of the high-pressure oil on the closing circuit is insufficient, and the closing operation cannot be completed.

According to the on-site inspection, during the operation of the circuit breaker corresponding to the opening coil fault, the disc spring displacements of the mechanisms are significantly larger than their normal values.

oil trace

Figure 3.   Oil trace on ejector rod.

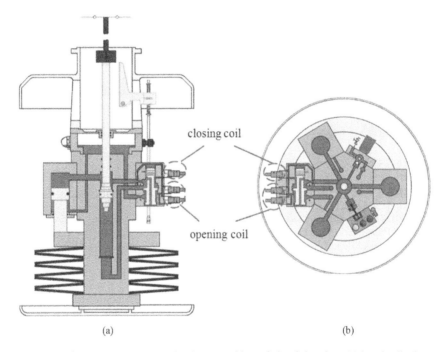

closing coil

opening coil

(a)   (b)

Figure 4.   Operating mechanism status in closed position of circuit breaker: (a) longitudinal cross section; (b) horizontal cross section.

The mechanism state for the circuit breaker's closing position is shown in Figure 4; the red elements represent high-pressure oil and the blue elements represent low-pressure oil.

The defective opening coils were disassembled to further analyze the causes of the ram ejector rod and oil traces. After disassembly, it was found that there was more grease on the connecting lever. Although the outer seal of the coil was tight, there was no special 'internal seal' between the mechanical and electrical parts of the coil. The overuse of grease had caused some of the grease to enter the electrical portion of the coil along the ejector rod, resulting in an increase in friction between the ejector rod and the core. When the coil is energized, the electromagnetic force is large enough to overcome the frictional force to drive the core and the ejector movement. Once the coil is de-energized, the ejector rod mainly relies on the return spring to return to position; however, the return spring has less force, and this is insufficient to overcome the frictional force and drive the ejector rod to return to its original position.

The overuse of grease in the connecting lever section is caused by poor process control by the manufacturer. According to the manufacturer's standard process requirements, it is only necessary to use a small amount of grease at the contact areas between the connecting lever, the stem and the ejector rod. However, it was found that grease had been applied everywhere on the connecting lever. A brush is used to apply grease to the connecting lever in the factory but it is difficult to apply a small amount of grease evenly using this type of tool. In addition, under the influence of the heater inside the mechanism, the internal temperature is higher and lowers the viscosity of the grease on the connecting lever. Together, the overuse of grease and the higher ambient temperature cause the 'internal leak' of the grease.

6 CONCLUSIONS

A circuit breaker's opening/closing coil current contains a lot of mechanical information about the circuit breaker. By detecting and analyzing the current waveform of the opening and closing coil, a mechanical defect in the circuit breaker can be effectively found. This paper introduces the operating principle of a hydraulic mechanism and the structure of the opening/closing coil. Based on the current-detection technology of the opening/closing coil, the reason for the frequent closing or reclosing failures of circuit breakers in a converter station is analyzed.

At present, the detection technology of the opening/closing coil current is rarely applied in power grid inspection. The conclusions of this paper can provide a reference for the wider application and popularization of this technology in the power grid.

REFERENCES

Bian, H.W. (2012). *Study on on-line monitoring and fault diagnosis system for high voltage circuit breaker.* Jiangsu, China: Engineering, Yang Zhou University.

Chen, T. (2009). *The software development of vacuum circuit breaker's mechanical characteristics testing based on LabWindows/CVI.* Sichuan, China: Power Electronics and Power Drives, Xi Hua University.

Dai, Q.W. & Jiao, F. (2011). Research on online monitoring of high voltage circuit breaker mechanical characteristic. *Power System Technology, 7,* 5–9.

Meng, Y.P., Jia, S.L. & Rong, M.Z. (2006). On-line monitoring method for mechanical characteristics of vacuum circuit breakers. *High Voltage Apparatus, 42*(1), 31–34.

Mu, T., Zhou, S. & Wang, X. (2013). Data acquisition in the mechanical condition monitoring of circuit breaker. *Low Voltage Apparatus, 3,* 26–30.

Rao, H.L. (2012). Cause analysis and preventive measurement of GIS open/close coil fault. *Hubei Electric Power, 36*(4), 14–15.

Rao, H.L., Yao, Q.X., Zhou, G.Y., Rao, L., Chen, F., Zhang, Y. & Liu, X. (2012). Reason analysis and preventive measurement to three-phase discrepancy of GIS 5053 circuit breaker in Yidu converter station. *Central China Electric Power, 25*(2), 47–49.

Song, J., Xu, C.Q. & Zhu, T.L. (2011). Coil current monitoring of SF_6 circuit breakers based on waveform identification. *Electric Power Construction, 32*(3), 65–68.

Xu, G.Z., Zhang, J.R., Qian, J.L. & Huang, Y.L. (2000). *The principle and application of high-voltage circuit breaker* (pp. 439–532). Beijing: Tsinghua University Press.

Yuan, J.L., Li, K., Guo, Z., Yue, D.W. & Wang, Y. (2011). Mechanical failure diagnosis of high voltage circuit breaker based on SVM and opening/closing coil current parameters. *High Voltage Apparatus, 47*(3), 26–30.

Zhang, Y.K., Zhao, Z.Z., Feng, X. & Guo, X. (2013). Mechanical fault diagnosis of high voltage circuit breakers based on opening/closing coil current parameters. *High Voltage Apparatus, 49*(2), 37–42.

Zhong, J.X., Li, B.Q. & Li, Y.H. (2011). Exploration and practices of mechanical state diagnosis and monitoring techniques for high voltage circuit-breaker. *High Voltage Apparatus, 47*(2), 53–60.

Automatic Control, Mechatronics and Industrial Engineering – He & Qing (Eds)
© 2019 Taylor & Francis Group, London, ISBN 978-1-138-60427-8

A new design for a reference voltage source

X.J. Li & Y.R. Yu
Zhonghuan Information College, Tianjin University of Technology, Tianjin, China

ABSTRACT: The main purpose of this design is to provide a stable reference voltage for a switch magnetism-sensing circuit. Because of the inherent characteristics of the circuit, the reference voltage is set to 3.3 V. In this paper, the reference voltage source is designed to have a wide input range, a stable output and a good temperature characteristic. The simulation results of the reference voltage and the complete design layout are described. The simulation results based on a 4 µm 25 V bipolar process verify the correctness of the theoretical analysis and the feasibility of the proposed methods.

Keywords: Reference voltage, constant current, temperature coefficient

1   INTRODUCTION

In recent years, the techniques around analog Integrated Circuits (ICs) have undergone big change. This places high demands on the voltage, power consumption, precision and response speed in an Application-Specific Integrated Circuit (ASIC) design. It is becoming increasingly difficult for the circuit structure of a traditional reference voltage source to meet the design requirement (Allen & Holberg, 2005; Gray & Meyer, 1993; Hastings, 2004). At present, there are three main types of design for a reference voltage source: the buried Zener diode, the Extra Implanted Field-Effect Transistor (XFET), and the bandgap reference voltage source. The buried Zener diode and the XFET reference voltage sources feature high precision and good stability. However, the buried Zener diode has a high power supply voltage and high power consumption, and XFET reference voltage source manufacturing is complex and is not compatible with the bipolar process. The traditional bandgap reference voltage source can be compatible with the bipolar process, and a circuit with high precision and low temperature coefficient (Qin et al., 2006). By adjusting the parameters of the emitter area and resistance, an output voltage with zero temperature coefficient can be achieved. However, the voltage is fixed at 1.25 V and cannot meet the switch magnetism-sensing circuit's reference voltage circuit requirements; at the same time, the reference voltage will be affected by the operational amplifier input offset voltage (Bilotti et al., 1997; Huang et al., 2015).

On the basis of this analysis, we propose a novel reference voltage source structure. Our circuit has a small area, low temperature coefficient and stable performance; it does not require changes to the original process (i.e. the original circuit can be smoothly upgraded).

2   ANALYSIS OF THE TRADITIONAL REFERENCE VOLTAGE SOURCE

2.1   *Circuit analysis*

Figure 1 shows a traditional reference voltage source circuit. The circuit is composed of differential amplifiers, P-type Metal-Oxide-Semiconductors (PMOSs) M_1 and M_2, bipolar transistors Q_1 and Q_2, and resistors R_1–R_3. From Figure 1, we can obtain the equation $V_{\text{ref\_conventional}} = V_{be1} + (R_2 V_T \ln n)/R_3$. The reference voltage with zero temperature coefficient is

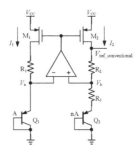

Figure 1. A traditional reference voltage source circuit.

calculated by taking the derivative with respect to temperature: $V_{\text{ref\_conventional}} = V_{be1} + 17.2\,V_T \approx$ 1.25 V, $\partial V_{be}/\partial T \approx -1.5$ mV/K, $\partial V_T/\partial T \approx 0.087$ mV/K. It can be seen that the traditional reference circuit $V_{\text{ref\_conventional}}$ is locked in at 1.25 V.

2.2 Defect analysis

The traditional reference voltage source circuit contains an operational amplifier, and its symmetry is not good. Therefore, the operational amplifier is affected by the input offset voltage, resulting in an error in the reference voltage. The input offset voltage can be quantified as V_{OS}, and the new equations $V_{be1} \pm V_{OS} \approx V_{be2} + R_3 I_{c2}$ and $V_{\text{ref\_conventional}} = V_{be2} + (R_3 + R_2)$ I_{c2} can be derived. So, $V_{\text{ref\_conventional}} = V_{be2} + (1 + R_2/R_3)(V_T \ln n \pm V_{OS})$. As can be seen from the equation, V_{OS} is amplified by a factor of $(1 + R_2/R_3)$. Thus, it is inevitable that an error exists in $V_{\text{ref\_conventional}}$. More importantly, V_{OS} changes with temperature, which increases the temperature coefficient of the conventional reference voltage source. The previous analysis shows that the reference voltage $V_{\text{ref\_conventional}}$ is locked in at 1.25 V. Therefore, it cannot satisfy the reference voltage demand of the switch magnetism-sensing circuit.

3 A NEW DESIGN FOR A REFERENCE VOLTAGE SOURCE

We can obtain the following equation from Figure 2: $V_{be1} = V_{be2} + I_o R_e$. Q_1 and Q_2 have the same parameters; I_o is the output reference current. On the basis of the base-emitter voltage formula of the triode, $V_{be} = V_T \ln(I_c/I_s)$, it is easy to derive Equation 1:

$$I_o = \frac{V_T \ln \dfrac{I_{c1}}{I_{s1}} - V_T \ln \dfrac{I_{c2}}{I_{s2}}}{R_e} = \frac{V_T \ln \dfrac{I_{c1}}{I_{s1}} \times \dfrac{I_{s2}}{I_{c2}}}{R_e} \tag{1}$$

where V_T is thermal voltage ($V_T = kT/q$, where k = 1.38×10^{-23} J is the Boltzmann constant, q is the electron charge and T is absolute temperature); I_c is the collector current and I_s is the inverse saturation current. The parameters of triodes Q_1 and Q_2 are the same ($I_{s1} = I_{s2}$). According to Figure 2, I_{c1} of collector current can be replaced by I_{ref} if we overlook the function of the base current. Similarly, I_{c2} can be replaced by I_o and we know that:

$$I_o = \frac{V_T \, Lambert\,W(R_e I_{ref})}{R_e} \tag{2}$$

I_o is the required current for our method and we can see that I_o is constant. Therefore, a new reference voltage source circuit is designed based on this source circuit; the circuit is composed of a reference voltage source circuit, start-up circuit, isolating circuit, feedback loop and switch magnetism-sensing circuit, as shown in Figure 3. The switch magnetism-sensing circuit is not our focus, so we will not describe it further in this paper.

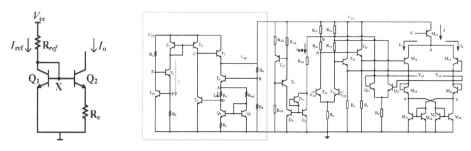

Figure 2.   The current source.   Figure 3.   The new reference voltage source circuit.

3.1   Zero temperature coefficient reference voltage source circuit

The reference voltage source circuit is composed of Q_1, Q_2, T_4, T_5, R_3, R_e, R_{ref}, R_6 and R_8. From Figure 3 we know that the voltage at point A is:

$$I_o R_3 + V_{be\_T_5} = I_{ref} R_{ref} + V_{be\_Q_1} \tag{3}$$

where V_{be_T5} and V_{be_Q1} approximately equal V_{be} (on). So:

$$I_o = \frac{R_{ref}}{R_3} I_{ref} \tag{4}$$

Substituting Equation 2 into Equation 4, we can obtain:

$$\frac{V_T Lambert W (R_e I_{ref})}{R_e} = \frac{R_{ref}}{R_3} I_{ref} \tag{5}$$

From Figure 3 we can see that $V_A = I_{ref} R_{ref} + V_{be1}$, $V_B = V_A + V_{be4}$. Because $V_{ref} = V_B (R_6 + R_8)/R_8$, V_{ref} can be expressed as:

$$V_{ref} = \frac{R_3 V_T ln\left(\dfrac{R_3}{R_{ref}}\right) + V_{be1} R_e + V_{be4} R_e}{R_e R_8}(R_6 + R_8) \tag{6}$$

The ratio of each resistance and zero temperature coefficient voltage of V_{ref} can be obtained from the derivative with respect to temperature. At the same time, the range of reference voltage is changed by adjusting the resistances of R_6 and R_8.

3.2   Start-up circuit

The function of the start-up circuit is to make the circuit avoid the zero state when the circuit starts, and enter the working state without disturbing normal function. The start-up circuit includes transistors T_1, T_2, T_3, T_9 and T_{10} and resistors R_1 and R_2. When the circuit is powered, the voltage at point E rises up and turns on the T_9. T_9 provides a path for the current mirror. When the current mirror composed of T_1 and T_2 is turned on, the voltage at point C rises and T_3 turns on. The reference voltage regulator circuit (inside the red dotted box in Figure 3) starts to work.

3.3   Isolating circuit

Once the reference voltage regulator circuit is working, the voltage at point D rises and T_5 turns on. Then the voltage at point C is pulled low and T_3 turns off. The start-up circuit is separated from the reference voltage regulator circuit. The reference voltage can be set independently of the supply voltage.

4 SIMULATION

Based on the performance indexes of the relevant modules and the relative parameters of the existing commercial chips, the main technical indexes of the overall circuit and the reference voltage source circuit are determined, as shown in Tables 1 and 2. The key point of the test is whether the design will affect the switch magnetism-sensing circuit when the supply voltage varies between 4.5 and 24 V.

In this paper, the circuit design provides a reference voltage for the switch magnetism-sensing circuit. Therefore, we first observe whether the overall circuit functions correctly.

The simulation results are shown in Figure 4, and the magnetic signal is simulated with a sinusoidal signal. When the magnetic field intensity is higher than the upper limit threshold, the circuit output will be HIGH, and when the intensity is lower than the lower limit threshold, the circuit output will be LOW. The performance index conforms to design requirements. The output rise time t_r is the time interval for the response curve to increase from 10% to 90%: $T_{M0} - T_{M2} \approx (1.91809-1.91794)$ ms = 0.05 μs, as shown in Figure 4a. The result is less than 0.2 μs, which conforms to the requirements of the design. The output fall time t_f is the time interval for the response curve to fall from 90% to 10%: $T_{M1} - T_{M3} \approx (1.91809-1.91794)$ ms = 0.05 μs, as shown in Figure 4b. Again, the result is less than 0.2 μs, which conforms to the requirements of the design.

Next, we need to verify whether the design will be affected by the reference voltage source circuit and the switch magnetism-sensing circuit when the supply voltage varies from 4.5 to 24 V.

The output voltage stability is an important basis for reference circuit performance. According to the above analysis, by substituting relevant design parameters into Equation 6, we can see that the reference voltage value is approximately 3.3 V. At a temperature of 25°C, the simulation result for reference voltage V_{ref} is shown in Figure 5b (DC scan of the reference voltage). The output voltage only changes by 87.0304 mV when the input voltage changes from 4.32 to 24 V, and the output error is 0.47%, which is far lower than 1%: thus, it meets the performance requirements.

Another important indicator of the reference source is its stability over a wide temperature range. Temperature Coefficient (TC) is an important parameter in measuring the temperature change of the output voltage of a bandgap reference voltage source. The simulation

Table 1. Overall performance indicators of circuit.

	Magnetic field	In-phase
Ambient temp. = 25°C	Supply voltage	4.5–24 V
	Output rise time	0.2 μs
	Output fall time	0.2 μs

Table 2. Performance indicators of reference voltage source.

Supply voltage	4.5–24 V
Reference voltage stability	±1%
Temperature coefficient (TC)	20 ppm/°C

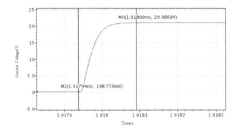 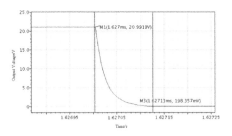

Figure 4. Simulation results: (a) the output rise time; (b) the output fall time.

120

result is shown in Figure 5a. The reference voltage changes 9.23102 mV when the temperature changes from 0 to 150°C and $V_{DD} = 10$ V. The TC of the voltage reference source is 18.65 ppm/°C, which meets the design requirement.

As shown in Figure 6, with power supply voltages of 5, 15 and 21 V, the curve of the reference voltage with temperature change is obtained. Under a constant supply voltage, the temperature is changed from 0 to 150 °C and the error in the reference voltage is lower than 27 mV. Under constant temperature, the supply voltage is changed from 5 to 21 V and the error in the reference voltage is lower than 90 mV.

In this paper, the design of the chip layout is completed using a BCD 4 μm 25 V bipolar process. Figure 7 shows the chip layout and the location of the reference voltage source. The layout is connected by a single layer of metal wire, with an area of 1.225692 mm² (length × width = 1,164 μm × 1,053 μm). In this chip, the layout area of the reference voltage source is 635 μm × 236 μm, and the decoupling capacitance is connected to the ground at the output end of the reference voltage. A test circuit and test environment is built for the sample.

The chip test results are shown in Table 3 and demonstrate that the simulation and experimental results have good consistency. The test results show that as the supply voltage changed

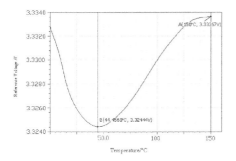 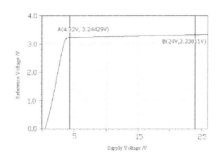

Figure 5. Simulation results for reference voltage: (a) at 25°C; (b) as the temperature changes from 0 to 150°C.

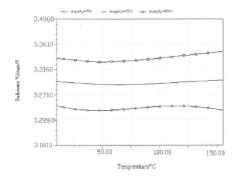 

Figure 6. Plot of reference voltage vs temperature.   Figure 7. Chip layout.

Table 3. Chip test results.

Ambient		Result		
Supply voltage (V)	Magnetic field (GS)	B_{OP} (GS)	B_{RP} (GS)	B_{HY} (GS)
5	0–500	149.7	100.03	49.67
10	0–500	150.4	100.01	50.39
15	0–500	150.3	100.02	50.28
20	0–500	149.0	100.08	48.92
25	0–500	150.2	100.05	50.15

from 5 to 25 V, the operating point (B_{OP}), release point (B_{RP}) and hysteresis (B_{HY}) of the chip remained about the same. That is to say, the chip can work normally under a 5–25 V supply voltage.

Compared with the traditional reference, we can see that the improved chip supports a wider supply voltage range, its layout area is smaller, and its reference voltage is higher (Bilotti et al., 1997; Huang et al., 2015; Zeng et al., 2018).

5 CONCLUSION

This design provides a stable reference voltage for the switch magnetism-sensing circuit. The simulation results show that the output voltage only changes 87.0304 mV when the input voltage changes from 4.32 to 24 V, and the output error is only 0.47%. At the same time, the reference voltage changes 9.23102 mV when the temperature changes from 0 to 150°C, and the temperature coefficient is 18.65 ppm/°C.

REFERENCES

Allen, P.E. & Holberg, D.R. (2005). *CMOS analog circuit design* (2nd ed.). Beijing: Publishing House of Electronics Industry.

Bilotti, A., Monreal, G. & Vig, R. (1997). Monolithic magnetic Hall sensor using dynamic quadrature offset cancellation. *Solid-State Circuits*, 3(6), 829–836.

Gray, P. & Meyer, R. (1993). *Analysis and design of analog integrated circuits*. New York: Wiley.

Hastings, A. (2004). *The art of analog layout*. Beijing: Tsinghua University Press.

Huang, H.Y., Wang, D.G. & Xu, Y. (2015). A monolithic CMOS magnetic Hall sensor with high sensitivity and linearity characteristics. *Sensors*, 15(10), 27359–27373.

Qin, B., Jia, C. & Chen, Z.L. (2006). A 1V MNC bandgap reference with high temperature stability. *Chinese Journal of Semiconductors*, 27, 2035–2038.

Zeng, Y.C., Xia, J.Y. & Cui, J.J. (2018). Design of a high performance CMOS bandgap voltage reference with piecewise linear compensation. *Electronic Components & Materials*, 3, 73–77.

The algorithm design for alignment and foreign object detection of the improved balanced coil used in wireless charging systems

J.J. Miao & H. Chang
State Grid Hebei Electric Power Co. Ltd., Beijing, China

B. Zhou & Z.K. Yue
School of Electrical Engineering, Shandong University, Shandong, China

Q. Liu
Shandong Luneng Intelligent Technology Co. Ltd., Shandong, China

ABSTRACT: Based on the improved balanced coil, the detection algorithm that considers the wireless charging alignment and foreign object detection is proposed. The detection method can solve the problem of alignment detection and metal detection before and during the process of wireless charging. We have designed a software algorithm, including the algorithms for the charging process and the comprehensive inspection process. In conclusion, the experiment shows that the algorithm is valid.

Keywords: foreign object detection, improved balanced coil, wireless charging system, algorithm

1 INTRODUCTION

As a new charging method, wireless charging technology is receiving more and more people's favor. However, a new problem has arisen. On the one hand, transmission power and transmission efficiency are affected by the alignment accuracy. On the other hand, part of power is consumed on metal due to the eddy current effect when metal is mixed in the energy transmission region, which greatly increases the charging time. More seriously, lots of heat is generated due to the metal eddy current effect. The sharply rising temperature can easily lead to serious safety accidents, which could cause great harm to people's lives and property. Therefore, the problem of alignment and foreign object detection in the wireless charging system urgently needs to be solved (Zhou, 2018; Wang, 2010; Ma et al., 2017; Bai et al., 2017).

A set of detection methods is proposed in this paper, which completely solves the problem of alignment detection, metal foreign object detection, both before charging and during charging. The software algorithm is designed to include the detection algorithm in each charging process and the complete detection flow chart.

2 STRUCTURE OF THE IMPROVED BALANCED COIL

In this paper, an improved balanced coil structure is proposed, which can be used in metal detection during the process of wireless charging. It takes into account both the wireless charging alignment and metal foreign object detection.

The coil adopts a figure of eight structure, including part A and part B, as shown in Figure 1. Ideally, when there is no foreign matter in electromagnetic coupling area, the electromagnetic fields passing through the two detection coils A and B are completely the same. According to the principle of electromagnetic induction, the electromotive force induced by

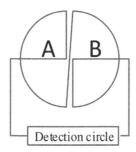

Figure 1. The improved balanced coil structure.

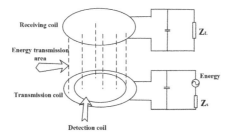

Figure 2. Balanced coil placed in magnetic field.

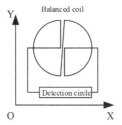

Figure 3. The diagram of alignment detection.

the A and B coils are equal in magnitude and opposite in direction. Since the electromotive forces of the A and B coils are cancelled, the total output voltage of the balanced coil is zero. However, when foreign objects are mixed in the charging system, the magnetic fields passing through A and B are not the same and the induced electromotive force is not equal. Therefore, the balance coil terminal voltage is not zero. According to this terminal voltage signal, we can judge that metal foreign metal matter has been mixed in the magnetic field coupling region (Yang et al., 2015; S. Yamazaki et al., 2002).

The balanced coil is laid on the surface of the transmitting coil, coupled with the transmitting coil and the receiving coil without occupying extra space. As shown in Figure 2, the structure has the advantages of simple device and high sensitivity.

When the balanced coil is used for alignment detection, the principle is similar. As shown in Figure 3, when the transmitter and receiver coils are offset in the X-axis direction, the terminal voltage of the balanced coil is not zero. By detecting the voltage value of the balance coil, it is determined whether the wireless charging system is accurately positioned in the X-axis direction.

3 THE DETECTION ALGORITHM

Alignment and foreign object detection circuit includes: rectification and filtering circuit; voltage amplification circuit; comparator circuit; Arm processor. Figure 4 shows the process flow of detection signal.

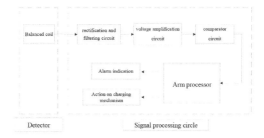

Figure 4. The flow chart of detection signal processing.

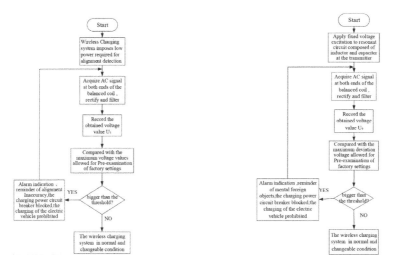

Figure 5. The flow chart of alignment detection.   Figure 6. Metal detection chart before charging.

3.1 *Alignment detection*

In order to solve the problems that the unbalanced voltage caused by the alignment is much larger than the imbalance voltage caused by metal foreign metal objects, as charging system applies low-power operation during alignment detection. In the comparison of the coil voltage and the alignment threshold, if it is greater than the maximum voltage allowed for the alignment, a command to stop charging is issued and a warning for re-alignment is issued to the user (Qu, 2016; Low et al., 2010). The detection flow is as shown in Figure 4.

3.2 *Metal foreign objects detection*

For metal detection in wireless charging, this article studies two phases of foreign object detection, including metal detection before charging and metal detection during charging. The following two separate metal detections are separately described.

Before the start of charging, only the transmitter coil exists. In the experiment, a fixed AC voltage was applied to the transmission coil to record the terminal voltage of the balance coil when there are no metal foreign objects. When there is metal placed over the transmitting coil, the terminal voltage of the balance coil is different from the situation when there is no metal foreign object. When the voltage change value is greater than the threshold, the processor considers that there is metal foreign matter mixed in, and therefore it will alert the user, whilst at the same time, the charging mechanism will be locked to be in a non-chargeable state. The detection process is shown in Figure 6.

In order to solve the problem of unbalanced voltage caused by coil winding asymmetry and alignment inaccuracy, the feedback voltage $-U2$ is introduced. First, the unbalanced voltage is completely cancelled out by the feedback voltage to ensure that the balance coil terminal voltage is zero at the charging start time, and then the metal detection in the charging process can be performed.

Through experimental observation, the power fluctuations in normal charging cause the little sudden changes in the balance coil terminal voltage. Therefore, the voltage is collected at intervals of one second in each sampling period, and the average value is acquired three times in succession. This value is compared with the value measured in the next cycle, so that the influence of power fluctuation on the detection can be avoided. In addition, large power bursts occur due to the switching time of the charging phase. At the moment of phase transition, foreign object detection is stopped. After waiting for a smooth transition to the new charging stage, restart the foreign object detection. The detection flow chart is shown in Figure 7.

An improved balance coil detection method is used to simultaneously handle two issues of wireless charging alignment and foreign matter detection. This article proposes innovatively a set of alignment and metal foreign object detection solutions, which are divided into three phases. The specific solution is as follows.

The first stage is metal detection before charging. At the factory, a fixed AC voltage is applied to the transmitting coil to record the terminal voltage as a factory value without metal foreign matter. When there is metal placed over the transmit coil, the balance coil terminal voltage changes. Because the transmission magnetic field is weak and the metal foreign body causes a small change in the balance coil terminal voltage, an amplifier is used to amplify the voltage of the collected balance coil by ten times. When the voltage change value is greater than the threshold, the processor considers that there is a metal foreign object mixed in to alert the user and at the same time the charging mechanism is locked to be in a non-chargeable state. If it is determined that there is no foreign matter mixed in, the charging system will be in a chargeable state and the device waits to enter the second stage.

The second stage is alignment detection. If it is determined in the first stage that there is no foreign matter mixed in, the electric device will be allowed to enter the charging area for alignment. Applying low power to charge after alignment, the balanced coil balancing voltage is collected in this low-power charging situation. At this time, the unbalanced voltage on both sides of the coil is compared with the allowable threshold voltage of the alignment. If the voltage is greater than the threshold voltage, the charging instruction will be stopped and will give the user a reminder of re-alignment. If the alignment is normal, it will enter the third stage.

The third stage is metal detection during charging. After the alignment detection, the device starts charging. The unbalanced voltage at this time is first recorded, and the feedback is introduced to cancel out the unbalanced voltage caused by coil winding asymmetry and alignment inaccuracy. Subsequently, the balance coil terminal voltage is collected once every second, the voltage values are collected three times in succession during each sampling period to obtain the mean value, and the recorded mean value is compared with the value measured in the next cycle. If the voltage change is larger than the voltage change threshold caused by the normal power fluctuation, it is considered that there is foreign matter mixed in until the end of the charging

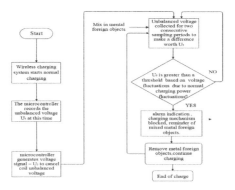

Figure 7. Metal detection process during charging.

process. In addition, foreign object detection is stopped at the transition of the charging stage. After waiting for a smooth transition to the new charging stage, foreign object detection restarts.

4 EXPERIMENT

4.1 *Alignment detection experiment*

The general flow chart of the wireless charging process with both alignment and metal foreign objects detection function is shown in Figure 8.

4.2 *Alignment detection experiment*

In the experiment, metal pieces with different diameters and materials were used for experiments. And metal foreign object detection experiments before and during charging were performed respectively.

When the transmitter coil voltage is 79.6 V, the balanced coil induction terminal voltage is high-frequency sinusoidal AC voltage. And the voltage amplitude is 2.5 V. The waveform is shown in Figure 8.

Different diameters, shapes, and materials were used for the detection experiments. The experimental results are shown in Table 2.

Table 1. Alignment detection experimental result.

Offset distances/mm	The terminal voltage of balanced coil/V	Experiment times	Number of alarms
0	2.0	20	0
30	3.1	20	0
50	4.2	20	0
80	6.5	20	19
100	8.5	20	20
120	11	20	19
140	15	20	20

Table 2. The experiment results before charging.

Metal sheets	The terminal voltage of balanced coil/mV	Experiment times	Number of alarms
Diameter 26 mm round iron sheet	54	20	19
Diameter 40 mm round iron sheet	81	20	20
Diameter 40 mm semicircle iron sheet	120	20	20
Diameter 40 mm round copper sheet	45	20	18

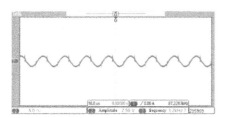

Figure 8. The voltage waveform of balanced coil.

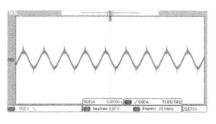

Figure 9. The waveform of balanced coil voltage.

127

Table 3. The experiment results during charging.

Metal sheets	The terminal voltage of balanced coil/mV	Experiment times	Number of alarms
Diameter 26 mm round iron sheet	200	20	18
Diameter 40 mm round iron sheet	150	20	19
Diameter 40 mm semicircle iron sheet	310	20	20
Diameter 40 mm round copper sheet	210	20	18

4.3 Metal foreign body detection experiment during charging

When the charging power is 500 W, the terminal voltage of balanced coil is high-frequency sinusoidal AC voltage. And the terminal voltage amplitude is 6.6 V. The waveform is shown in Figure 9.

Different diameters, shapes and materials were used for the detection experiments. The experimental results are shown in Table 3. The experiment verified that the algorithm is feasible and practical.

5 CONCLUSION

Firstly, an improved balanced coil structure was introduced that can take into account alignment and foreign objects detection. Then, a set of detection methods was proposed for the detection problems, which completely solved the problems of detection of the alignment and detection of metal foreign objects before and during charging. The detection algorithm in the charging process, as well as a comprehensive inspection flow chart, is designed. Finally, the experiment verified that the algorithm is feasible and practical. Limitations still exist in this study. Owing to the limited time and labor, only 11 sets of weights were simulated, which may be insufficient for the uncertainty analysis. A more detailed study, such as supported by the Monte Carlo simulation, should be further conducted to quantify the uncertainty of eco-corridor construction, which may lead to a deviation of positioning ecological engineers.

REFERENCES

Bai, S. & Dong, C. (2017). High precision algorithm of metal detector based on balance coil. *Journal of Shandong University (Engineering Science), 47*(4), 83–88.

Low, Z.N., Casanova, J.J., Maier, P.H., Taylor, J.A., Chinga, R.A. & Lin, J. (2010). Method of load/fault detection for loosely coupled planar wireless power transfer system with power delivery tracking. *IEEE Transactions on Industrial Electronics, 57*(4), 1478–1486.

Ma, Z., Liao, C. & Wang, L. (2017). Analysis of metal foreign object setting on electric vehicle wireless power transfer system. *Advanced Technology of Electrical Engineering and Energy, 36*(2), 14–20.

Qu. X. (2016). Research on wireless charging system of electric vehicle based on magnetic resonance. Shandong University.

Wang, Q. (2010). Design of metal detector based on balance coil technique [D]. Shandong University.

Yamazaki, S., Nakane, H. & Tanaka, A. (2002). Basic analysis of a metal detector. *IEEE Transactions on Instrumentation and Measurement, 51*(4), 810–814.

Yang, Q., Zhang, P., Zhu, L., Xue, M., Zhang, X. & Li, Y. (2015). Key fundamental problems and technical bottlenecks of the wireless power transmission technology. *Transactions of China Elecrtrotechnical Society, 30*(05), 1–8.

Zhou, B. (2018). Research on alignment and foreign object detection in wireless charging system. Shandong University.

Automatic Control, Mechatronics and Industrial Engineering – He & Qing (Eds)
© *2019 Taylor & Francis Group, London, ISBN 978-1-138-60427-8*

A synthetic assessment model for the condition of major equipment in nuclear power plants

T.C. Pan, J.F. Shen, X.M. Mao & M. Qu
Suzhou Nuclear Power Research Institute, Suzhou, Jiangsu Province, China

ABSTRACT: Maintaining healthy operating conditions of major equipment in nuclear power plants is an important foundation for safeguarding nuclear safety. Focusing on characteristics of equipment operation, management requirements and equipment status information, an evaluation method suitable for assessment of the health condition of major equipment is proposed. Taking full advantage of existing equipment monitoring data, this method is composed of a system for multi-level indicator establishment, and an architecture model for this system based on equipment failure phenomena. In the meantime, a method combining variable indicator weight distribution and fuzzy synthetic evaluation is proposed, which applies an analytic hierarchy process to recognize the deterioration in key indicators for major equipment in nuclear power plants. This method is shown to be simple and practical for evaluating the health status of main pump shaft systems.

Keywords: nuclear power equipment, analytic hierarchy process, fuzzy synthetic assessment

1 INTRODUCTION

Major equipment is the core and main resource for nuclear power plant production as well as an important guarantee of nuclear safety. During the normal operational period of primary circuit equipment, operators cannot enter nuclear islands to conduct repairs, and managers need to promptly establish intervention programs, such as contingency or outage plans, based on equipment health conditions to ensure the reliability of nuclear island equipment. Before overhaul, managers should also comprehensively assess the health condition of nuclear island equipment to arrange maintenance plans and tasks properly. In recent years, with more serious ramifications attaching to insufficient or excessive maintenance, research and application of condition-based maintenance have gradually become widespread. Compared to preventive maintenance, condition-based maintenance can extend equipment life, ensure safe and reliable operation of equipment, and significantly improve equipment reliability and economy (Lin, 2016; Li et al., 2012; Ahmad & Kamaruddin, 2012). According to the equipment health assessment results, condition-based maintenance comprehensively assesses the equipment health condition by obtaining data that characterizes the operational status of the equipment, and predicts equipment-condition trends and failure risk in order to develop the corresponding condition-based maintenance schedules (Yu, 2015; Lin et al., 2014).

Synthetic evaluation models of equipment have mainly involved such methods as principal component analysis, the Analytic Hierarchy Process (AHP), fuzzy mathematics, artificial neural networks or gray clustering. Because of their applicability to the assessment of complex equipment conditions, a combination of fuzzy synthetic evaluation and AHP has been widely applied. Meanwhile, more and more scholars have devoted time to combining various evaluation algorithms to develop new synthetic evaluation models and theories. Yu Qian, who has been conducting a great deal of research on electrical equipment health assessment, established a method for assessing the health condition of major equipment such as transformers (Yu & Li, 2013; Li, 2015; Zhang et al., 2013). Based on AHP and fuzzy synthetic evaluation, Fu Xuewei developed a multiple-experts cloud-model fuzzy AHP method and achieved remarkable application results (Fu, 2011). Li Rong proposed a synthetic assessment of the health condition of bridges based on the AHP and fuzzy synthetic evaluation methods, and applied a genetic algorithm to optimize weight value selection (Li, 2007). Luo Li established an assessment index system and evaluation theories for railway tamping vehicles (Luo, 2014). Wang Chuan established an evaluation system for equipment vibration indices and verified the indicators' validity with examples (Wang, 2011). Many scholars have conducted a great deal of research on theoretical systems and algorithms for major equipment health assessments in various industries. However, because of the sensitivity and complexity of nuclear power equipment, there has not been any in-depth theoretical and synthetic evaluation research on the health condition of nuclear equipment to date.

In this paper, a health condition assessment method for major equipment at nuclear power plants is presented, and a comprehensive, balanced and practical model for equipment health condition evaluation is established to provide a reference for the health management and maintenance of nuclear plant equipment.

2 EVALUATION INDEX SYSTEM

Condition evaluation indices are key to establishing equipment evaluation models. However, because most of the major equipment at nuclear power plants involves complex functional systems and structures with many influencing factors, it is necessary to establish the evaluation index model based on existing equipment monitoring systems at nuclear power plants.

Equipment-condition monitoring parameters, the most direct parameters to characterize equipment health conditions, are the main source of equipment evaluation indices. In reality, the health condition of an indicator is not only related to its parameter value, but also its change trend and type. Taking main pump vibration in a nuclear island as an example, in the process of analyzing the health condition of the axis displacement index, besides shaft displacement, attention should also be paid to axial displacement ratio, ratio variation, deviation value, deviation variation and trend analysis results for both axes. If each program is taken as a monitoring task, the indicator evaluation result can be judged by multiple monitoring tasks.

A monitoring task often corresponds to different fault modes and fault phenomena. Therefore, when selecting equipment indicators and monitoring tasks, it is necessary to analyze equipment fault modes first, then identify equipment fault phenomena to formulate monitoring tasks. Based on the importance of different failure modes, repeated monitoring tasks should be avoided under the premise of guaranteeing complete and comprehensive reflection of the equipment health condition, and the evaluation indicator is identified by the filtered monitoring tasks to establish the indicator models shown in Figure 1.

In order to fully reflect the equipment health condition, besides condition monitoring parameters, fault and defect indicators, safety indicators, reliability indicators and environmental protection indicators should also be established according to equipment functions and characteristics.

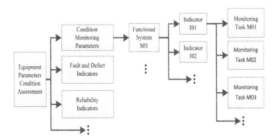

Figure 1.   Nuclear power plant major equipment indicator model.

3 EQUIPMENT HEALTH EVALUATION MODEL

A great number of phenomena in the objective world are fuzzy because their development represents a spectrum from quantitative changes to qualitative changes and clear boundaries cannot be found between them (He & Zheng, 2013). By quantifying the uncertainty of research objects, fuzzy theory uses computers to handle the information and thus, quantitatively describes the uncertainty of the objects (Shen, 2015). Based on fuzzy mathematics, fuzzy synthetic assessment is a method that quantifies the unclear, non-quantitative factors to synthetically assess the grade of membership (Li, 2013).

3.1 *Identification of index weight*

At present, the commonly used subjective weight determination methods are expert evaluation and AHP, of which AHP can express and process the experts' subjective judgments in the form of quantity and calculate the weights of equipment through mathematical methods using strict logic. When analyzing complex systems with very many indices, AHP helps confirm weights appropriately, which is difficult for the expert evaluation method. An importance-scale criterion should be selected according to the characteristics of equipment indices when using AHP to establish weighting models. The traditional criterion of importance scale divides the importance of indicators into ten levels, which makes it difficult to make accurate judgments in actual use and increases the risk of inconsistent judgment matrices. In order to improve applicability, based on the classification management practice of equipment at nuclear power plants, the importance scale is divided into four levels: *economic*, *common*, *important* and *key*, from which the determined economic-level index is directly excluded as redundant because it has little impact on equipment. The importance-scale values of common-, important- and key-level indices correspond to one, two and three, respectively.

After classifying the index importance, the judgment matrix C needs to be established according to the importance-scale values. The matrix is as follows:

$$C = \begin{bmatrix} C_{11} & C_{12} & \cdots & C_{1n} \\ C_{21} & C_{22} & \cdots & C_{2n} \\ \vdots & \vdots & \vdots & \vdots \\ C_{n1} & C_{n2} & \cdots & C_{nn} \end{bmatrix} \tag{1}$$

where $C_{ij} = {m_i}/{m_j}$, m_i is the importance-scale value.

After establishing the judgment matrix C, the maximum eigenvalue λ_{max} and the corresponding eigenvector V_{max} can be calculated. The weights of each index can be obtained after normalization processing of V_{max}:

$$W = \frac{V_{max}}{|V_{max}|} \tag{2}$$

Generally, the weight distribution models established with this method do not need to be verified for the consistency of the judgment matrix. If verification is needed for several indicators, it can be conducted on the basis of the traditional verification method presented by Saaty (1980).

3.2 *Fuzzy synthetic assessment model*

By constructing multi-level fuzzy subsets, the fuzzy synthetic evaluation quantifies various influencing factors of the evaluated object according to the fuzzy indicators, and then uses the fuzzy transformation principle to obtain comprehensive indices. In addition to index sets and index weights, the fuzzy synthetic evaluation also needs to establish a comment set and the membership functions of the indices.

The establishment of a comment set is mainly based on equipment evaluation requirements and industry characteristics. According to the usual practice of plant health assessment at nuclear power plants, the equipment health condition can generally be divided into four levels: *health*, *attention*, *warning* and *alarm*, which, respectively, correspond to four colors: green, white, yellow and red (Yuan, 2016). In other words, the comment set H can be expressed as H = [green, white, yellow, and red].

Construction of the index comment membership function should be determined based on major equipment management habits in nuclear power plants in order to match the evaluation result with the actual situation and enhance the value of on-site application. According to the characteristics of nuclear power plant equipment management, after the major equipment index parameters reach pre-warning or alarm values, all specialized and operations departments should analyze the equipment situation or implement corresponding intervention measures and contingency plans. Therefore, according to the requirements of existing equipment management systems of nuclear power plants, in order to guarantee the principle that the evaluation result matches the on-site equipment management, the membership function can be simulated by rectangular distribution when constructing the membership function of the yellow and red states. The white and green evaluation reflects an initial degradation of equipment health, which does not involve mandatory management requirements, so the initial state can be determined on the basis of the equipment health degradation mechanism. Most equipment failure processes continuously develop and accumulate, which means the equipment health deterioration is a process ranging from zero to one, and the relationship between parameter value and equipment condition is relatively fuzzy. However, although the equipment does not significantly deteriorate when the indicator value deviates from a historically normal range, the equipment condition is apparently more likely to be abnormal.

Based on the above principle, when establishing the membership function of the evaluation set, fitting can be performed according to the distribution curve shown in Figure 2.

In Figure 2, G_h, G_l, W_h, W_l, Y_h, Y_l, R_h and R_l are deterministic boundaries of corresponding color state and their values can be selected from the Electric Power Research Institute (EPRI) guidelines (Yuan, 2016). With adequate historical sample data, a 95% confidence interval for the sample data distribution during a normal historical period can be selected as the values of G_h and G_l, and a 99.7% confidence interval can be selected as the values of W_h and W_l. In normal operational processes, exceeding W_h and W_l is considered to be a low-probability event, and the membership degree that belongs to white is one. According to Figure 2, the membership function of the corresponding evaluation sets of red, yellow, white and green states is shown as:

$$\mu_R(x) = \begin{cases} 1 & x \le R_l \text{ or } x \ge R_h \\ 0 & x \ge R_l \text{ or } x \le R_h \end{cases} \tag{3}$$

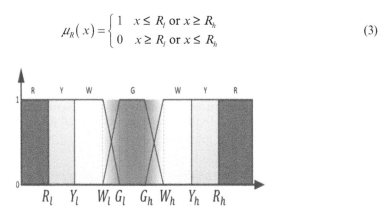

Figure 2.   Membership distribution curve of the evaluation set.

$$\mu_Y(x) = \begin{cases} 1 & x \le Y_l \text{ or } x \ge Y_h \\ 0 & x \ge Y_l \text{ or } x \le Y_h \end{cases} \tag{4}$$

$$\mu_W(x) = \begin{cases} 1 & x \le W_l \text{ or } x \ge W_h \\ \dfrac{G_l - x}{G_l - W_l} & W_l \le x \le G_l \\ \dfrac{x - G_h}{W_h - G_h} & G_h \le x \le W_h \\ 0 & x \ge G_l \text{ or } x \le G_h \end{cases} \tag{5}$$

$$\mu_G(x) = \begin{cases} 1 & G_l \le x \le G_h \\ \dfrac{x - W_l}{G_l - W_l} & W_l \le x \le G_{ll} \\ \dfrac{W_h - x}{W_h - G_h} & G_h \le x \le W_{hh} \\ 0 & x \le W_l \text{ or } x \ge W_h \end{cases} \tag{6}$$

The indicator evaluation results should be determined through monitoring tasks, and each monitoring task is actually a different monitoring and evaluation method of the corresponding indicators. The evaluation result vector of indicator I should be equal to the most deteriorative assessment results in the related monitoring tasks.

In reality, weighted-average principles can be used to determine the relative degree of deterioration for each monitoring task. The evaluation rating is regarded as a kind of relative position in the weighted-average principle. In order to process quantitatively, the numbers zero, one, three and nine can be used to localize each indicator level and define the rank of each grade. Each rank in the monitoring task result vector is weighted and summed to obtain the relative position of the evaluated object. The monitoring task result with the highest relative position A is selected as the indicator evaluation result, which is expressed by:

$$A = \frac{\mu_G^k \cdot U_G + \mu_W^k \cdot U_W + \mu_Y^k \cdot U_Y + \mu_R^k \cdot U_R}{\mu_G^k + \mu_W^k + \mu_Y^k + \mu_R^k} \tag{7}$$

In the formula, k is the amplification coefficient, which is applied to control effects caused by the larger membership degree. Generally, one or two indicates desirable; U is the rank of corresponding level.

After obtaining the evaluation vector of one index, the fuzzy relation matrix $Ri(r_{i1}, r_{i1}, ..., r_{in})$ of indicator set B to evaluation set H needs to be established after transforming the evaluation vector into the synthetic health condition of the equipment, where $0 \le rij \le 1$ means that factor Ri belongs to Hj membership degree. If the indicator number is m and the evaluation level number is n, the evaluation matrix is expressed as:

$$R = (r_{ij})_{m \times n} = \begin{bmatrix} r_{11} & r_{12} & \cdots & r_{1n} \\ r_{21} & r_{22} & \cdots & r_{2n} \\ \vdots & \vdots & \vdots & \vdots \\ r_{m1} & r_{m2} & \cdots & r_{mn} \end{bmatrix} \tag{8}$$

Using a fuzzy evaluation matrix and factor set weight vector, the fuzzy operator is selected to perform the fuzzy transformation operation and obtain the fuzzy evaluation result vector:

$$M = W \, o R = (\omega_1, \omega_2 ... \omega_m) o (r_{ij})_{m \times n} = (m_1, m_2 ... m_n) \tag{9}$$

133

where o is the fuzzy operator.

The commonly used fuzzy operators include the main factor outstanding type, the fully constrained type and the weighted-average type. The weighted-average fuzzy operator is appropriate to the health assessment of major equipment because it can calculate the contribution to the evaluation target according to the weight of each evaluation index, and can comprehensively balance the impact of different indicators. When using a weighted-average fuzzy operator, the element m_j in the evaluation result vector M is calculated as:

$$m_j = \sum_{i=1}^{m} \omega_i r_{ij} \tag{10}$$

3.3 Variable weight theory

Due to the complex system design of major equipment and the large numbers of key indicators affecting equipment health, each key indicator is only assigned a very small weight. As a result, the overall impact of each indicator on the equipment is limited, meaning that although the equipment may be abnormal or even broken, the health grading retains an illusion of good health. Apparently, it is difficult to apply such a synthetic evaluation result to the actual project. The comprehensive variable weighting theory can reflect the equilibrium of various elements and states in integrated decisions. The physical meaning of the indicator weight reflects the degree of attention paid to the indicator during the equipment health assessment. In fact, when the equipment indicator is only slightly degraded, it will not significantly improve the assessor's attention to the index. Only if the index has a significant degree of degradation, such as an indicator's color changing to yellow or red, will the assessor's attention be rapidly raised, which means the weight of the index increases. Similarly, there is no need to change the weight for non-critical indicators, because of the limited impact on equipment of their serious deterioration.

In application, first, it is necessary to select the index that needs to change the variable weight, according to the influence of the indicator on equipment following serious deterioration. Then, calculate the evaluation vector of each index according to the principle of maximum membership before calculating the synthetic evaluation result; multiply a given coefficient K_ω according to the evaluation of red or yellow status, where K_ω is proportional to the total number of indicators n, the degree of indicator deterioration m, and the extent of impact on overall assessment results K_m under this degree of deterioration. Normalize the newly obtained indicator after the weighting is adjusted, and update the weights in Equation 10 to obtain the new equipment health condition after variable weighting:

$$K_\omega = f(n, m, K_m) \tag{11}$$

3.4 Evaluation result

Through fuzzy synthetic evaluation, the fuzzy vectors of equipment health condition can be obtained. In order to get a clear evaluation conclusion, this paper adopts a method of calculating evaluation results using the fuzzy vector and standard vector closeness of each clear state.

The clear standard vectors for selecting green, white, yellow and red are [1, 0, 0, 0], [0 1 0 0], [0 0 1 0], [0 0 0 1]. The Euclidean distance approach method is applied to the approximation calculation model, calculated as follows:

$$N(\mathbf{A}, \mathbf{B}) = 1 - \frac{1}{n} \sqrt{\sum_{i=1}^{n} (a_i - b_i)^2} \tag{12}$$

Calculate the closeness between the evaluation result's fuzzy vector and the standard vector, and select the state with the highest degree of closeness to the evaluation result.

4 HEALTH CONDITION EVALUATION OF A NUCLEAR ISLAND MAIN PUMP

The main pump is the power source for a nuclear island primary circuit flow and the heart of the nuclear island. The main pump health condition is crucial to keeping the nuclear power plant running steadily and safely. Therefore, it is necessary to periodically evaluate the health state of the main pump and ensure its reliable operation. At present, the main pumps of nuclear power units in China are almost all shaft seal pumps, meaning the pump shaft seal system serves as a pressure boundary of the primary circuit. Safe and reliable operation is an important guarantee of the nuclear safety of nuclear power plants. In this paper, the health assessment of the main pump shaft seal system is taken as an example to illustrate the rationality of synthetic evaluation.

Figure 3 shows the shaft seal system structure, which is composed of three shaft seal systems. Through analysis of fault modes and fault phenomena, the main evaluation indicators of the main shaft seal system are obtained, as shown in Figure 4.

The evaluation model for each monitoring task is established according to the membership function model and parameter selection method described in Section 2.2: the relevant model parameters are shown in Table 1.

In order to establish the evaluation model according to the indicator weight described in Section 2.1, this article takes the weight vector of each indicator $W = [0.3, 0.3, 0.1, 0.2, 0.1]$, and selects nuclear power plant main pump data from August 2017 for evaluation. The corresponding index data are the worst values during the statistical period, and the evaluation vector results for each monitoring task are calculated as:

$$I_{01} = \begin{bmatrix} 0.7 & 1 & 1 & 1 \\ 0.3 & 0 & 0 & 0 \\ 0 & 0 & 0 & 0 \\ 0 & 0 & 0 & 0 \end{bmatrix} I_{02} = \begin{bmatrix} 1 & 0.8 & 1 \\ 0 & 0.2 & 0 \\ 0 & 0 & 0 \\ 0 & 0 & 0 \end{bmatrix} I_{03} = \begin{bmatrix} 1 \\ 0 \\ 0 \\ 0 \end{bmatrix} I_{04} = \begin{bmatrix} 0 \\ 1 \\ 0 \\ 0 \end{bmatrix} I_{05} = \begin{bmatrix} 0.87 & 0.22 & 1 \\ 0.13 & 0.78 & 0 \\ 0 & 0 & 0 \\ 0 & 0 & 0 \end{bmatrix}$$

(13)

Select amplification factor $k = 1$, calculate the relative position A of each monitoring task according to Equation 7, select the monitoring task with the largest relative position as the index evaluation result, and form an evaluation matrix R as:

$$R = \begin{bmatrix} 0.7 & 0.3 & 0 & 0 \\ 0.8 & 0.2 & 0 & 0 \\ 1 & 0 & 0 & 0 \\ 0 & 1 & 0 & 0 \\ 0.22 & 0.78 & 0 & 0 \end{bmatrix}$$

(14)

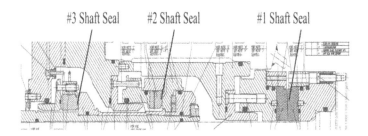

Figure 3. Main pump shaft seal system structure.

135

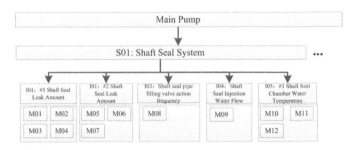

Figure 4. Main pump shaft seal system evaluation indicators.

Table 1. Monitoring tasks evaluation model.

No.	Task name	Model parameters
M01	#1 Shaft Seal Leakage (L/h)	$G_h = 600$, $G_l = 400$, $W_h = 700$, $W_l = 350$, $Y_h = 800$, $Y_l = 300$, $R_h = 900$, $R_l = 250$
M02	#1 Shaft Seal Leak Changes (L/h)	$G_h = 100$, $W_h = 150$
M03	Shaft Seal Leak Step Change	$G_h = 0$, $W_h = 1$
M04	Trend Analysis Deteriorates	$G_h = 0$, $W_h = 1$
M05	#2 Shaft Seal Leakage (L/h)	$G_h = 15$, $W_h = 20$, $Y_h = 30$, $R_h = 40$
M06	#2 Shaft Seal Leak Changes (L/h)	$G_h = 3$, $W_h = 5$
M07	Trend Analysis Deteriorates	$G_h = 0$, $W_h = 1$
M08	Valve Action Frequency (Times/Week)	$G_h = 5$, $G_l = 3$, $W_h = 6$, $W_l = 2$, $Y_h = 8$, $Y_l = 2$, $R_h = 14$, $R_l = 1$
M09	Shaft Seal Injection Water Flow Changes (m³/h)	$G_h = 0.05$, $W_h = 0.1$, $Y_h = 0.2$, $R_h = 0.4$
M10	#1 Shaft Seal Chamber Water Temperature (°C)	$G_h = 40$, $G_l = 55$, $W_h = 38$, $W_l = 58$, $Y_h = 60$, $Y_l = 35$, $R_h = 65$, $R_l = 30$
M11	Temperature Changes (°C)	$G_h = 3$, $W_h = 5$
M12	Trend Analysis Deteriorates	$G_h = 0$, $W_h = 1$

According to the operational requirements and equipment characteristics, indicators I_{01} and I_{02} are selected as key indicators, and the others are non-key indicators. Because I_{01} and I_{02} indicators have not deteriorated severely, there is no need to calculate the variable weight coefficients. The synthetic evaluation results of the shaft seal system calculated via Equation 9 are:

$$M = WoR = \begin{bmatrix} 0.572 & 0.428 & 0 & 0 \end{bmatrix} \qquad (15)$$

On the basis of the resulting fuzzy vectors, the lattice closeness of the remaining states' color standard vectors is calculated. The lattice closeness of the result vector to the standard vector of green, white, yellow and red states are 0.85, 0.8, 0.69 and 0.69, respectively. This result state is closest to the green state, followed by the white state. This means the color of the device is determined as green, but engineers still need to pay more attention to the equipment parameters, especially the change of shaft seal injection flow and the #1 shaft seal leakage changes.

If the flow of #1 shaft seal seriously deteriorates to 800 L/h or more, the evaluation result of M01 becomes yellow at this time and the other parameters are unchanged. If the indicator weight remains, the synthetic evaluation result will turn to [0.362 0.338 0.3 0]. According to the European proximity method, the synthetic evaluation state of the shaft seal system is still green, which obviously does not match the practical situation. Therefore, it is necessary to change the weight of the key indicators when they deteriorate, as per Section 2.3. If the variable coefficient is taken as $k_\omega = 8$, the subsequent results with variable weight evaluation are [0.117 0.109 0.774 0]. According to the European approach method, the synthetic evaluation result of the shaft seal system is yellow, which requires the operations and maintenance department engineers to intervene, conforming to the actual situation on site.

5 CONCLUSIONS

Aiming at the fuzziness, complexity and characteristics of equipment management during the health condition assessment of major equipment in nuclear power plants, this paper establishes a model of a health condition assessment system for major equipment in nuclear power plants, which is verified as suitable for nuclear engineering applications by applying the health condition assessment model to the main pump shaft seal system in a nuclear island.

Establishing a hierarchical indicator system based on monitoring tasks can effectively utilize the existing monitoring technology systems in nuclear power plants, and is beneficial to the application of the index system. The fuzzy synthetic evaluation method contains weighting method, the variable weighting method and multi-monitoring task is suitable for evaluation model of the health evaluation, which objectively and effectively evaluates the health condition of major equipment in a nuclear power plant. The application of the proposed model to evaluate the health condition of a main shaft seal system is simple and practical, and the evaluation results are comprehensive and accurate, proving the model suitable for the state evaluation of nuclear power plant equipment.

REFERENCES

Ahmad, R. & Kamaruddin, S. (2012). An overview of time-based and condition-based maintenance in industrial application. *Computers & Industrial Engineering, 63*(1), 135–149.

Fu, X. (2011). *The research and application of fuzzy AHP*. Harbin, China: Harbin Institute of Technology.

He, R. & Zheng, S. (2013). Independent modal variable structure fuzzy active vibration control of thin plates laminated with photostrictive actuators. *Chinese Journal of Aeronautics, 26*(2), 350–356.

Li, D. (2013). *Research on an equipment health management method based on spot inspection and supporting software*. Changsha, China: Central South University.

Li, M. (2015). *The synthetic evaluation of AC and DC distribution system based on improved analytic hierarchy process*. Beijing: North China Electric Power University.

Li, M., Han, X., Wang, Y., Guo, Z. & Liu, G. (2012). Decision-making model and solution of condition based maintenance for substation. *Proceedings of the CSEE, 32*(25), 196–202.

Li, R. (2007). *The research and application of the fuzzy synthetic assessment method for health condition evaluation of bridges based on AHP*. Changsha, China: Hunan University.

Lin, D. (2016). *Research on condition evaluation and condition-based maintenance for distribution equipment based on multi information source*. Hangzhou, China: Zhejiang University.

Lin, P., Gu, J. & Yang, M. (2014). Intelligent maintenance model for condition assessment of circuit breakers using fuzzy set theory and evidential reasoning. *IET Generation, Transmission & Distribution, 8*(7), 1244–1253.

Luo, L. (2014). *Study of synthetic evaluation on equipment status of railway tamping machine*. Changsha, China: Hunan University.

Saaty, T.L. (1980). *The analytic hierarchy process*. New York: McGraw-Hill.

Shen, Y. (2015). *Research on fuzzy evaluation method for the running state of transmission cable based on multi-level index system*. Guangzhou, China: South China University of Technology.

Wang, C. (2011). *The equipment status evaluation system construction and its key technology research*. Zhengzhou, China: Zhengzhou University.

Yu, Q. (2015). *Research on the model of state evaluation and risk assessment for distribution equipment*. Changchun, China: Jilin University.

Yu, Q. & Li, W. (2013). Application of fuzzy set pair analysis model to power transformer condition assessment. *Journal of Central South University (Science and Technology), 2*, 598–603.

Yuan, J. (2016). The supervision of nuclear power plant system based on AP-913 INPO framework. *Nuclear Science and Engineering, 5*, 671–676.

Zhang, C., Wang, D., Wang, A. & Liu, X. (2013). Establishment of judgment matrices for the condition assessment of the power transformer based on fuzzy AHP. *Journal of Northeastern University (Natural Science), 2013*(3), 317–321.

Automatic Control, Mechatronics and Industrial Engineering – He & Qing (Eds)
© 2019 Taylor & Francis Group, London, ISBN 978-1-138-60427-8

Optimization decision method for neutral grounding mode of distribution network considering electric shock risk

W.W. Zhang & M.B. Yang
Guizhou Power Grid Co. Ltd., Guiyang Power Supply Bureau, Guiyang Guizhou, China

X.Y. Yang, L. Long, J.X. Ouyang & G.T. Ma
State Key Laboratory of Power Transmission Equipment and System Security and New Technology, Chongqing University, Chongqing, China

ABSTRACT: Aiming at the problem that the current neutral grounding decision method lacks quantitative measurement of electric shock risk, this paper proposes an optimization decision method. Firstly, according to the influence factor of the neutral grounding method, the decision target set of electric shock risk, over voltage risk, life cycle cost and reliability is constructed. Secondly, based on minimizing the inconsistency between each scheme and decision makers, the decision problem is transformed into the linear programming problem, and the weights of each target are obtained. Then, use the weighted Euclidean distance to rank the neutral grounding schemes. Finally, the effectiveness of the method is verified by a practical example. The method, which combined the influence of decision makers, relevant standards and objective factors on the choice of neutral point grounding method, has strong practical significance.

1 INTRODUCTION

Reasonable neutral grounding method should balance the safety, economy and reliability of the distribution network. Selection decisions on the neutral grounding method have been studied in considerable literature. The weighting coefficient of the neutral grounding method is obtained by expert scoring method, and the largest weighted is selected as the best grounding method (Fu et al. 2010). The corresponding function is established based on the influencing factors of the neutral grounding method, and establishing the neutral grounding method by the fuzzy principle (Gan et al. 2013). Combining the reliability evaluation of the distribution network and the economic operation cost for safety factors, to calculate the cost of the distribution network under different neutral grounding modes. And the reliability index is the constraint condition to achieve the decision (Lu et al. 2006). But, the methods mentioned in do not consider the corresponding standards or specifications, so the results obtained may conflict with the standards. And they neglected the impact of different neutral grounding methods on personal safety when constructing the index evaluation system.

Based on the multi-dimensional preference linear programming analysis method, this paper proposes an optimization decision method for the neutral grounding method. Firstly, it build the four goals of personal electric shock risk, over voltage risk, life cycle cost and reliability; Secondly, aiming at the minimum degree of inconsistency between each program and decision makers, transforming the decision problem into the linear programming problem. Thirdly, using the weighted Euclidean distance to rank the neutral grounding schemes, and the optimal neutral grounding method is obtained. Finally, the effectiveness of the method is verified by an example.

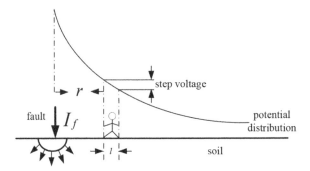

Figure 1. Potential distribution characteristic of ground fault point.

2 PERSONAL ELECTRICAL SHOCK RISK ASSESSMENT INDICATORS

Due to the grounding fault current of the large current grounding system is so large and the duration is short, while the grounding fault current of the small current grounding system is small and has a long duration, it is necessary to combine both of them to measure the risk of personal electric shock in different grounding methods.

Figure 1 shows the potential distribution characteristics of the ground fault point. Radial distance between step voltage and ground fault point can be computed:

$$U_s = \frac{\rho I_f l}{2\pi r(r+l)} \frac{R_b}{R_b + 6\rho_1} \tag{1}$$

where U_s corresponds to step voltage; ρ means soil resistivity around the fault point; I_f means earth fault current, l means the human stride distance; r means the radial distance from the ground fault point; R_b means the body resistance; ρ_1 means resistivity of the surface.

According to relevant regulations, the step voltage shall not be exceeded:

$$U_{s1} = \frac{174 + 0.7\rho_1 C_s}{\sqrt{t}} \tag{2}$$

U_{s1} means requirement of step voltage when ground fault can be quickly removed; C_s means surface attenuation coefficient, which related to the reflection coefficient of soil with different resistivity and surface soil thickness.

It can be known from equation (1–2) that the ground faults in different neutral grounding modes corresponds to a circular danger zone where a step voltage electric shock may occur, and the size of the circular danger zone is determined by the fault current and the fault duration. Defining personal electric shock risk ϕ is

$$\phi = \int r \times g(r)\, dr \tag{3}$$

where $g(r)$ corresponds to its probability density. ϕ represents the mathematical expectation of a hazardous area where different neutral grounding systems may experience striding voltage shocks.

3 DECISION MODEL DESIGN BASED ON LINMAP

Assume that the solution set of the multi-objective decision problem of the finite scheme is $A = \{A_1, A_2, \dots A_n\}$, the target set is $B = \{B_1, B_2, \dots B_m\}$, the attribute value of A_i for the target B_j is x_{ij}. Then the decision matrix is $X = (x_{ij})_{n \times m}$. Normalize decision matrix X and the normalized decision matrix is $Y = (y_{ij})_{n \times m}$. Decision makers get ordered sets Λ by comparing pairs of

140

programs in pairs, The ideal solution for the most preferred solution is $y^* = \{ y_1^*, y_2^*, \ldots, y_m^* \}$, Then the weighted Euclidean distance is

$$S_i = \sum_{i=1}^{m} \omega_j \left(y_{ij} - y_j^* \right)^2 \qquad (4)$$

where ω_j represents the weight of the target B_j.

According to the weighted decision result, the degree of inconsistency with the decision maker's preference is the smallest, and ω_j is solved by linear programming that:

$$\min J = \sum_{(k,l) \in \Lambda} G(k,l) \qquad (5)$$

$$s.t \begin{cases} \sum_{j=1}^{m} \omega_j \left(y_{kj} + y_{ij} - 2y_j^* \right) \left(y_{ij} - y_{kj} \right) + G(k,l) + \delta \geq 0 \\ G(k,l) \geq 0 \\ \sum_{j=1}^{m} \omega_j = 1 \\ \omega_j \geq \omega_c \\ (k,l) \in \Lambda \end{cases} \qquad (6)$$

where (k,l) is an ordered pair, indicating that when comparing A_k and A_l decision makers prefer k scheme. Recorded as $A_k \succ A_l$; $G(k,l) = \max(0, S_k - S_l - \delta)$ indicates the degree of inconsistency in order to (k,l), δ corresponds to a small positive number introduced to avoid an undifferentiated set of scenarios (Liu et al. 2010), ω_c corresponds to the lower limit of each target weight.

4   NEUTRAL GROUNDING METHOD DECISION TARGET SET

According to the principle that the evaluation indicators are as comprehensive as possible, the neutral point grounding decision target set is divided into safety indicators, reliability indicators and cost indicators.

The safety index comprehensively reflects the impact of different neutral grounding methods on system and personal safety. The cost indicator refers to the full life cycle cost of each neutral point grounding method. The reliability index reflects the adaptability of each neutral grounding method to the distribution network

4.1   *Decision target calculation*

4.1.1   *Safety indicators*
1. Personal risk of electric shock
The personal electric shock risk indicator (B_1) considers the fault point current and the fault duration. Assuming that the single-phase ground fault current is evenly distributed the risk of personal electric shock under different neutral grounding modes can be obtained by equation (1–3).

Ungrounded system personal electric shock risk ϕ_1 is computed:

$$\phi_1 = \frac{8}{3I_C K_1} r_1^3 + \frac{2l}{I_C K_1} r_1^2 \qquad (7)$$

$$\begin{cases} r_1 = \frac{1}{2} \left(\sqrt{l^2 + \frac{0.64 \rho I_C l R_b}{(50 + 0.2 \rho_1 C_s)(R_b + 6\rho_1)}} - l \right) \\ K_1 = \frac{0.64 \rho l R_b}{(50 + 0.2 \rho_1 C_s)(R_b + 6\rho_1)} \end{cases} \qquad (8)$$

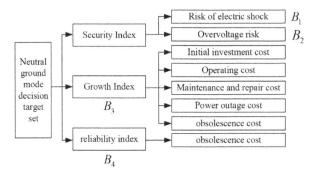

Figure 2.   Decision target set.

where I_C is corresponds to system capacitive current.

Arc suppression coil grounding system personal electric shock risk ϕ_2 is computed:

$$\phi_2 = \frac{8}{3\nu I_c K_2} r_2^3 + \frac{2l}{\nu I_c K_2} r_2^2 \tag{9}$$

$$\begin{cases} r_2 = \frac{1}{2}\left(\sqrt{l^2 + \dfrac{0.64\rho\nu I_c l R_b}{\left(50 + 0.2\rho_1 C_s\right)\left(R_b + 6\rho_1\right)}} - l \right) \\[4mm] K_2 = \dfrac{0.64\rho\nu l R_b}{\left(50 + 0.2\rho_1 C_s\right)\left(R_b + 6\rho_1\right)} \end{cases} \tag{10}$$

where ν is corresponds to Detuning degree of arc suppression coil.

When feeding line zero—sequence protection I segment action, the fault duration is t_1, and the fault current range is $\left[I_{0,set}^I, I_{f\_max} \right]$. Where I_{f_max} is corresponds to maximum single-phase ground fault current, $I_{f\_max} = E/R_0$, where E is corresponds to system rated phase voltage RMS, R_0 means neutral point resistance, $I_{0,set}^I$ means feeding line zero sequence protection I section setting value. The personal risk of electric shock ϕ_{31} is

$$\phi_{31} = \frac{8}{3I_{f\_max} K_{31}}\left(r_{31}^3 - r_{32}^3 \right) + \frac{2l}{I_{f\_max} K_{31}}\left(r_{31}^2 - r_{32}^2 \right) \tag{11}$$

$$\begin{cases} r_{31} = \frac{1}{2}\left(\sqrt{l^2 + \dfrac{0.64\rho I_{f\_max}\sqrt{t_1} l R_b}{\left(174 + 0.7\rho_1 C_s\right)\left(R_b + 6\rho_1\right)}} - l \right) \\[4mm] K_{31} = \dfrac{0.64\rho\sqrt{t_1} l R_b}{\left(174 + 0.7\rho_1 C_s\right)\left(R_b + 6\rho_1\right)} \\[4mm] r_{32} = \frac{1}{2}\left(\sqrt{l^2 + \dfrac{0.64\rho I_{0,set}^I \sqrt{t_1} l R_b}{\left(174 + 0.7\rho_1 C_s\right)\left(R_b + 6\rho_1\right)}} - l \right) \end{cases} \tag{12}$$

2. Overvoltage risk

The overvoltage risk (B_2) is measured by the arc grounding overvoltage multiple. According to the power frequency arc extinction theory, the range of the arc grounding overvoltage multiple is $\zeta_1, \zeta_2 \in (2.5, 3.5)$ (in the ungrounded system and the arc suppression coil grounding system), ζ_1, ζ_2 respectively indicate the arc overvoltage multiple of the ungrounded system and the arc suppression coil grounding system. For small resistance grounded systems, the arc overvoltage multiple η_3 can be estimated as

$$\zeta_3 = 1.5 \times \sqrt{1 + \left(1/K + 1/\sqrt{3} \right)^2} \tag{13}$$

142

where $K = 1/\omega C_s R_N$, usually K is 1.5~3. C_s represents the total capacitance of the system to ground, and R_N represents the resistance of the neutral point resistance.

4.1.2 Cost indicators

The cost indicator (B_3) reflects the sum of all the costs of the neutral point grounding method from the initial investment to the operation, maintenance, failure and abandonment of the project. It is a quantitative indicator, and the specific calculation method of the second level indicator is

$$
\begin{cases}
C_I = (1+\alpha)^\beta p \\
C_O = \dfrac{(1+\alpha)^\beta - 1}{(1+\alpha)-1} q \\
C_M = \gamma \times C_I \\
C_F = \dfrac{(1+\alpha)^\beta - 1}{(1+\alpha)-1}\left(EENS \times c_{price} \right) \\
C_D = \chi \times C_I
\end{cases}
\tag{14}
$$

where C_I means the initial investment cost; C_O means the operating costs; C_M means the maintenance cost; C_F means failure cost; C_D is the cost of abandonment; α is the annual interest rate; β is the service life; p is the fixed investment cost; q is the operating cost; γ is the maintenance and repair factor,; $EENS$ is the lack of power; c_{price} is the unit power failure loss; χ is the waste cost factor.

4.1.3 Reliability index

The neutral grounding mode selection needs to meet the user's requirements for reliable power supply, and its impact on the distribution network is mainly the line trip rate under the single-phase ground fault of the line. So, the reliability index (B_4) can be measured by the line trip frequency caused by the single-phase ground fault. For the feeder i, the failure rate is λ_i, the probability that the fault is a single-phase ground fault is η_i, and the probability that the single-phase ground fault is a permanent fault is σ_i.

1. Neutral point is not grounded
Probability of feeder i tripping P_1

$$
P_1 = P_{1\_i} + (n_1 - 1) P_{1-ji}
\tag{15}
$$

$$
\begin{cases}
P_{1\_i} = \left(\lambda_i \eta_i \sigma_i + \lambda_i \eta_i (1-\sigma_i)(1-\mu_1) \right) \varepsilon_1 \\
P_{1\_j} = P_{1\_i} \\
P_{1\_ji} = (1-\varepsilon_1)\left(1 - \dfrac{1}{n_1-1} \right) \dfrac{P_{1\_j}}{\varepsilon_1}(e_1 + e_2 - e_1 e_2)
\end{cases}
\tag{16}
$$

where P_{1_i} means the probability of a single-phase ground fault in the feeder line; P_{1_ji} means the probability that the fault will expand to make the line trip when the feeder fails. μ_1 is the probability of arc extinction; ε_1 represents the line accuracy rate in the ungrounded mode; $P_{1\_j}(j \neq i)$ is the probability that a single-phase ground fault occurs on the feeder and eventually trips; e_1 is the probability of over-voltage accident expansion; e_2 is the probability of expansion of the arc fire accident.

2. Arc suppression coil grounding method
Probability of feeder i tripping P_2

$$
P_2 = P_{2\_i} + (n_1 - 1) P_{2\_ji}
\tag{17}
$$

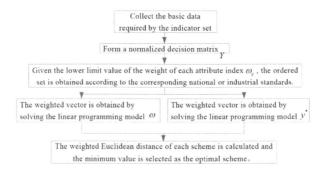

Figure 3. Decision flow chart.

$$\begin{cases} P_{2\_i} = \left(\lambda_i \eta_i \sigma_i + \lambda_i \eta_i \left(1 - \sigma_i \right) \left(1 - \mu_2 \right) \right) \varepsilon_2 \\ P_{2\_j} = P_{2\_i} \\ P_{2\_ji} = \left(1 - \varepsilon_2 \right) \left(1 - \frac{1}{n_1 - 1} \right) \frac{P_{2\_j}}{\varepsilon_2} \left(e_1 + e_2 - e_1 e_2 \right) \end{cases} \tag{18}$$

The meaning of each parameter in the formula is similar to the above
3. Neutral point through small resistance grounding method
Probability of feeder i tripping P_3

$$P_3 = \lambda_i \eta_i \sigma_i + \lambda_i \eta_i \left(1 - \sigma_i \right) \left(1 - \varepsilon_3 \right) \tag{19}$$

where ε_3 represents the success rate of the reclosing.

4.2 *Decision process*

Figure 3 is a flow chart of the neutral point grounding method decision based on the LINMAP method:

5 ANALYSIS OF SIMULATION RESULTS

The proposed method is applied to the modification of a city distribution network, selecting a substation to be transformed as the decision object. The basic information is shown in Table 1. The scheme of the neutral point transformation of the substation is as follows:

Scheme A_1: adopt the neutral point ungrounding method, configure the corresponding line selection device, and the line selection accuracy rate shall not be lower than 80%;

Scheme A_2: The neutral point is grounded by the arc suppression coil, and the corresponding line selection device is configured. The ground fault current is compensated to within 10 A, and the line selection accuracy is not less than 80%;

Scheme A_3: Neutral point is grounded by small resistance, and small resistance is taken as 10 Ω. Zero-sequence over-current protection is configured. The zero-sequence I segment of the feeder is set to 60 A, and the action time is 0.5 s. The zero-sequence of the feeder is set to 25 A, and the action time is 1.0 s.

Scheme A_4: The neutral point is grounded by a small resistor, and the small resistor is 16 Ω. Zero-sequence over-current protection is configured. The zero-sequence I segment of the feeder is set to 60 A, and the action time is 0.5 s. The zero-sequence of the feeder is set to 25 A, and the action time is 1.0 s.

Table 2 lists the basic parameters of the security indicators, reliability indicators, and economic indicators used for calculation and analysis.

Table 1. Basic parameters of feeder.

Feeder number	Line information						Capacitive current (A)	Current status of grounding mode
	Cable length (km)	Cable type	Length of overhead line (km)	Overhead line model	The power supply area	Load (MW)		
1	2.3	YJV22-300	6.2	JKLYJ-150	urban area	10.76		
2	6.6	YJV22-300	8.9	JKLYJ-150	urban area	9.64		
3	2.1	YJV22-300	2.7	JKLYJ-150	urban area	8.32		Arc suppression
4	2.5	YJV22-185	3.1	JKLYJ-120	urban area	7.68	75.6	coil grounding,
5	4.8	YJV22-300	5.7	JKLYJ-150	urban area	8.75		the model is
6	4.0	YJV22-185	5.4	JKLYJ-120	urban area	7.47		XHDCZ-600/10
7	2.4	YJV22-150	3.0	JKLYJ-120	urban area	5.33		
8	4.0	YJV22-150	4.6	JKLYJ-120	urban area	5.62		

Table 2. Basic parameters indicator.

Parameter	Evaluation	Parameter	Evaluation	Parameter	Evaluation
ρ	$200\ \Omega \cdot m$	ρ_1	$1000\ \Omega \cdot m$	σ_i	0.3
l	1 m	C_s	0.8	ε_2	0.8
R_b	$1000\ \Omega$			σ_i	0.3
α	5%	c_{price}	2.8	ε_2	0.8
β	15	χ	0.1	e_1	0.05
γ	0.2	T	1 h	e_2	0.05
λ_i	5	μ_1	0.5	e_1	0.05
η_i	0.7	ε_1	0.85	e_2	0.05

According to the basic parameters of Table 2, the safety index, reliability index and cost index of the above scheme $A_1 \sim A_4$ are calculated. And the calculation result is formed into a decision matrix X.

$$X = \begin{array}{c} \\ A_1 \\ A_2 \\ A_3 \\ A_4 \end{array} \begin{matrix} B_1 & B_2 & B_3 & B_4 \\ \begin{bmatrix} 0.5065 & 3.50 & 1043.4 & 2.48 \\ 0.1837 & 3.20 & 692.59 & 1.42 \\ 0.7002 & 1.89 & 813.33 & 1.79 \\ 0.5102 & 1.91 & 813.33 & 1.79 \end{bmatrix} \end{matrix} \quad (20)$$

Since the four indicators in the target set are all "cost-based" indicators, they are normalized to obtain a matrix.

$$Y = \begin{bmatrix} 0.38 & 0 & 0 & 0 \\ 1 & 0.49 & 1 & 1 \\ 0 & 1 & 0.66 & 0.65 \\ 0.37 & 0.97 & 0.66 & 0.65 \end{bmatrix} \quad (21)$$

The ordered set of decision makers combined with relevant standards is $\Lambda = \{(2,1),(3,1),(4,1)\}$, That is to say, the arc suppression coil grounding scheme and the small resistance grounding scheme are superior to the ungrounded scheme, and the order of the arc suppression coil grounding scheme and the small resistance grounding scheme cannot be determined.

Assume $\delta = 0.005$, $\omega_c = 0.05$, then, the weighted vector $\omega = \{0.05\ 0.05\ 0.05\ 0.85\}$ and the ideal solution point $y^* = \{1\ 1\ 1\ 1\}$ can be obtained by the multidimensional preference linear programming analysis method described in this paper.

Then the weighted Euclidean distance of each scheme is: $S_1 = 0.97$, $S_2 = 0.013$, $S_3 = 0.16$, $S_4 = 0.13$.

According to the weighted Euclidean distance sorting, the order of the advantages and disadvantages of each scheme is: $A_2 \succ A_4 \succ A_3 \succ A_1$, then the scheme A_2 is the best way to use the arc suppression coil grounding method. Therefore, the substation only needs to re-verify the capacitor current to ensure that the ground fault current is compensated within 10 A. At the same time, configure the appropriate line selection device.

6 CONCLUSION

Neutral grounding method decision affects the safety, economy and reliability of the distribution system and it is a multi-objective decision-making problem of limited scheme. Aiming at the problem that the current neutral grounding decision method lacks quantitative measurement of electric shock risk, this paper proposes an optimization decision method for neutral point grounding in distribution network considering the risk of electric shock. The method minimizes the inconsistency between each scheme and the decision makers' preferences, combines the influence of decision makers, relevant standards and objective factors on decision-making, and transforms decision-making problems into linear programming problems. The target set structure covers personal electric shock risk, over voltage risk, life cycle cost and reliability, which comprehensively reflects the influence of neutral grounding method on the power distribution system, and the effectiveness of the proposed method, is verified by a practical example.

REFERENCES

Cao Zhenchong, He Jiansheng, Yang Xuechang. 2007. Interval Analysis about Influence on Reliability of Power Distribution Lines with Various Grounded Neutrals [J]. Guangxi Electric Power, 30(5):5–7, 15.

Cao Zhenchong, Wang Wenli, Yang Xuechang. 2007. Decision Arithmetic of Neutral Grounding Mode Based on Interval Multi-attribute Weighted Grey Target Theory [J]. Guangxi Electric Power, 30(6):1–5.

Chen Bobo, Qu Weifeng, Yang Hongyu. 2016. Research on single phase grounding arc model and line selection for neutral ineffectively grounding system [J]. Power System Protection & Control, 44(16):1–7.

Dong Lei, He Lin, Pu Tianjiao. 2013. Effect of neutral grounding mode on reliability of distribution network [J]. Power System Protection & Control, 41(1):96–101.

Fu Xiaoqi, Xu Liangzhen, Zhao Baoli. 2010. Discussion on the technology and application of 10 kV distribution network neutral grounding through small resistance [J]. Power System Protection & Control, 38(23):227–230.

Fu Yingshuang, Wang Zhenggang, Kuang Shi. 2006. Medium-voltage grid grounding comprehensive selection method [J]. Power System Technology, 30(15):101–102,106.

Gan Yaosheng, Tang Qinghua, Fang Qiong. 2013. Analysis of Low-resistance Neutral Grounding in Urban Medium-voltage Power Grid [J]. Proceedings of the Chinese Society of Universities for Electric Power System & Its Automation, 25(3):138–141.

Lu Yanchao, Zhang Caiqing. 2006. Assessment of power supply structure optimization by using kernel PCA-LINMAP [J]. East China Electric Power, 34(8):88–91.

Leng Hua, Tong Ying, Li Xinran. 2017. Comprehensive evaluation method research of the operation state in distributed network [J]. Power System Protection & Control, 45(1):53–59.

Liu Jian, Rui Jun, Zhang Zhihua. 2018. Smart grounding power distribution systems [J]. Power System Protection & Control, 46(8):130–134.

Li Jinglu, Zhou Yusheng. 2004. Study on neutral grounding modes of distribution system [J]. Electric Power Automation Equipment, 24(8):85–86, 94.

Liu Yi, Nie Yixiong, Peng Xiangang. 2010. Application research on the mode selection of neutral grounding with fuzzy reasoning method [J]. Power System Protection & Control, 38(7):32–36.

Su Jifeng. 2013. Research of neutral grounding modes in power distribution network [J]. Power System Protection & Control, 41(8):141–148.

Xu Yunzhi, Guo Xijin, Zhang Qian. 2008. Characteristics of earth faults in electrical distribution networks with a compensated neutral and resistance earthing [J]. Power System Protection & Control, 36(11):85–87.

Xiong Xiaofu, Liu Hengyong, Ouyang Jinxin. 2014. Neutral point grounding mode decision algorithm for medium voltage distribution network [J]. Journal of Chongqing University (Natural Science Edition), 37(6):1–9.

Industrial engineering

Automatic Control, Mechatronics and Industrial Engineering – He & Qing (Eds)
© 2019 Taylor & Francis Group, London, ISBN 978-1-138-60427-8

Numerical simulation study of lithium battery electrode crack pattern based on mixed-mode cohesive interface model

L. Chen
China Special Equipment Inspection and Research Institute, Beijing, China

Y.P. Feng
China Academy of Civil Aviation Science and Technology, Beijing, China

K.Q. Ding
China Special Equipment Inspection and Research Institute, Beijing, China

ABSTRACT: The capacity, cycle-life and life time of Si Lithium-Ion batteries are closely dependent on the ability of its electrodes to avoid fracture failure brought by de/litigation during charging/discharging cycles. In this paper we study the effect of film thickness, on the formation of crack patterns systematically. Applying the Mixed-mode Cohesive Zone interface model on the film/substrate system, a developed finite element numerical model has been carried out under plane strain condition, the whole fracture process and crack pattern of silicon film with different thickness is documented. With the growth of the Si film thickness, the average crack spacing length is increase, which is in good qualitative agreement with former experimental results and other theoretical model. The results shown that two-dimensional film-substrate system model can capture the essential features of cracking patterns of electrode materials and effectively characterize the relationship between crack pattern of thin film Lithium-Ion Battery electrodes and the thickness of Si thin film.

Keywords: numerical simulation, crack pattern, mixed-mode cohesive interface model

1   INTRODUCTION

In lithium-ion batteries, anode materials are one of the important factors that affect the capacity and performance of batteries. Silicon and lithium can form various alloys with high capacity (up to 4200 mAh/g) and low reactivity with electrolyte. Silicon is a promising anode

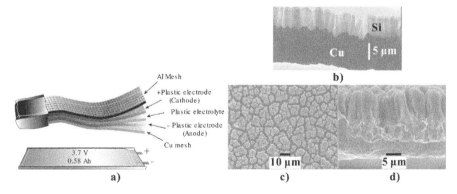

Figure 1.   (a) Side view of thin film lithium-ion battery; (b) cross-sectional TEM image; Surface (c) and cross-sectional (d) SEM images for the 6-μm film electrode after 250 cycles (Yin et al., 2006).

material for lithium-ion batteries because of its abundant reserves and low cost. However, in the charge- discharge process, the deintercalation lithium reaction of silicon will be accompanied by large volume changes (~300%), resulting in the destruction of the material structure, resulting in the separation of electrode materials, and then loss of electrical contact, resulting in rapid capacity degradation and deterioration of cycle performance. How to improve the cycling performance of silicon-based anode materials while obtaining high capacity is the research focus of silicon-based materials. In recent years, thin film materials have developed rapidly, with high specific capacity and good cycling performance, so silicon film electrode materials have also been paid attention to (Tarascon and Armand, 2001).

The two-dimensional film-substrate system model of thin film Lithium-Ion Battery electrodes is conducted through microscopic observations of Si thin film electrodes and crack pattern analysis after electrochemical litigation and delithiation cycles tests. The relationship between crack pattern of thin film Lithium-Ion Battery electrodes and the thickness of Si thin film is investigated by adopting mixed-mode cohesive interface model. Because of the characteristics of cohesion model, it can also simulate the whole process of crack initiation and propagation with loading.

2 THE MIXED-MODE COHESIVE INTERFACE MODEL

In the present investigation, a mixed-mode cohesive interface model developed by Turon (Turon et al., 2004, Turon et al., 2006) will be used to describe the damage evolution of interface. Figure 2 shows the scheme of the mixed-mode cohesive interface model within the space of both traction and separation displacements, and shows the relationship between the mixed-mode cohesive model with the pure normal separation cohesive model as well as with the pure shear separation cohesive model. Referring to Figure 2, the both triangle $O-T_1-\delta_1^f$ and $O-T_s-\delta_s^f$ (O is coordinate origin) are the bilinear responses in pure normal and pure shear modes, respectively. Any point located on the $O-\delta_s^f-\delta_1^f$ plane will correspond to a mixed-mode interface separation process. Subscripts '1' and 's' are used to represent the pure separation mode and pure shear mode, respectively. The critical relative

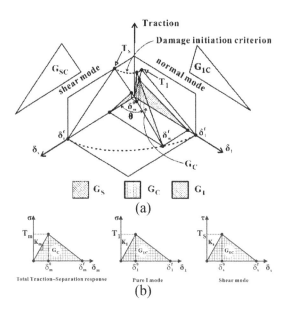

Figure 2.   Illustration of mixed-mode cohesive interface model (a) and bilinear T–S response (b).

displacements corresponding to the initiation of damage are identified with the superscript '0'. The limit relative displacements corresponding to failure state are identified with the superscript 'f'.

The damage is assumed to be initiate when the following quadratic relation is satisfied (Mi et al., 2008):

$$\left(\frac{\langle \sigma_1 \rangle}{T_1}\right)^2 + \left(\frac{\tau_s}{T_S}\right)^2 = 1 \tag{1}$$

where σ_1 and τ_s are the normal and shear stresses on the interface, respectively. T_1 and T_S are the limit separation and shear tractions, respectively.

The mixed-mode fracture criterion is described as follows (Mi et al., 2008):

$$\frac{G_1}{G_{1C}} + \frac{G_S}{G_{SC}} = 1 \tag{2}$$

where G_1 and G_S are the current fracture energies for the normal and shear cases, respectively, G_{1C} and G_{SC} are the fracture energies for pure separation mode and pure shear mode, respectively. $G_C = G_1 + G_S$ is the total current fracture energy for mixed-mode case when above condition is satisfied.

3 THEORETICAL MODEL OF CRACK PATTERN

In order to simplify the problem, a square pattern electrode with length of a_{cr} and height of h is shown in Figure 3. On delithiation, the Si film undergoes a tensile stress of σ_Y^{Si}. However, the Si film cannot yield or fracture freely because of the constrain from the substrate and the friction at the Si film/substrate interface τ_{cr}^{int}. According to the force balance (Li et al., 2011) of the Si film:

$$\tau_{cr}^{int} a_{cr} \frac{a_{cr}}{2} = \sigma_Y^{Si} a_{cr} h \tag{3}$$

The one-dimensional (1-D) critical size for crack initiation can be calculated by

$$a_{cr} = \left(2\frac{\sigma_Y^{Si}}{\tau_{cr}^{int}}\right) h \tag{4}$$

The linear relationship formula (4) represents is consistent with the result of experimental statistics.

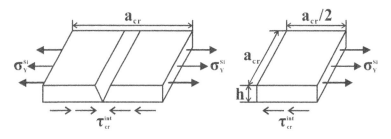

Figure 3.   Schematic of a fractured piece of Si film during delithiation.

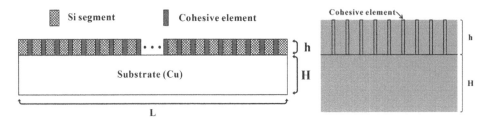

Figure 4. 2D finite element model and finite element mesh.

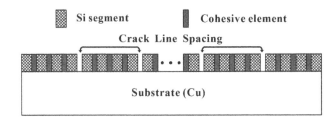

Figure 5. Schematic diagram of crack line spacing.

4 FINITE ELEMENT MODEL

Finite element method (FEM) is used in numerical simulations for crack pattern. The 2D finite element model and finite element mesh are shown in Figure 4. The mixed-mode cohesive interface model was used to describe the interface. The plain strain element (CPE4) was adopted in Si film and substrate which constitutive relation is elastic. The cohesive element COH2D4 was adopted in interface. The under surface of substrate is fixed.

As shown in Figure 5, the width of continuous Si segment is the crack line spacing. The number of interfaces between Si segments with complete failure is called the number of cracks.

5 RESULTS AND DISCUSSIONS

The finite element simulation results of crack patterns of Si films with different thicknesses were shown in Figure 6. During the calculation, the thickness of Si film (h) is changed form 1 to 5 $\mu\varepsilon$, whereas the substrate thickness (H = 5 $\mu\varepsilon$) remains unchanged. As shown in Figure 6, from top to bottom, the crack pattern simulation results of h/H = 0.2, 0.4, 0.6, 0.8 and 1 are given respectively. It can be seen that as the thickness of the Si film decreases, more and more islands are formed and the crack spacing (island width) becomes smaller and smaller. This is consistent with experimental phenomena.

The quantitative relationship between the number of cracks and the thickness of Si film was shown in Figure 7. With the increase of thickness of Si film, the number of cracks decreases gradually, which means that the average crack spacing (island width) increases.

As shown in Figure 8, the comparison between the simulation results of the average crack spacing and that predicted by formula (4) for Si films with different thickness after fragmentation was given. The formula (4) shows the linear relationship between the average crack spacing and the thickness of Si film. It can be seen that the simulation results are in good agreement with the formula derivation and the linear relationship observed experimentally.

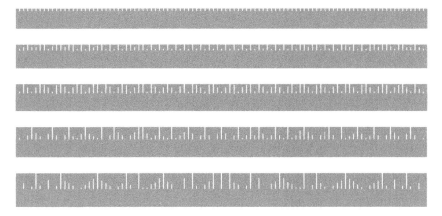

Figure 6.   Simulation results of crack pattern of Si films with different thicknesses.

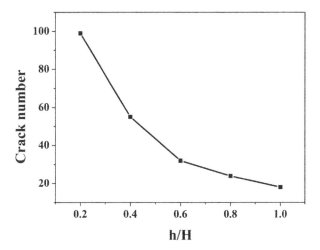

Figure 7.   Relationship between crack number and thickness of Si film after fragmentation.

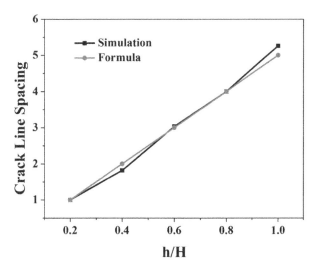

Figure 8.   Comparison between simulation results and formula prediction results.

6 CONCLUSIONS

The two-dimensional film-substrate system model of thin film Lithium-Ion Battery electrodes is conducted through microscopic observations of Si thin film electrodes and crack pattern analysis after electrochemical lithiation and delithiation cycles tests. The relationship between crack pattern of thin film Lithium-Ion Battery electrodes and the thickness of Si thin film is investigated by adopting mixed-mode cohesive interface model. With the increase of thickness of Si film, the average crack spacing increases. There is a linear relationship between the thickness of Si film and the average crack length. The present results shown that two-dimensional film-substrate system model can capture the essential features of cracking patterns of electrode materials and effectively characterize the relationship between crack pattern of thin film Lithium-Ion Battery electrodes and the thickness of Si thin film.

ACKNOWLEDGMENT

This research was supported by National key technologies Research & Development program through 2017YFC0805103.

REFERENCES

Li, J., Dozier, A.K., Li, Y., Yang, F. & Cheng, Y.T. 2011. Crack Pattern Formation in Thin Film Lithium-Ion Battery Electrodes. Journal of the Electrochemical Society, 158: A689–A694.

Mi, Y., Crisfield, M.A., Davies, G.A.O. & Hellweg, H.B. 2008. Progressive Delamination Using Interface Elements. Delamination Behaviour of Composites, 32: 367–386.

Tarascon, J.M. & Armand, M. 2001. Issues and challenges facing rechargeable lithium batteries. Nature, 414: 359–367.

Turon, A., Camanho, P.P., Costa, J. & D Vila, C.G. 2006. A damage model for the simulation of delamination in advanced composites under variable-mode loading. Mechanics of Materials, 38: 1072–1089.

Turon, A., Camanho, P.P., Costa, J. & Davila, C.G. 2004. An Interface Damage Model for the Simulation of Delamination Under Variable-Mode Ratio in Composite Materials. Recercat Home.

Yin, J., Wada, M., Yamamoto, K., Kitano, Y., Tanase, S. & Sakai, T. 2006. Micrometer-Scale Amorphous Si Thin-Film Electrodes Fabricated by Electron-Beam Deposition for Li-Ion Batteries. Journal of the Electrochemical Society, 153: A472–A477.

Automatic Control, Mechatronics and Industrial Engineering – He & Qing (Eds)
© *2019 Taylor & Francis Group, London, ISBN 978-1-138-60427-8*

Developmental status and analysis of agricultural electric drainage and irrigation system

K. Huang & C. Liu
Nari Technology Co. Ltd., Nanjing, Jiangsu, China

K.J. Lu, P. Chikangaise & X.Y. Zhu
Research Centre of Fluid Machinery Engineering and Technology, Jiangsu University, Zhenjiang, Jiangsu, China

ABSTRACT: The paper states the research status of agricultural electric drainage and irrigation system at home and abroad. The research also further expounds the meaning of electric energy replacement and agricultural electric irrigation and its significance. It introduces two advanced field irrigation methods, namely sprinkler irrigation and micro irrigation. In view of the strong seasonal characteristics of agricultural electric drainage and irrigation load, the research theoretically analyzed and solved the problems existing in the current agricultural electric irrigation and drainage system. This will be crucial in introducing a better role of the replacement of electric energy in agricultural irrigation projects.

Keywords: Agricultural irrigation and drainage, Electric energy substitution, Sprinkler irrigation, Micro irrigation, Load characteristics

1 INTRODUCTION

Electric energy has the advantages of being clean, safe and convenient. Through directly consuming electric energy and large-scale centralized conversion, it can increase fuel using efficiency and reduce pollutant emissions, to improve the terminal energy structure and promote environmental protection (Han 2018). Presently, energy saving and emission reduction issues has been prioritized to national strategic plan. The electric energy substitution included in the plan, has become a research frontier of energy conservation and environmental protection as a major measure to realize energy strategy transformation.

The State Grid Corporation is also promoting the implementation of electric drainage and irrigation in an all-round way. Many irrigation and drainage areas have built platform transformers to provide electric power for irrigation and drainage equipment. However, due to the wide range of drainage and irrigation areas and strong seasonal characteristics, the current average annual utilization rate of platform transformers is low, resulting to poor quality of electric energy. Therefore, it is imperative to optimize the power grid for platform transformers to realize the prosperous development of electric energy substitution.

2 RESEARCH STATUS OF AGRICULTURAL ELECTRIC IRRIGATION AND DRAINAGE ENGINEERING

In order to reduce the low utilization efficiency and serious pollutant emission caused by direct consumption of fossil fuels, many countries have applied electromechanical technology to agricultural drainage and irrigation projects. For instance, Holland, Japan and the United States and other countries are developing rapidly in this respect (Cheng 2005).

In Dutch, low-lying drainage problems are very prominent. Agricultural drainage and irrigation pumping stations are characterized by low head and large flow. The Netherlands

Emotton drainage station has a maximum head of 2.3 m, a single flow of 37.5 m³/s and the total drainage capacity of 150 m³/s. In recent years, Japan has built a large number of drainage pumping stations for farmland irrigation and water logging problems. A total of 6 tubular pumps with a diameter of 4.2 m were installed at the New River Estuary Pump Station. The head is 2.6 m and the flow rate of a single pump is 40 m³/s. Foreign agricultural electric drainage and irrigation projects are characterized by advanced technology, equipment, excellent engineering quality and high degree of automated operation. They also pay attention to the maintenance and update of electromechanical engineering equipment in time.

In China, the electromechanical drainage and irrigation industry have developed rapidly recently. In 2016, the State Development and Reform Commission and other eight ministries jointly issued the "*Guiding Opinions on Promoting Electric Energy Replacement*". This clearly pointed out that it is necessary to speed up power supply of pumping station combining with the construction of high-standard farmland and the promotion of agricultural water-saving irrigation (NDRC 2016). In Gansu Province, the electricity price of agricultural drainage and irrigation is designed to guide the application of electric drainage and irrigation system (Zhang 2011). The Chinese Academy of Electrical Sciences has designed an adaptive load distribution transformer which can realize voltage tap-changer adjustment and automatic switching of rated capacity operation mode according to the actual situation of system voltage and load without cutting off load (Wang et al. 2014). Nanjing Nari Group are also developing an integral distribution transformer and power quality control equipment to simplify the whole platform structure and facilitates installation.

3 CONCEPT AND SIGNIFICANCE OF AGRICULTURAL ELECTRIC DRAINAGE AND IRRIGATION

3.1 *Concept and significance of electric energy substitution*

Electric energy substitution refers to the use of electric energy to replace the energy consumption mode of coal and oil burning in the terminal energy consumption link. Electric energy substitution is of great significance for promoting the energy consumption revolution, implementing the national clean energy strategy. From the perspective of drainage and irrigation, promoting the "*Guiding Opinions on Promoting Electric Energy Substitution*", is to speed up the power supply of pumping station in an all-round way as well as building an efficient and energy-saving agricultural electric drainage and irrigation system.

3.2 *Concept and significance of agricultural electric irrigation and drainage system*

Drainage and irrigation project refer to the engineering and technical measures to use power machinery to drive water pump, drain and then irrigate the field. The electromechanical irrigation and drainage project applied to farmland irrigation is called agricultural electric irrigation and drainage. The agricultural irrigation and drainage project plays an irreplaceable role in improving agricultural production conditions, building stable and high-yield farmland and solving high water supply demands. The in-field project application of agricultural power irrigation and drainage mainly involves advanced water-saving irrigation methods such as sprinkler irrigation and drip irrigation (Lan et al. 2010, Yuan et al. 2017, Zhu et al. 2018, Li et al. 2018).

Sprinkler irrigation is shown in Figure 1. It makes use of pump pressure or natural water gravity to supply water to the field through pressure pipeline. Water is sprayed into the air through the nozzle to form fine water droplets, which are evenly sprayed on the farmland to provide necessary water requirements for normal growth of crops. The practice has proved that spraying has the advantages of increased production, water saving, labor saving, soil conservation, fertilizer retention, strong adaptability, easy mechanization and automation (Xu et al. 2018, Tang et al. 2018, Lu et al. 2018).

Microirrigation, as shown in Figure 2, is another new type of water-saving irrigation technology that can transport the water and nutrients required for crop growth with a small flow

Figure 1.   Sprinkler irrigation.

Figure 2.   Micro irrigation.

rate according to crop water requirements, through a piping system and a special emitter installed on the final pipeline. Compared to traditional full-surface moist ground irrigation and sprinkler irrigation, micro-irrigation only wets part of the soil near the root zone of the crop with a small amount of water. Micro-irrigation makes it very convenient to apply water to the soil near each plant maintaining low water stress to meet crop growth needs.

4   EXISTING PROBLEMS AND OPTIMIZATION METHODS

Pumping stations can create a good environment for irrigation and drainage of farmland as well as playing a great role in ensuring the safety of grain production. It can promote the development of clean energy more effectively by optimizing the traditional drainage and irrigation pumping stations into the electromechanical systems.

4.1   *Problems in agricultural electric drainage system*

Adopting electric energy substitution and upgrading the traditional pumping station into an electric drainage and irrigation system is conducive to promote the energy revolution, but in the process of pumping station renovation, the following problems need to be solved.

1. Low power factor
The main load of drainage and irrigation pumping station comes from asynchronous motor, which has the characteristics of strong seasonality, large single motor capacity and relatively concentrated power consumption time. Because of its simple structure, low price, durability and high efficiency, asynchronous motors are widely used in small and medium pumping stations to drive pumps. However, when the motor is running, it needs to absorb a great deal of inductive reactive current from the grid. For drainage and irrigation pumping stations, the operation under low power factor not only increases the power supply and distribution loss, but also probably be fined by the power supply department so that the economic and social benefits cannot be brought into full play. Therefore, the reactive power compensation for the motor can greatly reduce the line voltage, power loss, and improve economic and social benefits (Dai, Hu 2009).

2. Large change of power load rate

Agricultural electricity, especially in the agricultural electric drainage system, relies more on seasons throughout the year. The power load is very low in the winter, while very high in the summer. For instance, when not pumping water, the transformer is only used for electricity consumption by villagers, with low load rate and high line loss rate. But in the busy agricultural season, the load increases sharply resulting to an overload or even the transformer burned out. Therefore, forecasting the agricultural electricity drainage and irrigation load and making the power grid supply at high load rate is of great interest to improve the utilization efficiency (Yang, Cheng 2018).

4.2 *Optimization research*

1. Reactive power compensation technology

① *Compensation methods*

Parallel capacitor banks are usually used to compensate the reactive power, which can be divided into two modes: local compensation and centralized automatic compensation. Due to the use of automatic compensation devices and shunt circuit breakers to switch capacitor banks, the investment of centralized automatic compensation is large and it is generally not used in pumping stations. Local reactive power compensation has been widely used in pumping stations because it can synchronously switch with load and no special circuit breaker is needed.

② *Compensation requirements*

According to the regulation of "*Technical Guidelines for Voltage and Reactive Power in Power Systems*", the power factors of power users should meet the following requirements: The power factors of industrial users with high voltage power supply should be above 0.9; the power factor of other power users (whose power factor are above 100kVar) and large-sized or medium-sized power irrigation and drainage stations should be more than 0.85. The necessary reactive power compensation devices should be installed if the above requirements are not met.

③ *Compensation capacity*

In order to prevent the self-excited overvoltage generated when the motor is out of operation, the compensation capacity is generally not greater than the no-load reactive power of the motor (Dai, Hu 2009), which can be calculated as formula (1):

$$Q_c = (0.95 \sim 0.98) \sqrt{3} U_N I_0 \tag{1}$$

where: Q_c = the capacity of the compensation capacitor, kVar; U_N = rated voltage, kV; I_0 = the no-load current of motor, A.

For motors with large mechanical inertia of the motor for drain irrigation, the compensation capacity can be appropriately increased. The motor for drain irrigation is often powered off with the mechanical load of the pump. Therefore, the motor speed drops sharply and the self-excited overvoltage does not occur even if the compensation capacity is slightly larger than the no-load reactive load of the motor. For ordinary motors for irrigation and drainage, the compensation capacity can be determined by the following formula (2):

$$Q_c = (0.5 \sim 0.6) P_N \tag{2}$$

where: P_N = the rated active power of the motor for drain irrigation, kW.

2. Analysis of power load

One of the characteristics of electric power production is that the power cannot be stored in large scale. The use of electricity for irrigation and drainage in agriculture has strong seasonality. The change of power load within the day is relatively small, but during the month, especially during the quarter and year, the load changes greatly. Therefore, it is necessary to establish evaluation indexes and prediction method for the electrical load of agricultural electric drainage and irrigation systems.

Proper description and accurate estimation of load variation characteristics can be achieved through the analysis and calculation of some specific index parameters.

Load rate
The daily load rate and daily minimum load rate are used to describe the characteristics of daily load curve and to characterize the daily imbalance. Higher load rate is conducive to the economic operation of drainage and irrigation power system. For the agricultural electric irrigation and drainage system, the daily electric load change is small, so the daily load rate and the daily minimum load rate are generally small.

Monthly load rate
The monthly load rate is mainly related to electricity consumption structure, and seasonal changes. For the agricultural electric irrigation and drainage project, it accounts for a large proportion of rural electricity consumption, and thus has a greater impact on the monthly load rate. Since the irrigation and drainage project is seasonal, it will be affected by natural rainfall. And the natural rainfall is uneven. This imbalance also leads to the imbalance of electricity consumption in agricultural irrigation and drainage.

Seasonal load rate
The seasonal load rate, also known as the seasonal imbalance factor, reflects the seasonal variation of the power load. It is mainly affected by the seasonal configuration of the electrical equipment, the annual overhaul of the equipment.

Annual load rate
The annual load rate is a comprehensive indicator. It is the ratio of annual average load to annual maximum load.

The peak-to-valley difference
The peak-to-valley difference is defined as the difference between the highest load and the lowest load. It directly reflects the peaking capacity required by the grid. For agricultural irrigation and drainage, seasonal changes have a great impact on peak-to-valley difference.

Annual maximum load utilization hours
Annual maximum load utilization hours are mainly used to measure the time utilization efficiency.
② *Mathematical forecasting methods for power load*
Least squares
The trend of load development can be predicted by the least squares method which is to express the development trend of the load sequence by equation and then use the trend equation to predict the future trend development. The least squares method is used to determine the development trend curve, and the sum of the squares of the deviations of the actual values of the load sequence from the trend is required to be the smallest.

Regression analysis
The regression analysis method fits the past load records with random characteristics to obtain certain curve, and then extends the curve to the appropriate time to obtain the load prediction value at that time. The dependent variable of the regression equation is the power system load, and the independent variables are various factors that affect the load of the electric irrigation and drainage system, such as the pump station scale, the pump type, the operating conditions and the area size controlled by the pumping station.

Fuzzy control method
Power load forecasting is the use of past data to find out the law of load changes, and thus predict the trend and state of power load in the future. In actual forecasting, it is often carried out in the conditions that the historical load and the related environmental factors are not clear, so fuzzy mathematics provides an effective means for such problems. In recent years, more combined methods have emerged, such as fuzzy regression, fuzzy clustering, fuzzy neural network.

5 CONCLUSION

The characteristics of agricultural electricity are different from those of industrial and residential electricity consumption. Agricultural electricity consumption is very uneven during the year, but relatively balanced during a single day. The seasonality is the most important characteristic of agricultural drainage load. It is determined by the seasonality of agricultural production.

Finding the existing problems, considering various factors like project scale, pump type, drainage area and drainage operation conditions, analyzing the characteristics of agricultural irrigation and drainage systems and establishing electric load characteristic, only in these ways, can we build a cleaner and more efficient agricultural irrigation and drainage project, and promote the benign development of electric energy substitution.

ACKNOWLEDGMENTS

This research was funded by 2017 State Grid Scientific Project Research and application of typical scenario optimization interaction in electric energy substituting and power grid regulation and support technology, and the Key R&D Project of Yangzhou City (Modern Agriculture), No. YZ2017050.

REFERENCES

Cheng J. 2005. Development and operation management of pumping stations abroad [J]. *China Water Resources,* 2005(23):116.

Dai Z X, Hu G W. 2009. Application of the local reactive power compensation in the agriculture pump station [J]. *Journal of Agricultural Mechanization Research*, 2009, 31(09):194–196.

Han J. 2018. Promoting electric energy substitution and clean heating [J]. *Electric Power Equipment Management*, 2018(02):18.

Lan Y B, Thomson S J, Huang Y B, Hoffmann W C, Zhang H H. 2010. Current status and future directions of precision aerial application for site-specific crop management in the USA[J]. *Computers and Electronics in Agriculture*, 2010, 74(1): 34–38.

Li Y F, Liu J P, Li T, Xu J E. 2018. Theoretical model and experiment on fluidic sprinkler wet radius under multi-factor[J]. *Journal of Drainage and Irrigation Machinery Engineering*, 2018, 36(8): 685–689.

Lu M Y, Lu K J, Hu G, Zhu X Y. 2018. Experiment on hydraulic performance of type SD-03 pop-up sprinkler[J]. *Journal of Drainage and Irrigation Machinery Engineering*, 2018, 36(11): 1120–1124.

National Development and Reform Commission. 2016. Guiding Opinions on Promoting Electric Energy Replacement [J]. *Popular Utilization of Electricity*, 2016, 31(09):5–6.

Tang L D, Yuan S Q, Qiu Z P. 2018. Development and research status of water turbine for hose reel irrigator[J]. *Journal of Drainage and Irrigation Machinery Engineering*, 2018, 36(10): 963–968.

Wang J L, Sheng W X, Fang H F, Wang J Y, Yang H L, Wang L. 2014. Design of a Self-adaptive Distribution Transformer[J]. *Automation of Electric Power Systems*, 2014, 38(18):86–92.

Xu Z D, Xiang Q J, Waqar A Q, Liu J. 2018. Field combination experiment on impact sprinklers with aerating jet at low working pressure[J]. *Journal of Drainage and Irrigation Machinery Engineering*, 2018, 36(9): 840–844.

Yuan S Q, Darko R O, Zhu X Y, Liu J P, Tian K. 2017. Optimization of movable irrigation system and performance assessment of distribution uniformity under varying conditions[J]. *Int J Agric & BiolEng*, 2017, 10(1): 72–79.

Yang B Y, Cheng S J. 2018. Overview of electrical load forecasting researches and forecasting analysis [J]. *Sichuan Electric Power Technology*, 2018, 41(03):56–60 91.

Zhang H W. 2011. Study in the reform of Gansu agricultural electricity price of irrigation[J]. *China Electric Power(Technology Edition)*, 2011(01):63–67.

Zhu X Y, Chikangaise P, Shi W D, Chen W H, Yuan S Q. 2018. Review of intelligent sprinkler irrigation technologies for remote autonomous system[J]. *Int J Agric & BiolEng*, 2018, 11(1): 23–30.

Zhu X Y, Yuan S Q, Jiang J Y, Liu J P, Liu X F. 2015. Comparison of fluidic and impact sprinklers based on hydraulic performance[J]. *IrrigSci*, 2015, 33(5): 367–374.

Automatic Control, Mechatronics and Industrial Engineering – He & Qing (Eds)
© 2019 Taylor & Francis Group, London, ISBN 978-1-138-60427-8

Knowledge discovery of design rationale based on frequent-pattern mining

H. Jiang, W. Yang, J. Mei, R.L. Wu & L. Guo
The Netherlands Institute of Telecommunications Satellite, China Academy of Space Technology, Beijing, China

ABSTRACT: Knowledge of design rationale provides support for design reuse, innovative design, collaborative design and other design behaviors. It is difficult to guarantee the accuracy and universality of design knowledge in a single instance of design rationale, and it is difficult for designers to reuse the design knowledge directly. This paper proposes a design rationale knowledge-discovery method based on frequent-pattern mining. In order to facilitate the mining of frequent patterns from a fine-grained design rationale model, we first transform it into a label graph, then extract the frequent patterns from the label graph, and then transform these into a design rationale model. Then, through the design rationale model fragment integrity detection and inference method, the extracted design rationale model structure is perfected as a complete design rationale model fragment. Finally, the validity of this method is illustrated by way of an example of design rationale models for an automatic marking machine.

Keywords: Knowledge discovery, design rationale, frequent-pattern mining, design reuse, label graph

1 INTRODUCTION

According to modern design theory, design is a creative activity based on knowledge. The acquisition and application of new knowledge is the core of design, and is of great significance. Design rationale is an explicit expression of the systematic knowledge of the design process. The purpose of research into the acquisition, recording and modeling of design rationale knowledge is to realize the reuse of design rationality. Design rationale knowledge supports design reuse, innovative design, collaborative design and other design activities. Currently, the reuse of design rationality is mainly supported by design rationale retrieval and model processing.

DRed (Bracewell et al., 2004, 2009) builds a design rationale model node index through keywords. Based on the user entering the keywords and the semantic retrieval of their synonyms, the results are returned to the user after sorting. In addition, DRed realizes tunnel linking, which supports the retrieval of associated nodes in different charts. Hong-wei Wang and colleagues improved the original retrieval method of DRed and proposed a structure-based design rationale retrieval method (Wang et al., 2012). This method makes use of implicit structural relations in the DRed model, the designer's design situation information, and semantic information in the design node to realize design rationale retrieval. Luye Li and colleagues put forward the expression method of rationale design based on ontology, and a retrieval method based on ontology: keywords retrieval, natural language retrieval and retrieval based on design rationale record (Li et al., 2013a, 2013b). The Issue, Solution and Artifact Layer (ISAL) model (Liang et al., 2010) provides users with a series of interactive interfaces for design rationale retrieval. This approach enables users to perform multiple types of retrieval and displays the details of the problem layer, solution layer and product layer through the display interface. With more and more comprehensive and detailed knowledge of design rationality model description, the design rationality model becomes bigger and bigger. This large model makes it difficult for users to browse and retrieve, and to find the design knowledge they need accurately and quickly. Based on the knowledge reasoning of Web Ontology Language (OWL) description logic, Hu and colleagues (Hu, 2013; Jing et al., 2013) realized the segmentation and extraction of the design context of a fine-grained design rationality model. This method encapsulates and expresses the knowledge of design rationality in design rationality segments. Aiming at a design rationality model driven by intention, Shi-kai Jing and colleagues proposed a semantically based extraction method for design rationality fragments (Jing et al., 2014). It is difficult for designers to directly reuse design knowledge because the accuracy and universality of design knowledge in a single design rationale model is difficult to guarantee. In response to these challenges, we propose a design rationale knowledge-discovery method based on frequent-pattern mining. The basic idea is that a design knowledge or model mechanism with poor accuracy and universality is a small-probability event in a large number of design rationale model cases. In order to facilitate the mining of frequent patterns from fine-grained design rationale models, this paper first converts them into labeled graphs, then excavates the frequent patterns from the labeled graphs, and finally converts them into design rationale models. The ontology-based design rationality model is composed of multiple design fragments, combined according to certain logic. Each design rationality fragment describes a relatively complete design task. The frequently discovered design rationale model structure is often not a complete design fragment. In this paper, the structure of the excavated design rationality model is improved to form a complete design rationality model fragment.

2 DESIGN RATIONALE MODEL AND MAPPING MECHANISM OF A LABEL GRAPH

The fine-grained design rationale model mainly consists of the relationship between nodes. A label graph consists of vertices and edges. Mapping mechanism for fine-grained design rationale model into label the vertices in the graph, nodes in a fine-grained design rationale models of relationship between nodes into label the edges in the graph, at the same time generate model mapping table used to explain the excavated frequent subgraphs. Figure 1 shows a typical example of a fine-grained design rationale model, and Figure 2 shows the model's conversion into a label graph. Table 1 is the relational mapping table associated with this conversion. The standard code of a graph is a formal description of it. It is possible to determine whether a graph is isomorphic by inspecting this standardized coding. In this paper, the Depth-First Search (DFS) (Yan & Han, 2002) algorithm is used to realize the standardized encoding of the graph.

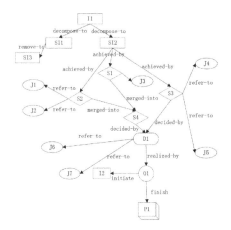

Figure 1. Fine-grained design rationale model.

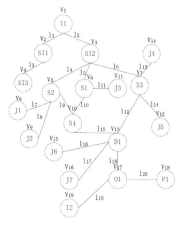

Figure 2. Labeled graph.

Table 1. Mapping table.

Node in design rationale model	Node in labeled graph	Node in design rationale model	Node in labeled graph
I1	V_1	J2	V_9
I2	V_{19}	J3	V_{11}
SI1	V_2	J4	V_{14}
SI2	V_3	J5	V_{12}
SI3	V_4	J6	V_{15}
S1	V_6	J7	V_{16}
S2	V_5	D1	V_{13}
S3	V_{11}	O1	V_{17}
S4	V_{10}	P1	V_{18}
J1	V_8		

Relationship	Edge	Relationship	Edge
decompose-to	l_1, l_2	decided-by	l_{12}, l_{15}
achieved-by	l_4, l_5, l_6	realized-by	l_{18}
remove-to	l_3	initiate	l_{19}
merged-into	l_9, l_{10}	finish	l_{20}
refer-to	l_7, l_8, l_{11}, l_{13}	refer-to	l_{14}, l_{16}, l_{17}

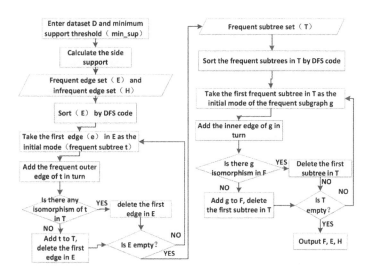

Figure 3. Incremental frequent-pattern mining algorithm based on DFS encoding.

3 INCREMENTAL FREQUENT-PATTERN MINING BASED ON DFS ENCODING

The label graph in this paper is transformed from the fine-grained design rationale model, which has the structural characteristics of the design rationale model. This label graph contains only a small number of rings.

For frequent-pattern mining of labeled graphs with a small number of rings, Li et al. (2007) proposed a mining algorithm that first mined frequent subtrees and then extended to frequent subgraphs. The algorithm selects a frequent edge as the initial node, and then expands the frequent subtree by adding frequent edges. The frequent inner edges of the subtrees are selected from frequent edge sets to produce frequent subtrees. The algorithm is shown in Figure 3.

The growth-type frequent-pattern mining algorithm based on DFS encoding mainly consists of three parts: the first manages the atlas, extracts necessary information, and generates frequent edge sets; the second generates frequent subtrees on the basis of frequent edge sets; the third extends the frequent subtree to the frequent subgraph.

The modification of the original algorithm in this paper is mainly reflected in that the infrequent edge set (H) is output through traversal and sorted according to the degree of support. Infrequent edge sets are important for the extension of typical design knowledge fragments. The inverse of the mined frequent subgraph needs to be mapped to the frequent design rationale model according to the mapping table.

4   DESIGN RATIONALE MODEL FREQUENT-PATTERN INTEGRITY DETECTION AND INFERENCE

The frequent structure of the design rationale model, which is mined by a growth-type frequent pattern based on DFS encoding, is a frequent pattern in the mathematical sense, regardless of its semantic constraints. Design rationale models generated directly from frequent subgraphs by inverse mapping are often only the combination of several nodes and their relations, and cannot represent complete design knowledge. Therefore, there is a need to excavate the design of the rationale structure of frequent integrity testing and reasoning, according to certain rules, to expand and tailor, to express an independent relatively complete design knowledge, and to ensure the integrity of its semantic level.

We divide the excavated design rationale frequent structure into three types: intention analysis fragments, intention realization fragments and scheme evaluation fragments. These three types of structural fragments correspond to three types of design knowledge mined from fine-grained design rationale models. The intention analysis fragment describes how the designer decomposes an abstract initial intention into several relatively independent sub-intentions. The intention realization fragment describes the designer's proposal for multiple possible design schemes for the design element intention, as well as the evaluation and implementation process of the design scheme. The scheme evaluation fragment considers the design basis for a given design scheme, and gathers its positive or negative aspects.

The design intention fragment excavated by the mining algorithm must include the initial design intention node and the sub-intention node. If the frequent patterns mined by a mining algorithm have initial intention nodes and sub-intention nodes, the integrity detection and inference of intention analysis fragments are carried out. If the frequent patterns mined by the mining algorithm contain a design intention node and at least one design scheme node, it can be extended by inference to complete the intention realization fragment. If the main structure of a frequent pattern is the design scheme or design decision node and its related design basis, it can be treated as a scheme evaluation fragment.

5   APPLICATION EXAMPLES

Taking the design rationale model for an automated marking machine as an example, this paper explains the design rationale knowledge-discovery method based on frequent-pattern mining proposed in this paper. First of all, we construct ten different models to describe the knowledge of the marking machine design rationale; Figure 4 shows one example.

After building multiple design rationale models, we transform them into label graphs and build relational mapping tables. All the label graphs converted from the marking machine design rationale model are imported into the graph set, and then multiple frequent subgraphs are mined, using a growth-type frequent-subgraph mining algorithm based on DFS encoding. Three representative frequent subgraphs were selected from multiple frequent subgraphs, as shown in Figure 5 (a1, a2 and a3). Frequent-pattern integrity detection and inference are carried out for the frequent subgraphs to generate relatively independent and complete fragments of the design rationale model, also as shown in Figure 5 (b1, b2 and b3).

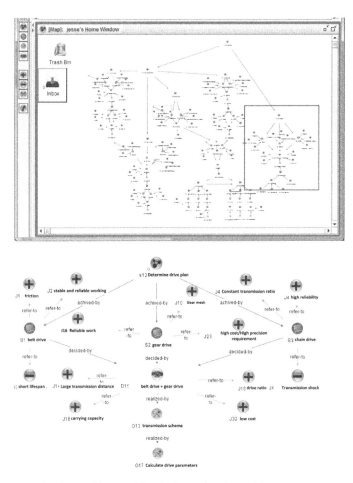

Figure 4.   An example of a marking machine design rationale model.

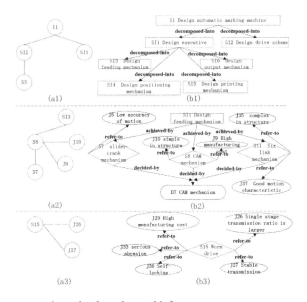

Figure 5.   Frequent-pattern integrity detection and inference.

6 CONCLUSION AND PROSPECTS FOR ADAPTATION

Aiming at the problem that the accuracy and reliability of design knowledge are difficult to guarantee in a single design rationale model example, this paper proposes a design rationale knowledge-discovery method based on frequent-pattern mining. In this paper, the design rationale model is transformed into a label graph, and the frequent patterns are found from the label graph set transformed by multiple design rationale models, through a growth-type frequent-subgraph mining algorithm based on DFS. In this paper, three kinds of design rationale knowledge fragments and their integrity detection and inference methods are proposed to realize the inverse mapping and interpretation of frequent subgraphs. Finally, the validity of this method is illustrated by its application to the design rationale model for a marking machine.

ACKNOWLEDGMENT

We acknowledge the support for this research of the National Natural Science Foundation of China (NSFC) (Grant no. 2015BAF18B01).

REFERENCES

Bracewell, R., Ahmed, S. & Wallace, K. (2004). DRed and design folders: A way of capturing, storing, and passing on knowledge generated during design projects. In *Proceedings of ASME 2004 Design Automation Conference* (pp. 235–246). doi:10.1115/DETC2004-57165.

Bracewell, R., Wallace, K., Moss, M. & Knott, D. (2009). Capturing design rationale. *Computer-Aided Design, 41*(3), 173–186.

Hu, X.-J. (2013). Study on the theory and method of multi-granularity modeling in design thinking process. Beijing: Beijing University of Aeronautics and Astronautics.

Jing, S.-K., Liu, J.-H. & Hu, X.-J. (2013). Design reuse oriented transformation of design rationale models. *Computer Integrated Manufacturing Systems, 19*(6), 1186–1193. Retrieved from http://www.cims-journal.cn/EN/abstract/abstract4443.shtml.

Jing, S.-K., Liu, J.-H., Wang, K. & Zhou, J.-T. (2014). Semantic-based design rationale segment extraction method. *Computer Integrated Manufacturing Systems, 20*(6), 1291–1299. doi:10.13196/j.cims.2014.06.jingshikai.1291.9.2014063.

Li, L., Qin, F.W. & Gao, S.M. (2013a). An approach to design rationale retrieval using ontology-aided indexing. In *ASME 2013 International Design Engineering Technical Conferences and Computers and Information in Engineering Conference* (vol. 2B). doi:10.1115/DETC2013-12522.

Li, L., Qin, F.W. & Gao, S.M. (2013b). An extended representation of design rationality that effectively supports design knowledge retrieval and reuse. *Journal of Computer-Aided Design and Graphics, 25*(10), 1514–1522.

Li, X.-T., Li, J.-Z. & Gao, H. (2007). An efficient frequent subgraph mining algorithm. *Journal of Software, 18*(10), 2469–2480.

Liang, Y., Lu, W.F., Liu, Y. & Lim, S.C.J. (2010). Interactive interface design for design rationale search and retrieval. In *ASME 2010 International Design Engineering Technical Conferences & Computers and Information in Engineering Conference* (vol. 5, pp. 73–82). doi:10.1115/DETC2010-28392.

Wang, H.W., Johnson, A. & Bracewell, R. (2012). The retrieval of structured design rationale for the re-use of design knowledge with an integrated representation. *Advanced Engineering Informatics, 26*(1), 251–266.

Yan, Y. & Han, J. (2002). gSpan: Graph-based substructure pattern mining. In *Proceedings of the 2002 International Conference on Data Mining (ICDM 2002), Maebashi City, Japan* (pp. 721–724). doi:10.1109/ICDM.2002.1184038.

Automatic Control, Mechatronics and Industrial Engineering – He & Qing (Eds)
© 2019 Taylor & Francis Group, London, ISBN 978-1-138-60427-8

Transfer alignment of large misalignment angles with a Quaternion algorithm combining Strong Tracking and Cubature Kalman Filtering (QSTCKF)

K. Liu, L. Yan, Z.G. Zhu & Y. Zhao
Beijing Institute of Aerospace Control Device, Beijing, China

ABSTRACT: Transfer alignment is usually applied to weapon systems that need a rapid launch and complex external environment. In actual conditions, the misalignment angles are usually not small and the error model cannot be processed simply by linear filtering for large misalignment angles. In this paper, a large misalignment angle error model based on the multiplicative quaternion is derived, and a quaternion strong-tracking fifth-order cubature Kalman filter is proposed. To keep the quaternion's norm unchanged during the filtering process, a quadratic Padé approximation method is used in the quaternion iterative decoding process. The adaptive fading factor from a strong tracking filter is added to the time update and measurement update equations from the fifth-order quaternion cubature Kalman filter; thus, the gain matrix can be adaptively adjusted, and the filtering divergence can be effectively prevented. Simulation results show that the proposed algorithm can solve the problem of transfer alignment for large misalignment angles, and the proposed algorithm can improve the transfer alignment accuracy and rate of convergence in large misalignment angle cases.

Keywords: large misalignment, transfer alignment, strong tracking filter, fifth-order cubature Kalman filter

1 INTRODUCTION

Initial alignment is a key link before an inertial device enters navigation. It can be said that the accuracy of initial alignment determines the final navigation accuracy. Conventional initial alignment methods include the compass loop method and optimal filtering method, but the compass loop method utilizing the compass method requires that the vehicle is under static conditions. The second method utilizes modern control theory to estimate the misalignment angles between the on-board Master Inertial Navigation System (MINS) and the low-accuracy Strapdown INS (SINS) of weapon systems.

It is impossible to guarantee that the initial misalignment angle is small when transfer alignment actually takes place. When the filtering algorithm adopts an Extended Kalman Filtering (EKF) algorithm, the EKF algorithm only provides a first-order approximation to the nonlinear object. When the system is highly nonlinear, the EKF algorithm will cause a large truncation error, resulting in filtering divergence (Forbes, 2010). Unscented Kalman Filtering (UKF) can obtain higher accuracy than EKF (Chang et al., 2013), but for nonlinear systems with higher dimensions, a covariant non-positive definite situation easily occurs. If UKF is adopted, it will cause filtering divergence and affect filtering accuracy. The Particle Filter (PF) based on a Monte Carlo sampling strategy requires a large amount of calculation, which leads to a time delay and particle degradation. Based on third-order spherical radial cubature rules, Cubature Kalman Filtering (CKF) is proposed (Arasaratnam & Haykin, 2009). Although the estimation speed is improved, the estimation accuracy is lower than PF.

In the literature (Sun & Tang, 2012; Gao et al., 2011, Guo et al., 2017), an error model based on large azimuth misalignment angles is proposed, but the horizontal misalignment angle is still assumed to be small. Using quaternions instead of attitude angle can solve the situation of level misalignment angle for a small angle. However, quaternions are rotation vectors and cannot be directly involved in the operation. Reference (Markley et al., 2007) based on Frononius norm calculating the weighted average of the quaternion. The algorithm obtains the weighted-average quaternions by calculating the extreme value of the attitude-matrix cost function. Based on this method, Quaternion UKF (QUKF) (Zhou et al., 2011) and Quaternion CKF (QCKF) (Chen et al., 2013) algorithms are proposed and applied to transfer alignment. Li and Sun (2013) proposed the combination of CKF and a Strong Tracking Filter (STF) algorithm, which effectively improved the tracking accuracy and robustness of filtering.

In this paper, large misalignment angles of transfer alignment based on a quaternion error model are established. The algorithm uses Padé approximation in the recurrence formula that guarantees the norm of quaternion invariance. In addition, a strong tracking algorithm combined with a fifth-order CKF algorithm is proposed. The simulation results show that the proposed algorithm (QSTCKF) is superior to both QCKF and QUKF in convergence speed and accuracy.

2 LARGE MISALIGNMENT ANGLE ERROR MODEL

2.1 *Attitude error equation based on multiplicative quaternions*

When the misalignment angle is small, the error model can be linearized in the derivation process, that is:

$$C_{n'}^n = I + [\Phi \times] = \begin{bmatrix} 1 & -\phi_z & \phi_y \\ \phi_z & 1 & -\phi_x \\ -\phi_y & \phi_x & 1 \end{bmatrix} \tag{1}$$

where n stands for the navigation coordinate, n' is the computational navigation coordinate, ϕ_x, ϕ_y and ϕ_z are the misalignment angles between n and n'.

However, when the misalignment angle is large, the above linear simplification cannot be made, and the nonlinear error model needs to be rederived. Based on the small misalignment angle model, the large misalignment angles based on the quaternion are derived.

In the large misalignment angles transfer alignment, the ideal attitude quaternion of the SINS to the ideal navigation coordinate q_s^n satisfies the differential equation:

$$\dot{q}_s^n = \frac{1}{2} q_s^n \otimes \omega_{ns}^s = \frac{1}{2} q_s^n \otimes (\omega_{is}^s - q_n^s \otimes \omega_{in}^n \otimes q_s^n) \tag{2}$$

where ω_{ns}^s is the projection of angular velocity of the carrier coordinate s relative to the navigation coordinate n under the coordinate S, and $\omega_{ns}^s = \omega_{is}^s - \omega_{in}^s$; ω_{is}^s is the projection of angular velocity of the carrier coordinate s relative to the earth centered inertial coordinate i under the coordinate s; ω_{in}^n is the projection of angular velocity of the navigation coordinate n relative to the earth centered inertial coordinate i under the coordinate n. $\otimes$ means quaternion multiplication.

According to the actual output of SINS, the attitude quaternion differential equation can be deduced as follows:

$$\dot{q}_s^{n'} = \frac{1}{2} q_s^{n'} \otimes \tilde{\omega}_{ns}^s = \frac{1}{2} q_s^{n'} \otimes (\tilde{\omega}_{is}^s - q_{n'}^s \otimes \tilde{\omega}_{in}^n \otimes q_s^{n'}) \tag{3}$$

where $q_s^{n'}$ is the SINS actual attitude quaternion, $\tilde{\omega}_{is}^s$ is the actual output of SINS gyros, and $\tilde{\omega}_{is}^s = \omega_{is}^s + \delta\omega_{is}^s$, $\delta\omega_{is}^s$ is the output error of SINS gyros; $\tilde{\omega}_{in}^n$ is obtained by calculation and has a calculation error.

According to the formula for quaternion multiplication, we get:

$$q_s^{n'} = q_n^{n'} \otimes q_s^{n} \tag{4}$$

The derivative of Equation 4 can be written as:

$$\dot{q}_s^{n'} = \dot{q}_n^{n'} \otimes q_s^{n} + q_n^{n'} \otimes \dot{q}_s^{n} \tag{5}$$

Substitution of Equations 2 and 3 into Equation 5 yields:

$$\frac{1}{2} q_s^{n'} \otimes (\tilde{\omega}_{is}^{s} - q_{n'}^{s} \otimes \tilde{\omega}_{in}^{n} \otimes q_s^{n'}) = \dot{q}_n^{n'} \otimes q_s^{n} + \frac{1}{2} q_n^{n'} \otimes q_s^{n} \otimes (\omega_{is}^{s} - q_n^{s} \otimes \omega_{in}^{n} \otimes q_s^{n})$$

$$\dot{q}_n^{n'} \otimes q_s^{n} = \frac{1}{2} q_s^{n'} \otimes (\tilde{\omega}_{is}^{s} - q_{n'}^{s} \otimes \tilde{\omega}_{in}^{n} \otimes q_s^{n'}) - \frac{1}{2} q_s^{n'} \otimes (\omega_{is}^{s} - q_n^{s} \otimes \omega_{in}^{n} \otimes q_s^{n}) \tag{6}$$

$$= \frac{1}{2} q_s^{n'} \otimes (\tilde{\omega}_{is}^{s} - \omega_{is}^{s}) - \frac{1}{2} \tilde{\omega}_{in}^{n} \otimes q_s^{n'} + \frac{1}{2} q_n^{n'} \otimes \omega_{in}^{n} \otimes q_s^{n}$$

When multiplying both sides of Equation 6 by q_n^{s} at the same time, it can be concluded that the differential equation of attitude quaternion is:

$$\dot{q}_n^{n'} = \frac{1}{2} q_s^{n'} \otimes (\tilde{\omega}_{is}^{s} - \omega_{is}^{s}) \otimes q_n^{s} - \frac{1}{2} \tilde{\omega}_{in}^{n} \otimes q_n^{n'} + \frac{1}{2} q_n^{n'} \otimes \omega_{in}^{n}$$

$$= \frac{1}{2} q_s^{n'} \otimes \varepsilon^{s} \otimes q_n^{s} - \frac{1}{2} \tilde{\omega}_{in}^{n} \otimes q_n^{n'} + \frac{1}{2} q_n^{n'} \otimes \omega_{in}^{n} \tag{7}$$

2.2 *Velocity error equation based on multiplicative quaternions*

The ideal velocity differential equation can be obtained from the inertial navigation specific force equation:

$$\dot{V}^{n} = f^{n} - (2\omega_{ie}^{n} + \omega_{en}^{n}) \times V^{n} + g^{n} \tag{8}$$

The actual output of the SINS accelerometer not only has its own error, but also the disturbed acceleration caused by the lever arm effect. Therefore, the actual output of the SINS accelerometer is:

$$\tilde{f}^{n'} = q_s^{n'} \otimes (f^{s} + f_r^{s} + \delta f^{s}) \otimes q_{n'}^{s} \tag{9}$$

where f_r^{s} is the disturbed acceleration due to the lever arm effect and δf^{s} is the output error of the SINS accelerometer.

The differential equation of actual velocity can be obtained by using the SINS specific force equation:

$$\dot{\tilde{V}}^{n'} = \tilde{f}^{n'} - (2\tilde{\omega}_{ie}^{n'} + \tilde{\omega}_{en}^{n'}) \times \tilde{V}^{n'} + \tilde{g}^{n'} \tag{10}$$

The velocity error is defined as $\delta \dot{V}^{n} = \dot{\tilde{V}}^{n'} - \dot{V}^{n}$, by substituting Equations 8 and 10, and ignoring the second-order small quantity:

$$\delta \dot{V}^{n} = \tilde{f}^{n'} - f^{n} - (2\tilde{\omega}_{ie}^{n} + \tilde{\omega}_{en}^{n}) \times \delta V^{n} - (2\delta \omega_{ie}^{n} + \delta \omega_{en}^{n}) \times V^{n} + \delta g^{n}$$

$$= q_s^{n'} \otimes (f^{s} + f_r^{s} + \nabla^{s}) \otimes q_{n'}^{s} - q_n^{n} \otimes f^{s} \otimes q_n^{n} - (2\tilde{\omega}_{ie}^{n} + \tilde{\omega}_{en}^{n}) \times \delta V^{n}$$

$$- (2\delta \omega_{ie}^{n} + \delta \omega_{en}^{n}) \times V^{n} + \delta g^{n} \tag{11}$$

$$= q_n^{n'} \otimes q_s^{n} \otimes f^{s} \otimes q_n^{s} \otimes q_{n'}^{n} - q_n^{n} \otimes f^{s} \otimes q_n^{s} - (2\tilde{\omega}_{ie}^{n} + \tilde{\omega}_{en}^{n}) \times \delta V^{n}$$

$$- (2\delta \omega_{ie}^{n} + \delta \omega_{en}^{n}) \times V^{n} + q_n^{n'} \otimes q_s^{n} \otimes (f_r^{s} + \nabla^{s}) \otimes q_n^{s} \otimes q_{n'}^{n} + \delta g^{n}$$

2.3 Error model of the inertial device and installation

The errors of the SINS gyroscope and accelerometer are modeled, and the error of the transfer alignment inertial device is a combination of constant error and Gaussian white noise, then the error model of the gyros is:

$$\delta \omega_{is}^s = \varepsilon_s + \varepsilon_w, \dot{\varepsilon}_s = 0 \tag{12}$$

where ε_s is the constant drift of the gyros, ε_w is the output noise of the gyros.
The error model of the accelerometers is:

$$\delta f^s = \nabla_s + \nabla_w, \dot{\nabla}_s = 0 \tag{13}$$

where ∇_s is the constant bias of accelerometers, ∇_w is the output noise of the accelerometers.
When the MINS is installed on the carrier, there is an installation error between SINS and MINS. The installation error is unchanged, the quaternion of installation error is:

$$\dot{q}_m^s = 0 \tag{14}$$

2.4 State equation

Because the system noise of the quaternion cannot be directly added and subtracted, the process noise and measurement noise in the error model are extended to the state vector. According to Equations 7, 11, 12, 13 and 14, the system state vector can be written as:

$$x_k = [(x_k^q)^T \quad (x_k^e)^T \quad w_k^T \quad v_k^T]^T$$

where x^q is the quaternion part of the state vector, and x^e is the non-quaternion part of the state vector, respectively:

$$x_k^q = [(q_n^{n'})^T \quad (q_s^m)^T], x_k^e = [(\delta V^n)^T \quad \varepsilon_s^T \quad \nabla_s^T]$$

w_k is the system noise vector, and is the zero mean Gaussian white noise, the variance is Q_k; v_k is the observed noise vector, and is the zero mean Gaussian white noise, and the variance is R_k.

2.5 Measurement equation

The matching method of transfer alignment is velocity matching and attitude quaternion matching. The velocity measurement is the difference of velocity between MINS and SINS; velocity of MINS should compensate the lever arm error. The velocity measurement is:

$$Z_v = \tilde{V}^{n'} - V^n - V_r^n \tag{15}$$

where $\tilde{V}^{n'}$ is the calculated velocity of SINS, V^n is the velocity of MINS and V_r^n is the error velocity of the lever arm effect.
The measurement of the attitude quaternion is:

$$Z_Q = q_s^{n'} \otimes q_m^s \otimes q_n^m \tag{16}$$

where $q_s^{n'}$ is the calculated attitude quaternion of SINS, q_m^s is the installation error quaternion between MINS and SINS and q_n^m is the attitude quaternion of MINS.
According to the definition, the nonlinear measurement equation is established as follows:

$$Z_k = \begin{bmatrix} Z_Q \\ Z_v \end{bmatrix} = \begin{bmatrix} q_{s,k}^{n'} \otimes q_{m,k}^s \otimes q_{n,k}^m \\ \tilde{V}_k^{n'} - V_k^n - V_{r,k}^n \end{bmatrix} + v_k = H(x_k) + v_k \tag{17}$$

3   RECURRENCE FORMULA OF QUATERNIONS

According to the quaternion differential equation, we can obtain:

$$\dot{q} = \frac{1}{2}Wq, q(t_0) = q_0 \tag{18}$$

where t_0 is the initial moment of carrier motion, q_0 is the initial attitude quaternion, and let ω_x, ω_y and ω_z be the angular velocity vectors of rotation carrier respectively, then:

$$W = \frac{1}{2}\begin{bmatrix} 0 & -\omega_x & -\omega_y & -\omega_z \\ \omega_x & 0 & \omega_z & -\omega_y \\ \omega_y & -\omega_z & 0 & \omega_x \\ \omega_z & \omega_y & -\omega_x & 0 \end{bmatrix} \tag{19}$$

Because the actual angular velocity output is discontinuous and gyros output is the angular increment within the time of sampling interval Δt, then:

$$\Delta\theta_x = \omega_x \cdot \Delta t, \Delta\theta_y = \omega_y \cdot \Delta t, \Delta\theta_z = \omega_z \cdot \Delta t$$

Equation 18 can be expressed as a recursive form:

$$q(t_k) = e^{\Delta\theta}q(t_0) \tag{20}$$

The matrix $\Delta\theta$ satisfies the following properties: $\Delta\theta^2 = -\Delta\theta_m I_{4\times4}, \Delta\theta_m = \frac{1}{4}(\Delta\theta_x^2 + \Delta\theta_y^2 + \Delta\theta_z^2)$, $I_{4\times4}$ is 4×4 unit matrix.

By using Taylor series expansion and special properties of matrix $\Delta\theta$, we can obtain:

$$e^{\Delta\theta} = [1 - \frac{1}{2}m + ... + (-1)^i\frac{1}{(2i)!}m^i]\cdot I_{4\times4} + [1 - \frac{1}{3!}m + ... + (-1)^i\frac{1}{(2i+1)!}m^i]\cdot\Delta\theta \tag{21}$$

According to the j-order diagonal Padé approximation, the recursive formula of the quaternion can be obtained as follows:

$$q_k = [\sum_{i=0}^{j}(-1)^i a_i\Delta\theta^i]^{-1}\cdot[\sum_{i=0}^{j}a_i\Delta\theta^i]\cdot q_{k-1} \tag{22}$$

where $a_i = \dfrac{(2j-i)! \ j!}{(2j)! \ i! \ (j-i)!}$.

According to the properties of the matrix $\Delta\theta$, we can obtain:

$$\sum_{i=0}^{j}a_i\Delta\theta^i = A(\Delta\theta_m)I_{4\times4} + B(\Delta\theta_m)\Delta\theta \tag{23}$$

where $A(\Delta\theta_m)$ and $B(\Delta\theta_m)$ represent the coefficient sum of matrix $I_{4\times4}$ and $\Delta\theta$, respectively. Therefore, Equation 23 can be expressed as:

$$q_k = [A(\Delta\theta_m)I_{4\times4} - B(\Delta\theta_m)\Delta\theta]^{-1}\cdot[A(\Delta\theta_m)I_{4\times4} + B(\Delta\theta_m)\Delta\theta]\cdot q_{k-1} \tag{24}$$

The inverse matrix in Equation 16 can be expressed as:

$$[A(\Delta\theta_m)I_{4\times4} - B(\Delta\theta_m)\Delta\theta]^{-1} = \frac{[A(\Delta\theta_m)I_{4\times4} + B(\Delta\theta_m)\Delta\theta]}{A^2(\Delta\theta_m) + B^2(\Delta\theta_m)\Delta\theta} \tag{25}$$

Substituting Equation 25 into Equation 24, the Padé approximation of the quaternion can be approximated as:

$$q_k = \frac{[(A^2(\Delta\theta_m) - B^2(\Delta\theta_m) \cdot \Delta\theta_m)I_{4\times4} + 2A(\Delta\theta_m)B(\Delta\theta_m)\Delta\theta]}{A^2(\Delta\theta_m) + B^2(\Delta\theta_m)\Delta\theta} \cdot q_{k-1} \qquad (26)$$

Take the second-order diagonal Padé approximation, and according to Equation 24, it can be obtained:

$$\begin{cases} A(\Delta\theta_m) = 1 - \dfrac{1}{12}\Delta\theta_m, \ B(\Delta\theta_m) = \dfrac{1}{2} \\ q_k = \dfrac{1}{1 + \dfrac{1}{12}\Delta\theta_m + \dfrac{1}{144}\Delta\theta_m{}^2}[(1 - \dfrac{5}{12}\Delta\theta_m + \dfrac{1}{144}\Delta\theta_m{}^2)I_{4\times4} + (1 - \dfrac{1}{12}\Delta\theta_m)\Delta\theta]q_{k-1} \end{cases} \qquad (27)$$

4 IMPROVED FIFTH-ORDER QCKF ALGORITHM

The CKF algorithm adopts the principle of spherical radial cubature, and the estimation accuracy is higher when the nonlinear model is established more accurately. However, the external environment of transfer alignment is relatively complex and the system model cannot be very accurate, which will lead to a large estimation error when filtering. Therefore, the STF is combined with the fifth-order CKF algorithm in the delivery of transfer alignment; the elimination factor obtained from the calculation in STF is substituted into the time update and measurement update equations. In addition, a new strongly tracking fifth-order cubature Kalman filter (QSTCKF) algorithm is proposed. The new algorithm can adapt to the system with strong nonlinearity, and can also adapt to the complex and variable environment effectively. The specific algorithm for QSTCKF is applied as follows:

1. Calculate the cubature points of QSTCKF:

$$x_{0i,k} = \hat{x}_k, \ x_{1i,k} = \sqrt{n+2}S_k \cdot e_i + \hat{x}_k, \ x_{2i,k} = -\sqrt{n+2}S_k \cdot e_i + \hat{x}_k \qquad (28)$$

where $i = 1, 2, ..., n$, e_i is the unit vector in the direction of coordinate axis, and $P_k = S_k \cdot S_k{}^T$.

$$\begin{cases} x_{3i,k} = \sqrt{n+2}S_k \cdot s_i^+ + \hat{x}_k, \ x_{4i,k} = -\sqrt{n+2}S_k \cdot s_i^+ + \hat{x}_k \\ x_{5i,k} = \sqrt{n+2}S_k \cdot s_i^- + \hat{x}_k, \ x_{6i,k} = -\sqrt{n+2}S_k \cdot s_i^- + \hat{x}_k, \ \text{where}: i = 1,2,...,n(n-1)/2. \end{cases} \qquad (29)$$

$$s_i^+ = \frac{1}{\sqrt{2}}(e_j + e_k), s_i^- = \frac{1}{\sqrt{2}}(e_j - e_k), (j < k, j,k = 1,2,...,n) \qquad (30)$$

2. Time update:

By using the calculated cubature points obtained in Step 1, the transmitted cubature points can be obtained through the nonlinear state equation:

$$\mathcal{X}_{mi,k+1/k} = f(x_{mi,k}), m = 0,1,...,6 \qquad (31)$$

According to Equation 31, the predicted state value and the covariance matrix of state error at the moment $k+1$ can be obtained, $\hat{x}_{k+1/k}^e$ represents the non-quaternion part of the system state, the attitude error quaternion $\hat{x}_{k+1/k}^q$ can be calculated from Equation 27:

$$\hat{x}_{k+1/k}^e = \omega_0 \mathcal{X}_{0i,k+1/k}^e + \omega_1 \sum_{j=1}^{n} (\mathcal{X}_{1i,k+1/k}^e + \mathcal{X}_{2i,k+1/k}^e) + \omega_2 \sum_{j=1}^{n(n-1)/2} (\mathcal{X}_{3i,k+1/k}^e + \mathcal{X}_{4i,k+1/k}^e + \mathcal{X}_{5i,k+1/k}^e + \mathcal{X}_{6i,k+1/k}^e) \qquad (32)$$

where $\omega_0 = 2/(n+2)$, $\omega_1 = (4-n)/[2(n+2)^2]$, $\omega_2 = 1/(n+2)^2$

$$\hat{x}_{k+1/k} = \begin{bmatrix} \hat{x}^q_{k+1/k} \\ \hat{x}^e_{k+1/k} \end{bmatrix} \tag{33}$$

$$P^x_{k+1/k} = \omega_0 \chi_{0i,k+1/k} \chi^T_{0i,k+1/k} + \omega_1 \sum_{j=1}^{n} (\chi_{1i,k+1/k} \chi^T_{1i,k+1/k} + \chi_{2i,k+1/k} \chi^T_{2i,k+1/k})$$

$$+\omega_2 \sum_{j=1}^{n(n-1)/2} (\chi_{3i,k+1/k} \chi^T_{3i,k+1/k} + \chi_{4i,k+1/k} \chi^T_{4i,k+1/k} + \chi_{5i,k+1/k} \chi^T_{5i,k+1/k} + \chi_{6i,k+1/k} \chi^T_{6i,k+1/k}) - \hat{x}_{k+1/k} \hat{x}^T_{k+1/k} \tag{34}$$

3. Measurement update:

The updated state cubature points $x_{k+1/k}$ are calculated according to the statistical characteristics of x_{k+1},

$$x_{0i,k+1/k} = \hat{x}_{k+1/k}, x_{1i,k+1/k} = \sqrt{n+2} S_{k+1/k} \cdot e_i + \hat{x}_{k+1/k}, x_{2i,k+1/k} = -\sqrt{n+2} S_{k+1/k} \cdot e_i + \hat{x}_{k+1/k} \quad (35)$$

where $i = 1, 2, ..., n$, e_i is the unit vector in the direction of the coordinate axis, and $P^x_{k/k+1} = S_{k/k+1} \cdot S_{k/k+1}{}^T$.

$$\begin{cases} x_{3i,k+1/k} = \sqrt{n+2} S_{k+1/k} \cdot s^+_i + \hat{x}_{k+1/k}, x_{4i,k+1/k} = -\sqrt{n+2} S_{k+1/k} \cdot s^+_i + \hat{x}_{k+1/k}, i = 1,2,...,n(n-1)/2 \\ x_{5i,k+1/k} = \sqrt{n+2} S_{k+1/k} \cdot s^-_i + \hat{x}_{k+1/k}, x_{6i,k+1/k} = -\sqrt{n+2} S_{k+1/k} \cdot s^-_i + \hat{x}_{k+1/k} \end{cases} \tag{36}$$

$$s^+_i = \frac{1}{\sqrt{2}}(e_j + e_k), s^-_i = \frac{1}{\sqrt{2}}(e_j - e_k), (j < k, j,k = 1,2,...,n) \tag{37}$$

The cubature points can be obtained by the measurement equation:

$$Z_{mi,k+1} = H(x_{mi,k+1/k}), m = 0,1,...,6 \tag{38}$$

According to Equation 38, the predicted value of the one-step measurement is obtained:

$$\hat{Z}^e_{k+1} = \omega_0 Z^e_{0i,k+1/k} + \omega_1 \sum_{j=1}^{n} (Z^e_{1i,k+1/k} + Z^e_{2i,k+1/k}) + \omega_2 \sum_{j=1}^{n(n-1)/2} (Z^e_{3i,k+1/k} + Z^e_{4i,k+1/k} + Z^e_{5i,k+1/k} + Z^e_{6i,k+1/k}) \tag{39}$$

The one-step measurement of the quaternion part $\hat{Z}^q_{k+1}$ can be calculated via Equation 27.

$$\hat{Z}_{k+1} = \begin{bmatrix} \hat{Z}^q_{k+1} \\ \hat{Z}^e_{k+1} \end{bmatrix} \tag{40}$$

$$P^{xz}_{k+1/k} = \omega_0 x_{0i,k+1/k} Z^T_{0i,k+1/k} + \omega_1 \sum_{j=1}^{n} (x_{1i,k+1/k} Z^T_{1i,k+1/k} + x_{2i,k+1/k} Z^T_{2i,k+1/k})$$

$$+\omega_2 \sum_{j=1}^{n(n-1)/2} (x_{3i,k+1/k} Z^T_{3i,k+1/k} + x_{4i,k+1/k} Z^T_{4i,k+1/k} + x_{5i,k+1/k} Z^T_{5i,k+1/k} + x_{6i,k+1/k} Z^T_{6i,k+1/k}) - \hat{x}_{k+1/k} \hat{Z}^T_{k+1} \tag{41}$$

$$P^{zz}_{k+1} = \omega_0 Z_{0i,k+1/k} Z^T_{0i,k+1/k} + \omega_1 \sum_{j=1}^{n} (Z_{1i,k+1/k} Z^T_{1i,k+1/k} + Z_{2i,k+1/k} Z^T_{2i,k+1/k})$$

$$+\omega_2 \sum_{j=1}^{n(n-1)/2} (Z_{3i,k+1/k} Z^T_{3i,k+1/k} + Z_{4i,k+1/k} Z^T_{4i,k+1/k} + Z_{5i,k+1/k} Z^T_{5i,k+1/k} + Z_{6i,k+1/k} Z^T_{6i,k+1/k}) - \hat{Z}_{k+1} \hat{Z}^T_{k+1} \tag{42}$$

4. Calculate the fading factor λ_{k+1}:

$$N_{k+1} = V_{k+1} - L_{k+1}Q_k L_{k+1}^T - \beta R_{k+1} \qquad (43)$$

where $V_{k+1} = \begin{cases} [Z_{k+1} - H(\hat{x}_{k+1/k})][Z_{k+1} - H(\hat{x}_{k+1/k})]^T, & k = 0 \\ \dfrac{\rho V_k + [Z_{k+1} - H(\hat{x}_{k+1/k})][Z_{k+1} - H(\hat{x}_{k+1/k})]^T}{1+\rho}, & k > 0 \end{cases}$, is the forgetting factor,

$0.95 \leq \rho \leq 0.995$. β is the softening factor.

$$M_{k+1} = L_{k+1}P_{k+1/k}^x L_{k+1}^T \qquad (44)$$

where, according to Equations 34 and 41, $L_{k+1} = P_{k+1/k}^{xz}(P_{k+1/k})^{-1}$ can be obtained.

Matrices N_{k+1} and M_{k+1} are obtained according to Equations 43 and 44, respectively, and the fading factors can be calculated:

$$\lambda_{k+1} = \max(1, \frac{tr(N_{k+1})}{tr(M_{k+1})}) \qquad (45)$$

5. Update measurement with fading factor:
Based on the elimination factor calculated by Equation 45, the one-step state error prediction covariance matrix is calculated:

$$P_{k/k+1} = \lambda_{k+1}P_{k/k+1}^x \qquad (46)$$

Then bring $P_{k/k+1}$ into Step 3 (Measurement update) and apply it again.

6. State update:
According to Step 5, the updated measurement error covariance matrix and one-step prediction cross-correlation covariance matrix calculate the filter gain matrix:

$$K_{k+1} = P_{k/k+1}^{xz}(P_{k+1}^{zz})^{-1} \qquad (47)$$

According to Step 5, the updated measurement predicted value calculates the state estimation value and state error covariance matrix:

$$\hat{x}_{k+1} = \hat{x}_{k+1/k} + K_{k+1}(Z_{k+1} - \hat{Z}_{k+1}) \qquad (48)$$

$$P_{k+1} = P_{k+1/k} + K_{k+1}P_{k+1}^{zz}K_{k+1}^T \qquad (49)$$

5   SIMULATION TEST AND ANALYSIS

To verify the effectiveness of the algorithm, a simulation experiment was carried out. In this simulation experiment, first, the motion trajectory of the carrier was established as a uniform linear motion and, in the process, the vehicle swing maneuver mode was rolled around the vertical axis of the carrier. Carrier of initial position for the east longitude 119°, latitude 39°, the speed of the uniform is 5 m/s, affected by the waves of the sea and in the process of driving pitch, roll and yaw wave motion by sine wave simulation, because the time is short, a simplified treatment for real waves do, with only a sine wave simulation:

$$\theta = \theta_m \cos(\omega_\theta t + \theta_0), \gamma = \gamma_m \cos(\omega_\gamma t + \gamma_0), \psi = \psi_m \cos(\omega_\psi t + \psi_0) + \varphi_0 \qquad (50)$$

where θ_m, γ_m and ψ_m are the maximum values of attitude angle; θ_0, γ_0 and ψ_0 are the initial angle; let $\theta_m = 7°$, $\gamma_m = 10°$, $\psi_m = 5°$; $\theta_0 = 10°$, $\gamma_0 = 15°$, $\psi_0 = 20°$; $\omega_\theta = 2\pi/15$ rad/s, $\omega_\gamma = \pi/7$ rad/s, $\omega_\psi = \pi/8$ rad/s, $\varphi_0 = 30°$.

It is assumed that there is no error in the output information of MINS during the alignment, and there is a large error in the accuracy of SINS. The assumption is that the gyro constant drift of SINS is $0.1°/h$, the intensity of Gaussian white noise of the gyroscope is $(0.01°/h)^2$, the accelerometer bias is $10^{-4} g$ and the intensity of Gaussian white noise is $(10^{-4} g)^2$. The simulation time is 480 s. In comparison with QUKF and QCKF, the new QSTCKF filtering algorithm simulation results are shown in Figures 1, 2 and 3.

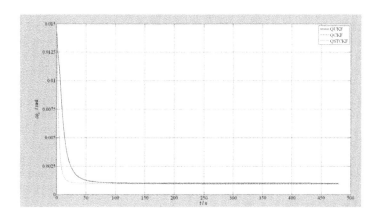

Figure 1.   Comparison of pitch angle error standard deviation estimates of QUKF, QCKF and QSTCKF.

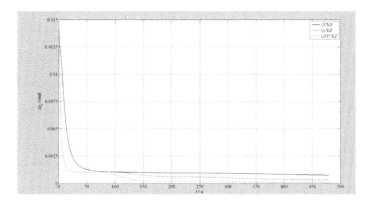

Figure 2.   Comparison of roll angle error standard deviation estimates of QUKF, QCKF and QSTCKF.

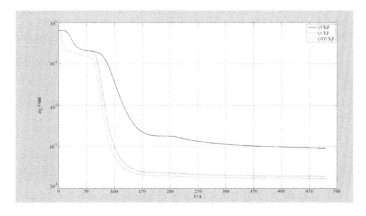

Figure 3.   Comparison of yaw angle error standard deviation estimates of QUKF, QCKF and QSTCKF.

It can be seen from the simulation diagrams that the newly proposed algorithm is superior to QUKF and QCKF in terms of convergence speed and convergence accuracy. The error curve of roll angle and pitch angle can be rapidly convergent within 40" by using fifth-order STCKF, and the yaw angle can be rapidly convergent within 5'.

6 CONCLUSION

The error equation based on the multiplicative quaternion is derived in detail. According to the system's strong nonlinearity, the proposed algorithm combines a strong tracking algorithm with the fifth-order CKF algorithm. The specific steps of the algorithm realization are derived in detail, which verifies the effectiveness of the algorithm. Meanwhile, QUKF and QCKF are compared and analyzed with QSTCKF.

Based on this research, we can derive the following conclusions: the error equation based on the quaternion avoids small-angle approximation and is no longer limited to the case of a small misalignment angle. The proposed algorithm combines a STF and fifth-order CKF. An adaptive fading factor is introduced into the calculation, adjusting the gain matrix in real time according to the residual, giving the filtering algorithm of the nonlinear system strong robustness. The simulation results show that under the condition of a complex peripheral environment, the proposed algorithm has strong adaptability, fast convergence speed and high convergence precision.

REFERENCES

Arasaratnam, I. & Haykin, S. (2009). Cubature Kalman filters. *IEEE Transactions on Automation Control*, *54*(6), 1254–1269.

Chang, L.B., Hu, B.Q., Li, A. & Qin, F. (2013). Transformed unscented Kalman filter. *IEEE Transactions on Automation Control*, *58*(1), 252–257.

Chen, Y., Zhao, Y. & Li, Q.S. (2013). QCKF based rapid transfer alignment for large misalignment angles. *Journal of Beijing University of Aeronautics and Astronautics*, *39*(13), 1624–1628.

Fernandez-Prades, C. & Vila-Valls, J. (2010). Bayesian nonlinear filtering using quadrature and cubature rules applied to sensor data fusion for positioning. *IEEE International Conference on Communications, Cape Town, South Africa*. doi:10.1109/ICC.2010.5502587.

Forbes, J.R. (2010). Extended Kalman filter and sigma point filter approaches to adaptive filtering. *AIAA Guidance, Navigation, and Control Conference, Toronto, Canada*. Reston, VA: American Institute of Aeronautics and Astronautics. doi:10.2514/6.2010-7748.

Gao, Q.-W., Zhao, G.-R. & Z. & Wu, F. (2011). Research on nonlinear error model of transfer alignment with large azimuth misalignment angle. *Control and Decision*, *26*(3), 402–406.

Gao, W., Ben Y.Y., Zhang, X., Li, Q. & Yu, F. (2011). Rapid fine strapdown INS alignment method under marine mooring condition. *IEEE Transactions on Aerospace and Electronic Systems*, *47*(4), 2887–2896.

Guo, S.L., Xu, J.N. & Li, F. (2017). Strong tracking cubature Kalman filter for initial alignment of inertial navigation system. *Journal of Chinese Inertial Technology*, *25*(4), 436–441.

Juier, S.J. & Uhlmann, J.K. (2004). Unscented filtering and nonlinear estimation. *Proceedings of the IEEE*, *92*(3), 401–422.

Li, Q.R. & Sun, F. (2013). Strong tracking cubature Kalman filter algorithm for GPS/INS integrated navigation system. *Proceedings of 2013 IEEE International Conference on Mechatronics and Automation Control* (pp. 1113–1117).

Markley, F.L., Cheng, Y., Crassidis, J.L. & Oshman, Y. (2007). Averaging quaternions. *Journal of Guidance, Control and Dynamics*, *30*(4), 1193–1197.

Sun, F. & Tang, L.J. (2012). Initial alignment of large azimuth misalignment angle in SINS based on CKF. *Chinese Journal of Scientific Instrument*, *33*(2), 327–333.

Sun, T. & Xin, M. (2014). Hypersonic entry vehicle state estimation using high-degree cubature Kalman filter. *2014 AIAA Atmospheric Flight Mechanics Conference*. Reston, VA: American Institute of Aeronautics and Astronautics. doi:10.2514/6.2014-2383.

Zhou, W.D., Ji, Y.R. & Qiao, X.W. (2011). Quaternions augmented unscented Kalman filter and its application to rapid transfer alignment under large misalignment. *Control Theory and Applications*, *28*(11), 1583–1588.

Automatic Control, Mechatronics and Industrial Engineering – He & Qing (Eds)
© 2019 Taylor & Francis Group, London, ISBN 978-1-138-60427-8

Overall satellite model composition research using Model-Based Systems Engineering (MBSE)

R.L. Wu, Z. Peng, X.H. Gao, Y.Q. Xia & H. Jiang
Institute of Telecommunication Satellite, China Academy of Space Technology, Beijing, China

ABSTRACT: Due to the increasing complexity of satellite systems, the disadvantages of the traditional Text-based System Engineering (TSE) method are becoming notable, such as non-unification of data sources, limited descriptive ability, development processes subject to breakpoint management, and so on. In this paper, Model-Based Systems Engineering (MBSE) has been used to solve the problems existing in present satellite development systems. An overall satellite model structure derived from MBSE has been generated according to the requirements analysis of different satellite development stages.

Keywords: satellite, MBSE, overall model

1 INTRODUCTION

Traditional satellite system development mainly relies on a variety of documents for overall design, manufacturing, testing and project management, that is, Text-based System Engineering (TSE) (Zhang et al., 2014). Often, these documents are developed separately by different staff within an organization, with specific disciplines responsible for different sections (Russell, 2012). The design and manufacturing parameters are scattered among various documents in the TSE method, while some parameters exist in more than one document. Several documents are needed to finish one design or manufacturing activity. The management of projects based on TSE also relies on documents, that determines the projects development mainly depend on manually. With the increasing scale and complexity of satellite systems, the TSE method can no longer satisfy satellite design and manufacturing needs.

In recent years, Model-Based Systems Engineering (MBSE) has become the focus of research and application of satellite system engineering. MBSE is the formal and standardized application of modeling methods in system engineering activities, which enables modeling methods to support system requirements in every development stage, such as design, analysis, verification and validation activities. These activities start from the conceptual design stage and continue through all life-cycle stages of product development progress (Han et al., 2014).

The MBSE method takes the model as the foundation of system description, and has to satisfy the needs of satellite design, manufacture and orbital operation. From the definition, there is no essential difference between the MBSE method and the TSE method in terms of basic theory and process. The difference is mainly in the form of design process management, work form and design result display. In the MBSE method, every activity can be described as a model instead of a document, and the parameters can be transmitted automatically on an appropriate platform. This means that as soon as all the parameters needed for a task are ready, the task leader will know that the task can be started without any command or reminder from others.

This paper proposes an innovative method based on the MBSE concept to support complex system characteristics for satellite design, simulation, integration, testing and in-orbit management. An MBSE-based structure for the overall satellite model and its method of implementation is created according to current satellite development processes.

2 REQUIREMENTS ANALYSIS

Due to increasing complexity, the design, manufacture and management of satellites has become a set of collaborative tasks shared among multidisciplinary, distributed teams. The overall satellite model should meet the requirements of the satellite's life-cycle development, including all kinds of data and models for the different stages of the development. The satellite development process can be divided into three stages: design; Assembly, Integration and Testing (AIT); in-orbit operation. Requirements analysis is carried out on the basis of these three stages.

2.1 Requirements analysis in the design stage

Overall satellite design, subsystem design and product design are carried out in the design stage:

a. The overall satellite design is the high-level design of the satellite system. User requirements are transformed into functions and performance parameters of systems, and systems are composed of several subsystems (Xu, 2003).
b. Subsystem design plays an important role in overall satellite design. The result of subsystem design should satisfy the design index of the subsystem and be compatible with other subsystems' functions, physics and program interfaces.
c. Product design is the basis of satellite system design. The product design results determine the final performance of the system.

Figure 1 shows the simplified data flow in the overall design process of the satellite. From the diagram, it can be determined that the model framework for the overall design stage should contain the following content:

a. All parameters required or generated during the overall design, subsystem design and product design phases that are needed to maintain the basic operation of the overall system. The amount of data is large and the data flow direction is complicated. Optimized data management is very important for efficient operation of the system.
b. Satellite requirements analysis model based on user requirements; large system constraints have to be created, and subsystem indices are distributed by satellite requirements analysis.
c. Subsystem models, which are the main components of the overall design model. Subsystem simulation analysis models (including a 3D model) are also needed, which can be further divided into a system requirements analysis model and subsystem simulation analysis model if necessary. Subsystem models need a lot of basic data and generate a lot of intermediate data while they are running.

2.2 Requirements analysis in the AIT stage

The AIT process is a vital phase in the progress of satellite development. Satellite assembly and testing are interspersed with each other throughout the AIT phase, and cannot be strictly segregated by development time or phase.

Figure 1. Simplified data flow of the overall satellite design process.

Satellite assembly includes the design of satellite assembly and its implementation. Satellite assembly design is the process of planning and promoting demands on satellite AIT requirements, and finally writing them into documents and drawings, according to requirements from system, subsystem, and overall integration tests. Satellite assembly implementation can be divided into component assembly, functional component assembly, cabin assembly and overall satellite assembly. The quality of satellite assembly will directly affect the satellite performance. According to statistics from the Russian Space Department, satellite assembly workload accounts for 35% of the total processing capacity of the satellite (Xu, 2003).

Satellite testing is a necessary part of satellite development. Through a comprehensive and systematic test of the satellite, the design faults and manufacturing technological defects of the satellite are fully exposed, and early failure detection of electronic components and/or lack of software are found to ensure the high quality of satellite products.

AIT is a stage of the satellite development process that involves interaction with physical models. The model construction involved in satellite assembly is shown in Figure 2:

a. An assembly design model that is an effective merger of structural design, thermal design and overall configuration can fully describe satellite assembly and integration. The model is the basis of the process design model, which can be directly used to guide workers to conduct assembly operation.

b. The assembly status data at various stages should be recorded, such as the grounding resistance of the products, the state of on-board plugs, and so on. The data not only records the status of the satellite assembly, but can also be compared with the design values, allowing assembly problems to be found in time, and could be used in assembly model updating.

Figure 3 shows the model composition and data flows for the testing of the subsystems and the overall satellite testing:

a. The subsystem or overall satellite test model (including the test program) can be constructed according to the simulation analysis model of each subsystem, which is built in the design phase. Several test models can be built if necessary.

b. A large amount of test data is produced at each stage, such as mechanical, thermal and electrical test data. All these data reflect the real performance of the satellite, and must be compared with the design index to confirm whether the satellite performance will meet the system requirements.

2.3 *Requirements analysis in the in-orbit operation stage*

During the in-orbit operation of the satellite, various engineering parameters of the satellite in orbit are transmitted through a Tracking Telemetry and Command (TT&C) subsystem. Instructions are sent to the satellite in order to control its attitude, orbit and working state.

Satellite in-orbit parameters reflect the true operating status of the satellite in orbit. The parameters can be compared with satellite design data to make sure that the satellite is

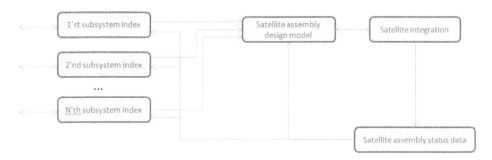

Figure 2.   Simplified satellite assembly model.

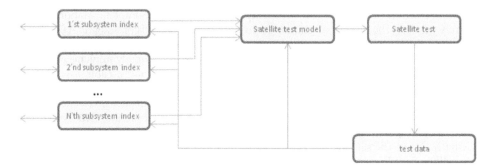

Figure 3.   Simplified satellite test model.

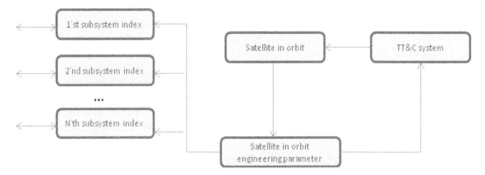

Figure 4.   Interactions between satellite in-orbit data and satellite model analysis data.

functioning normally, and can also be used in model updating for optimal satellite system design. Figure 4 represents the data flow directions between the in-orbit data and satellite model analysis data during the satellite's in-orbit operational phase.

3   OVERALL SATELLITE MODEL COMPOSITION

From the requirements analysis outlined in the previous section, it can be seen that the overall satellite model is divided into two main parts: a basic database and a system construction model (Graignic et al., 2013), described in more detail below.

3.1   *Basic database*

To perform a simulation of a satellite system, engineers have to gather numerous data about the system they have to analyze. The basic database covers the data required and produced by the satellite development process at various stages of design, manufacture, testing and in-orbit operation. Figure 5 shows the high-level composition of the satellite database, including all the data needed or produced in the process of satellite development:

» Data needed in the design phase, such as user indices, large system constraints, subsystem indices, on-board products' design and performance parameters, intermediate calculation values, and other related parameters necessary for subsystem design or overall design;
» Data produced in the AIT phase, such as assembly status data, test data and other process data;
» In-orbit data transmitted through the TT & C subsystem.

Each data item mentioned above exists uniquely and independently in the system with a specific identity, and can be used by any model in the overall satellite model system.

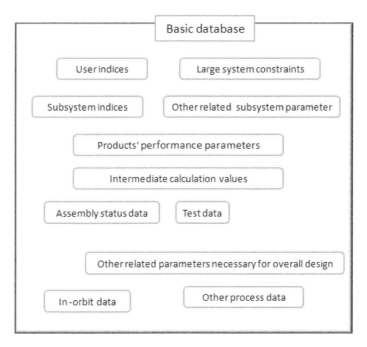

Figure 5.   Composition of the satellite database.

Data identification and management methods for this database are very important for this system, because they can determine the efficiency of the system's operation.

3.2   *Satellite system construction model*

According to the different application stages and application levels of the satellite model, the system architecture model can be divided into four levels in the MBSE method: requirements model, analysis model, functional model and physical model, as shown in Figure 6 (Estefan, 2008). With the exception of the physical model, each level of the model can be further divided into several sublevels:

a. Requirements model: includes the satellite requirements analysis model and subsystem requirements analysis model, which are the top level of the entire system model. System and subsystem indices are carried out through the requirements model and are important parameters for the simulation analysis model.
b. Analysis model: includes each subsystem simulation model, which is the important method and basis for overall detailed design. To a certain extent, the analysis models form a virtual satellite system and describe how the satellite works and how it should perform ahead of its manufacture.
c. Functional model: includes the satellite assembly design model and the test model, which is the level at which the models are associated with objects. The virtual model guides workers in how to assemble the products into a whole satellite and how to test the performance of the satellite.
d. Physical model: refers to the real objects, and takes the single on-board product as the smallest unit, according to the design and manufacturing requirements of the whole satellite. Workers assemble the products and test the satellite performance according to the functional model, obtaining the real status data and real performance data, and then feeding back the data to the model.

Figure 6 shows the relationship between the requirements model, analysis model, functional model and physical model. The four levels are in a progressive structure, which can

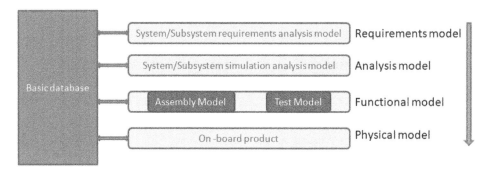

Figure 6. System composition model.

meet the requirements of satellite life-cycle development, and each model level interacts with the basic database directly. Such a model architecture has the following advantages:

a. The unified data source ensures consistency of data and reduces errors arising through data transmission.
b. The unified data source makes the maintenance and management of the data easier.
c. Each subsystem model and each level model are managed in a real-time interactive platform. Once all the parameters needed in a model are ready, the model can run automatically and produce the analysis results, which feedback to the database. Any other model in the system that needs the results will know the parameters are ready as soon as the results are produced. Such spontaneous driving of the model fundamentally changes the serial development model based on text, and can effectively improve development efficiency.
d. It is conducive to the integration of the knowledge engineering management method and the foundation for future intelligent satellite manufacturing.

4 CONCLUSION

An overall satellite model composition has been proposed in this paper, based on an MBSE approach. The problems existing in the original TSE method (such as data sources not being unified, limited descriptive ability, and so on) are solved via the system's operational mechanisms. However, the design and management platform appropriate to satellite development must, in future, be chosen carefully to promote the application of this modeling system.

REFERENCES

Estefan, J.A. (2008). *Survey of model-based systems engineering (MBSE) methodologies*. INCOSE MBSE Initiative. San Diego, CA: International Council on Systems Engineering. Retrieved from http://www.omgsysml.org/MBSE_Methodology_Survey_RevB.pdf.
Graignic, P., Vosgien, T., Jankovic, M., Tuloup, V., Berquet, J. & Troussier, N. (2013). Complex system simulation: Proposition of a MBSE framework for design-analysis integration. *Procedia Computer Science*, 16, 59–68.
Han, F., Lin, Y. & Fan, H. (2014). Research and practice of model-based systems engineering in spacecraft development. *Spacecraft Engineering*, 3(24), 119–125.
Russell, M. (2012). Using MBSE to enhance system design decision making. *Procedia Computer Science*, 8, 188–193.
Xu, F. (2003). *Introduction to satellite engineering*. Beijing: China Astronautic Publishing House.
Zhang, Y., Yang, L., Wang, P. & Chen, R. (2014). Discussion on application of model-based systems engineering method to human spaceflight mission. *Spacecraft Engineering*, 5(23), 121–128.

Automatic Control, Mechatronics and Industrial Engineering – He & Qing (Eds)
© *2019 Taylor & Francis Group, London, ISBN 978-1-138-60427-8*

Review of propulsion subsystems for a high-orbit Synthetic-Aperture Radar (SAR) satellite

H.F. Zhang & L. Jian
Beijing Institute of Spacecraft System Engineering, Beijing, China

ABSTRACT: High-orbit satellites with larger antennae as loads require large station-keeping velocity increments and rational propulsion systems. We compared the bipropellant propulsion system, hybrid propulsion system and ion bipropellant system in terms of system weight, configuration layout difficulty, and so on. The weight of the hybrid propulsion system is the lowest on the premise of the same velocity increment within a given weight range. Compared with the extant platform, the modification to a hybrid propulsion system is minor and the complexity is relatively low; plume effect and jet contamination of the arc-jet thruster are lower than others, and the thrust and station-keeping efficiency of hybrid propulsion systems with an arc-jet thruster are higher. In conclusion, compared with other systems, the hybrid propulsion system can better meet the requirements of high-orbit satellites with large antennae.

Keywords: High-orbit SAR satellite, hybrid mode propulsion system, arc thruster, system weight, configuration layout and complexity, plume effect and jet

1 INTRODUCTION

The Geosynchronous-orbiting Synthetic-Aperture Radar (GEOSAR) satellite is totally new in China. As a satellite with large antennae, the weight is relatively high, beyond the carrying capacity of a regular satellite platform, leading to a high station-keeping velocity increment, which places high requirements on the design of the propulsion system. It is difficult for the traditional bipropellant propulsion system to meet the requirements of this satellite. This review compared three system solutions—the traditional bipropellant propulsion system, a hybrid propulsion system based on bipropellant and arc, and a traditional chemical and ion propulsion system—in terms of the system weight, system inheritance and complexity, thruster plume effect, thruster-specific impulse and efficiency of use.

2 COMPARISON OF SYSTEM WEIGHT

According to the demand characteristics of satellite payload and in-orbit velocity increments, the total weight of the propulsion subsystem is important for the realization of the overall mission.

Table 1. Weight comparison of the three systems in configuration design.

	Bipropellant propulsion system	Bipropellant ion propulsion system	Hybrid propulsion system
Propellant demand (kg)	2560	2379	2436
Propulsion system dry weight (kg)	191	191 + 140 = 331	202
Total weight (system and propellant) (kg)	2751	2710	2638

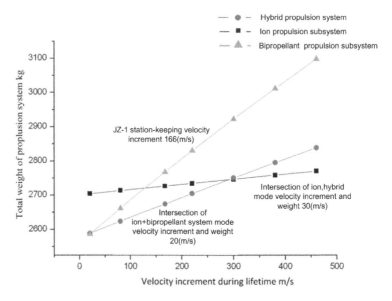

Figure 1.　Comparison of total weights at different speed increment requirements in the three systems.

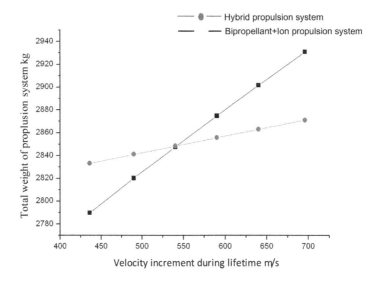

Figure 2.　Comparison of the total weights in the bipropellant + ion propulsion system and the hybrid-mode propulsion system of a foreign communication satellite orbit and station-keeping mode for various in-orbit velocity increments.

　　Taking an orbit design of a high-orbit satellite as an example, the satellite station-keeping velocity increment is 166.848 m/s. Based on this velocity increment and parameters for the propulsion system, we compared the weight requirement in a bipropellant propulsion system, ion bipropellant system and hybrid propulsion system, and found that the total weight of the hybrid-mode propulsion system is smallest (Table 1) (Smith, Aadland et al., 1997; Smith, Roberts et al., 1997).

　　According to the scale of the satellite, the total weights in bipropellant, ion and hybrid propulsion systems were calculated under the different requirements of in-orbit velocity increments (Figure 1). It can be seen that there is an advantage in the hybrid propulsion system unless the in-orbit velocity increment of the satellite is very high and exceeds a certain limit.

184

Table 2. Comparison of the system configuration of the two systems.

	Hybrid-mode propulsion system	Bipropellant + ion propulsion system
Chemical propulsion system configuration	74 sets of equipment (consistent with a bipropellant propulsion system)	74 sets of equipment (consistent with a bipropellant propulsion system)
Electric propulsion system configuration	Two arc thruster power supplies and four arc thrusters	Two helium cylinders, one set of high-pressure adjustment modules, four sets of flow adjustment modules, two sets of vector adjustment mechanisms, one ion thruster power supply and four ion thrusters

Based on experience in the application of foreign communication satellite electric propulsion, the ion propulsion system can be only used in North-South Station-Keeping (NSSK) because of the large volume and fixed installation position. Because of its small volume and convenient installation, an arc-jet thruster can be used for East-West Station-Keeping (EWSK) and NSSK. In Figure 2, there is a comparison of the total weights in the bipropellant and ion propulsion system and the hybrid propulsion system of a foreign communication satellite orbit and station-keeping mode for various in-orbit velocity increments.

3 COMPARISON OF STRUCTURE LAYOUT AND SYSTEM COMPLEXITY

Compared with the Chinese DFH-4 communications satellite platform, the hybrid-mode propulsion system has a gas–liquid supply system consistent with the traditional bipropellant propulsion system. The orbit transform motor and thruster are changed accordingly, which are on the periphery of the satellite and have no effect on its internal layout. In addition, two arc thruster powers are added. Compared with the traditional bipropellant propulsion system, the hybrid propulsion system involves fewer changes, is simpler, and can be directly loaded into the load compartment.

The bipropellant and ion propulsion system requires more modification and requires reconfiguration of the ion propulsion system, including a storage supply system, thruster, ion thruster power supply and vector adjustment structure, which is incompatible with the traditional chemical propulsion system, requiring more internal space in the satellite and making the whole system more complicated. At present, the satellite is built on the DFH-4 platform. The antenna takes one-third of the cabin space and the loaded elements are many and the weight is large. Therefore, there is no space in the cabin in which to place a storage and supply system. The specific configuration comparison is shown in Table 2 (Hotta et al., 1998; Morgan & Meinhardt, 1999; Zubeet al., 2003).

Only the hybrid-mode propulsion system can meet the requirements of the entire satellite in the case of satellites using electric propulsion.

4 COMPARISON OF THRUSTER PLUME AND JET MATTER

According to the current design of various domestic thrusters, the diameter of the 10 N thruster nozzle is 40 mm and the plume divergence angle is 30°. For the ion thruster, the nozzle diameter is 200 mm and the plume divergence angle is 15°. The nozzle diameter of the arc thruster is 12 mm and its plume divergence angle is 17°. All the parameters are listed in Table 3.

The mechanism of the arc thruster involves heating the decomposition products of mono-propellant anhydrous hydrazine (N_2H_4), which will not pollute the antenna. The spray of the ion thruster spray is plasma, which cannot be completely neutralized and will attach to the antenna mesh surface causing huge influences on the antenna (Stechman et al., 2000; Wu et al., 2001; Krismer et al., 2003).

Table 3. Comparison of the main parameters influencing the plume.

	Nozzle diameter (mm)	Plume divergence angle (°)	Jet
10 N thruster	40	30	NO_2, CH_4 etc.
Ion thruster	200	15	Helium plasma
Arc thruster	12	15	N_2, H_2, NH_3

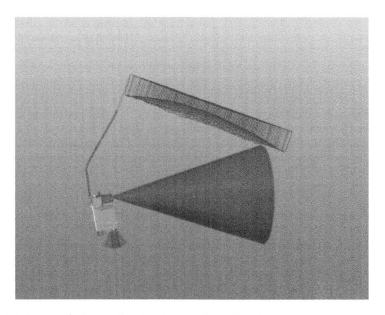

Figure 3. The impact of a 15° arc thruster plume angle on the antenna.

According to the analysis of the satellite plume and jet, the antenna can be protected from the plume only by using an arc thruster with a small nozzle diameter, a relatively small plume divergence angle and minimal spray contamination (Fisher & Meyer, 1998).

5 COMPARISON OF SYSTEM POSITION-RETENTION PERFORMANCE

In a satellite with large antennae, the in-orbit imaging task consumes a lot of power, the demand for heat dissipation is great and the mission takes a long time (loading work for 0.5 hour, heat dissipation for one hour). It is necessary to reduce position-retention times and increase the retention cycle, while saving the propellant in the orbital position (Zube et al., 1998).

The position-retention performance of the different thrusters is summarized in Table 4. The time and period of position retention are mostly determined by the magnitude and efficiency of the thrust. When the thrust of the arc thruster is 130 mN/kW or more, the use efficiency is 0.9. When the thrust of a 10 N thruster is 10 N, the use efficiency is 0.88. When the thrust of the ion thruster is 26 mN/kW, the use efficiency is 0.7.

For the station-keeping system, the 10 N thruster of a bipropellant propulsion system can be used for a single operation of 0.5 hour, once every two weeks in the maintenance of the inclination and regression accuracy. The arc thruster of a hybrid propulsion system can be used in a single operation for four hours, once a week, and the ion propulsion system can be used in a single operation for two hours, twice a day.

Table 4. Comparison of position-retention efficiency.

	Thrust power ratio (mN/kW)	Thruster efficiency	Station-keeping time*
10 N thruster	–	0.88	0.5 hours, once every two weeks
Ion thruster	26	0.7	2 hours, twice aday
Arc thruster	130	0.9	4 hours, once a week

*The station-keeping can only be carried out at the track ascending and descending points.

Based on the demands of the satellite and a working mode of load on for 0.5 hour and heat dissipation charging for one hour, only the electric propulsion with fewer working times (high thrust power ratio) can be used. Therefore, for a LZ-1 satellite, electric propulsion is the best choice (Skelly & Kay, 1997; De Grys et al., 2005).

6 SUMMARY

In this review, we compared the system weight, structural layout, system complexity and thruster plume of the bipropellant propulsion system, hybrid propulsion system and bipropellant and ion propulsion system in high-orbit SAR satellites. Because the arc thruster can provide greater speed increments for the same weight, the configuration layout is simpler, the plume and jet pollution are smaller and the thrust and efficiency are higher, the hybrid propulsion system with an arc thruster is more suitable for satellites with large antenna loads in high orbit.

REFERENCES

De Grys, K., Welander, B., Dimicco, J., Wenzel, S., Kay, B., Khayms, V. & Paisley, J. (2005). 4.5 kW hall thruster system qualification status. *41st AIAA/ASME/SAE/ASEE Joint Propulsion Conference & Exhibit, Tucson, AZ*. Reston, VA: American Institute of Aeronautics and Astronautics. doi:10.2514/6.2005-3682.

Fisher, J.R. & Meyer, S.D. (1998). The design and development of the MR-510 arcjet power conditioning unit. *34th AIAA/ASME/SAE/ASEE Joint Propulsion Conference & Exhibit, Cleveland, OH*. Reston, VA: American Institute of Aeronautics and Astronautics. doi:10.2514/6.1998-3630.

Hotta, N., Ichino, H., Toriyama, K., Kosugi, S. & Kanamori, Y. (1998). Development of system engineering model for data relay test satellite. *17th AIAA International Communications Satellite Systems Conference & Exhibit, Yokohama, Japan*. Reston, VA: American Institute of Aeronautics and Astronautics. doi:10.2514/6.1998-1219.

Krismer, D., Dorantes, A., Miller, S., Stechman, C. & Lu, F. (2003). Qualification testing of a high-performance bipropellant rocket engine using MON-3 and hydrazine. *39th AIAA/ASME/SAE/ASEE Joint Propulsion Conference & Exhibit, Huntsville, AL*. Reston, VA: American Institute of Aeronautics and Astronautics. doi:10.2514/6.2003-4775.

Morgan, O.M. & Meinhardt, D.S. (1999). Monopropellant selection criteria—Hydrazine and other options. *35th Joint Propulsion Conference & Exhibit, Los Angeles, CA*. Reston, VA: American Institute of Aeronautics and Astronautics. doi:10.2514/6.1999-2595.

Skelly, P. & Kay, R. (1997). RHETT/EPDM power processing unit. *25th International Electric Propulsion Conference, Cleveland, OH*. Retrieved from http://erps.spacegrant.org/uploads/images/images/iepc_articledownload_1988-2007/1997 index/7104.pdf.

Smith, R.D., Aadland, R.S., Roberts, C.R. & Lichtin, D.A. (1997). Flight qualification of the 2.2 kW MR-510 hydrazine arcjet system. *25th International Electric Propulsion Conference, Cleveland, OH*. Retrieved from http://erps.spacegrant.org/uploads/images/images/iepc_articledownload_1988-2007/1997index/7082.pdf.

Smith, R.D., Roberts, C.R., Aadland, R.S., Lichtin, D.A. & Davies, K. (1997). Flight qualification of the 1.8 kW MR-509 hydrazine arcjet system. *25th International Electric Propulsion Conference, Cleveland, OH*. Retrieved from http://erps.spacegrant.org/uploads/images/images/iepc_articledownload_1988-2007/1997 index/7081.pdf.

Stechman, C., Woll, P., Fuller, R. & Colette, A. (2000). A high performance liquid rocket engine for satellite main propulsion. *36th AIAA/ASME/SAE/ASEE Joint Propulsion Conference and Exhibit, Las Vegas, NV*. Reston, VA: American Institute of Aeronautics and Astronautics. doi:10.2514/6.2000-3161.

Wu, P.-K., Woll, P., Stechman, C., McLemore, B., Neiderman, J. & Crone, C. (2001). Qualification testing of a 2nd generation high performance apogee thruster. *37th AIAA/ASME/SAE/ASEE Joint Propulsion Conference, AIAA Paper 2001-3253, Salt Lake City, UT*. Reston, VA: American Institute of Aeronautics and Astronautics. doi:10.2514/6.2001-3253.

Zube, D.M., Fye, D., Masuda, I. & Gotoh, Y. (1998). Low voltage bus hydrazine arcjet system for geostationary satellites. *34th AIAA/ASME/SAE/ASEE Joint Propulsion Conference & Exhibit, Cleveland, OH*. Reston, VA: American Institute of Aeronautics and Astronautics. doi:10.2514/6.1998-3631.

Zube, D.M., Wucherer, E.J. & Reed, B. (2003). Evaluation of HAN-based propellant blends. *39th AIAA/ASME/SAE/ASEE Joint Propulsion Conference & Exhibit, Huntsville, AL*. Reston, VA: American Institute of Aeronautics and Astronautics. doi:10.2514/6.2003-4643.

Automatic Control, Mechatronics and Industrial Engineering – He & Qing (Eds)
© 2019 Taylor & Francis Group, London, ISBN 978-1-138-60427-8

A conceptual framework for product–service configuration based on customer demand in a processing-equipment Product–Service System (PSS)

C.B. Zhao & W. Cao
Fujian Provincial Key Laboratory of Special Energy Manufacturing, Xiamen Key Laboratory of Digital Vision Measurement, Huaqiao University, Xiamen, China

ABSTRACT: Currently, market competition is intensifying and customer demand is diverse. The traditional 'product and after-sales service' mode can no longer adapt to latest market demands. In order to increase profits and enhance competitiveness, companies are linking products and services to customers as a single entity, the Product–Service System (PSS). In this paper, a conceptual framework for the design of a processing-equipment PSS (pe-PSS) is proposed. The main functions of the framework and design process are illustrated. Finally, some key enabling technologies are described in detail, that is, identification and analysis of customer demand, feature matching between customer demand and service technology, division of product and service modules, configuration design and optimization of the pe-PSS.

Keywords: customer demand, pe-PSS, framework, design

1 INTRODUCTION

Due to improvements in raw materials, labor costs and global environmental protection, competition in the manufacturing industry has intensified and profit margins have shrunk. Low value-added industrial products can no longer meet the requirements of manufacturing enterprises and environmental development (Gaiardelli et al., 2014; Aurich et al., 2006; Cavalieri & Pezzotta, 2012). In the past, most companies simply sold material products, and this traditional business model has been unable to bring huge profits to enterprises. In order to achieve sustainable development, more and more enterprises are transforming from product manufacturers to product service providers. Based on customer demand, a series of customized services are provided to customers by enterprises. These value-added services include product operation, maintenance, upgrading, recycling, and so on.

In order to adapt to this strategic change, a new concept has been established—the Product–Service System (PSS). The idea is intended to provide customers with solutions, including products and services (Reim et al., 2015). With the continuous changes in market demand, customer requirements of processing equipment are becoming higher and higher. The processing equipment referred to here includes Computerized Numerical Control (CNC) machining equipment, such as CNC lathes, CNC milling machines, machining centers and so on. In addition, due to the professional, high-end precision and other characteristics of such processing equipment, there are high requirements around its design, use and maintenance. Typically, customers lack experience in the operation and maintenance of such processing equipment. The traditional model of 'product and after-sales service' cannot meet the customer's individual needs. To this end, processing equipment and related services are integrated by product manufacturers or service providers, and the end-customer is provided with a processing-equipment Product–Service System (pe-PSS).

Configuration design is considered to be a key aspect of the development of a PSS. The aim is to provide a solution composed of material products and services based on customer demand. When the scheme is designed, it is necessary to maximize the profit of the company while satisfying the customer's needs.

For a configuration design in pe-PSS, the performance, cost and other factors are mostly determined in the initial design phase. Once the scheme is determined, the performance and cost of the pe-PSS will not be controlled if a change occurs during system execution. Therefore, one of the issues that current research institutes should pay attention to is how to design a low-cost pe-PSS based on customer demand. In this paper, a conceptual framework for product–service configuration in pe-PSSs is proposed.

2 FRAMEWORK OF A PROCESSING-EQUIPMENT PRODUCT–SERVICE SYSTEM

2.1 Framework

The conceptual framework of the pe-PSS configuration design implementation model is shown in Figure 1. The framework consists of three layers, that is, the key technology layer at the bottom, the application method layer in the middle and the system layer on the top. The key technology layer includes four supported technologies that ensure the application layer can run smoothly. In the middle application layer, some main functions of the framework are included, and the design flow of a configuration scheme of a pe-PSS is given. The system layer supports the data and information of the whole design framework, including user information, a service module database, and a service process and resource database.

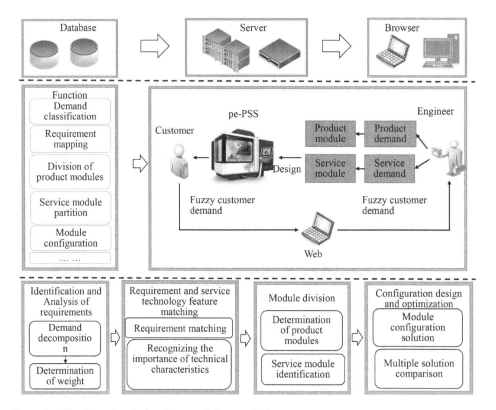

Figure 1. Configuration design framework for a pe-PSS.

190

2.2 *Design process*

In some cases, clearly understanding how a scheme of pe-PSS is designed, can help us identify some of the associated problems and difficulties. First of all, customer requirements are obtained through a Web terminal, and these requirements are often general. Therefore, it is necessary to identify, decompose and analyze the importance of these requirements, obtaining an operational requirement unit. Then, many requirement units with different weights are mapped and the corresponding technical features are obtained. According to the technical characteristics of these solutions, the product module and the service module are divided according to the customer usage preference. Then, the initial scheme of pe-PSS configuration is obtained through the solution of these modules. In the end, whether the customer is satisfied with the scheme indicates whether or not the configuration is successful.

3 SUPPORTED KEY TECHNOLOGIES

In this section, some key enabling technologies are described in detail to ensure that the framework operates smoothly, that is, identification and analysis of customer demand, feature matching between customer demand and service technology, division of product and service modules, configuration design and optimization of the pe-PSS.

3.1 *Identification and analysis of customer demand*

Given that customer demand is often reflected in a customer's production activities, ambiguity can be high and difficult to identify. Therefore, it is necessary to consider the customer's activities in different situations. Through deep understanding and analysis of customer activities, the needs of industrial products and services at different life-cycle stages are actively obtained. As shown in Figure 2, the fuzzy requirements of customers are refined into specific product requirements and corresponding service requirements. Then, these requirements are refined to obtain non-decomposable demand units.

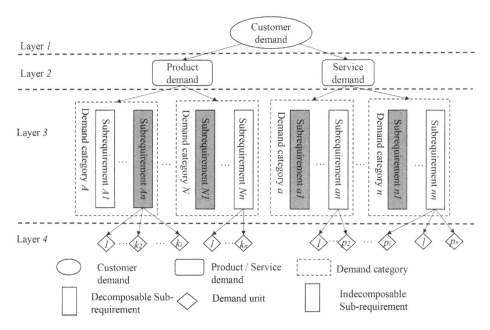

Figure 2.   Customer demand model.

The functions of the four layers in the model are as follows:

- Layer 1: This layer is the initial customer requirements layer. Different makes and models of products are associated with different customer needs (e.g. CK CNC machine tools, CJK CNC machine tools).
- Layer 2: This layer is a preliminary classification of customer needs. Combined with the idea of product–service, customer demand is decomposed into two parts: product requirements and service requirements.
- Layer 3: This layer contains different categories of requirements for service requirements and product requirements. Different requirements are analyzed and combined. The product requirement categories are identified by upper-case letters, the service requirements are identified by lower-case letters, and the sub-requirements are identified by numbers. For example, A_1 represents the first sub-requirement in the first category of product requirements and c_2 represents the second sub-requirement in the third category of service requirements. On the basis of this method, all of the different sub-requirements are numbered.
- Layer 4: This layer represents the ultimate non-decomposable requirement unit. According to the decomposition rules, the sub-requirements are decomposed and merged to obtain a series of requirement units. The coding rules for this layer are given. A product requirement unit is represented by the lower-case letter k and numbered from one. A service requirement unit is represented by the lower-case letter p, and numbered from one to n.

According to customer demand preference, the weights of technical characteristics of follow-up services are determined. Customer demand is also a measure of the quality of the subsequent product and service configuration scheme. When determining service requirements and their weights, the results are not unique because different experts have differing views. A method based on a standard weighting matrix can deal with these subjective and fuzzy service requirements effectively, and can achieve more accurate customer demand. Then, the key service requirements are identified to prepare for the subsequent scheme design and optimization. Figure 3 shows the flow chart of customer-demand importance analysis.

3.2 *Feature matching between customer demand*

Through analysis of customer demand and decomposition, several demand units are derived. However, these units are demands for products and services and cannot direct the configuration design of pe-PSS. In the configuration process, engineers also need to map these customer requirements to the technical features of products and services from a technical point of view.

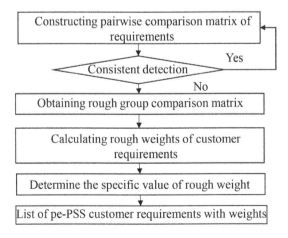

Figure 3.   Flow of customer-demand importance analysis.

In this paper, Quality Function Deployment (QFD) is used as a transformation tool to establish the relationship matrix between customer requirements and technical features. Through the transformation of a relational matrix, customer requirements are transformed into technical features of products. QFD is used to obtain the information needed in the process of product or service configuration and its structure is shown in Figure 4.

As can be seen from Figure 4, the structure contains the customer requirement unit and its requirement weight. The requirement unit is an independent and specific unit obtained from the decomposition of customer requirements. By resolving the importance of the requirement unit, the weight of each requirement unit can be obtained. Expectations are set up to represent the degree of customer demand for products or services. Each expectation corresponds to each requirement unit and has a mapping relationship with the technical target feature.

In configuration design, the technical characteristics determined by the engineer play a decisive role in the whole configuration design process. The content described by technical features includes two aspects: qualitative description and quantitative description. For example, the diameter of the spindle hole is 600 mm, in which 'the diameter of the spindle hole' is a technical feature, and 'φ 600 mm' is a quantitative description. The correlation between the customer demand unit and the technical feature is represented by r_{ij}. The degree of correlation between the units of items i and the technical feature of item j is usually expressed by r_{ij}. There are four relationships: strong correlation, intermediate correlation, weak correlation and uncorrelated. In order to quantitatively describe the correlation between customer demand and technical characteristics, a correlation evaluation table is established, which is shown in Table 1.

A correlation matrix is used to describe the relationship between customer requirement units and technical features. The matrix and the weight of the technical feature are used to describe its importance. The target value of the technical feature is the expected value of the possible target of the technical feature. It corresponds to the technical feature and provides the basis for the subsequent configuration design.

Some of the obtained technical features can be directly mapped from customer requirements. For some fuzzy requirement information, rough set theory is used to determine the

Customer demand unit	Demand weight	Desired value	Technical characteristics			
			TA_1	TA_2	$\cdots$	TA_n
CR_1	ω_1	D_1	Correlation Matrix between customer demand Unit and Technical Features			
CR_2	ω_2	D_2				
$\cdots$	$\cdots$	$\cdots$				
CR_n	ω_n	D_n				
	Technical feature weight		β_1	β_2	$\cdots$	β_n
	Technical feature target value		TB_1	TB_2	$\cdots$	TB_n

Figure 4.   Quality house structure of customer demand mapping.

Table 1.   Relative assessment between requirement unit and technical characteristics.

Correlation description	Strong correlation	Intermediate correlation	Weak correlation	Uncorrelated
Correlation score	1	2	3	4

corresponding technical features. Considering the potential conflicts between different service technology features, reasonable solutions are needed. For example, on the basis of requirements mapping, a certain service conflict resolution method is used to identify and analyze the feature conflict of service technology, and finally solve the problem.

3.3 *Division of product and service modules*

At present, due to the advantages of modular design, more and more attention is being paid to the design of a PSS. It should be noted that the division of product modules is a key element in modular design, and the partition of service modules is configuration-oriented. The purpose is to realize the rapid provision and application of services. The design of a service module should be oriented to a physical product, and the partition of a service module should be consistent with that of a physical product. The aim is to facilitate the implementation of service modules and the refinement of physical products.

Based on the analysis of customer demand, the whole life-cycle approach is used to divide the product modules from the point of view of structure and function. The structure attributes and function attributes are decomposed, and the module partition criterion for the whole life cycle is obtained. The correlation matrix of parts is established by analyzing the interrelations between parts. According to the constraints of the rules, a part relationship model is constructed to express the relative relationship between the parts. Finally, the fuzzy mean clustering algorithm is used to cluster each part and obtain the product module. Taking aCJK CNC machine tool product as an example, dividing the related module, is shown in Table 2.

From the point of view of manufacturing enterprises, the service provided to customers is also a product. The service module is a relatively independent service function carrier which is provided by the service provider for customers. It is the construction element of a product–service scheme. It is composed of a series of service components with interrelated relationships. Each service module contains the same function but a different performance module instance. A module example refers to the concrete implementation of a service module. Thus, 'system fault diagnosis, network feedback response within two hours' and 'engineer manual diagnosis, video and voice feedback within four hours' are two examples of a diagnostic service module. They are both for fault diagnosis, but their service performance is not the same. Therefore, the service module can be created on the basis of fuzzy tree graph theory; then, the module is selected according to the importance of the module to customer demand.

3.4 *Configuration design and optimization of pe-PSS*

A multi-objective optimization model for pe-PSS solution configuration optimization was established. This model can enhance the flexibility of industrial product service solution design and satisfy the customer's customization requirements quickly. The scheme is based on the degree of correlation between industrial product module examples and technical

Table 2. Physical module division for aCJK CNC machine tool.

No.	01	02	03	04	05	06	07
Name	Bed module	Guideway module	Principal axis module	Feed drive module	Lubrication module	Control system	Tool magazine
No.	08	09	10	11	12	13	14
Name	Tool holder	Cooling module	Hydraulic module	Electrical module	Chip removal	Remote control	Tailstock module

features, service cost and service response time. By solving the model, a set of configuration optimization solutions for a series of service schemes is obtained. Then, based on the criteria of customer satisfaction, the scheme in the service configuration optimization scheme set is further selected in order to obtain the best service scheme.

4 CONCLUSION AND FUTURE WORK

A framework of a pe-PSS based on customer demand is proposed. The function and design idea of the framework are expounded. In order to ensure the operation of the frame-work, some key support technologies are described, including customer requirement identification and analysis, customer demand and service technology feature matching, product and service module partition, configuration design and optimization of pe-PSS.

Further research on the relationship between product and service modules, configuration design algorithms will be carried out in the future.

ACKNOWLEDGMENTS

The research presented in this paper was supported by the Promotion Program for Young and Middle-aged Teachers in Science and Technology Research of Huaqiao University (no. ZQN-PY404) and the Subsidized Project for Postgraduates' Innovation Fund in Scientific Research of Huaqiao University (no.17013080046).

REFERENCES

Aurich, J.C., Fuchs, C. & Wagenknecht, C. (2006). Life cycle oriented design of technical product-service systems. *Journal of Cleaner Production*, *14*(17), 1480–1494.

Cavalieri, S. & Pezzotta, G. (2012). Product–service systems engineering: State of the art and research challenges. *Computers in Industry*, *63*(4), 278–288.

Gaiardelli, P., Resta, B., Martinez, V., Pinto, R. & Albores, P. (2014). A classification model for product-service offerings. *Journal of Cleaner Production*, *66*, 507–519.

Reim, W., Parida, V. & Örtqvist, D. (2015). Product–service systems (PSS) business models and tactics – A systematic literature review. *Journal of Cleaner Production*, *97*, 61–75.

Automatic Control, Mechatronics and Industrial Engineering – He & Qing (Eds)
© 2019 Taylor & Francis Group, London, ISBN 978-1-138-60427-8

An improved attitude estimation algorithm based on MPU9250 for quadrotor

Q.Q. Zhao, Y.G. Fu, Z.G. Liu, Y. Xu & X.Y. Liu
School of Port and Transportation Engineering, Zhejiang Ocean University, Zhoushan, China

ABSTRACT: To resolve the problem due to drift and noise in attitude estimation process of MEMS inertial measuring unit on quadrotor, an improved algorithm based on complementary filtering is presented. A test platform of quadrotor carried out with MPU9250 as attitude measuring unit is established. Both under static and dynamic conditions, attitude data obtained from different methods are collected and compared, including Kalman filtering algorithm for information integration, data fusion with traditional complementary filtering, and our improved filtering fusion algorithm. Experimental results show that the present improved attitude fusion algorithm has advantages in terms of higher estimation precision and lesser drift and noise errors of the final attitude angle in different conditions, and is easy to implement in low cost aircraft control system.

Keywords: quadrotor, complementary filtering, attitude fusion, MPU9250

1 INTRODUCTION

Unmanned aerial vehicle (UAV) is an aircraft with no pilot on board. UAV is now increasingly used as a cost effective and timely method of several applications, such as navigation, rescuing and reconnaissance. There exist many advantages of UAV technology, such as low cost, ecological operation, and most of all (Ammour et al. 2017). In general, there are four categories of UAV, namely single rotor helicopters, multi rotor-crafts, fixed wing planes and hybrid combinations (Kanellakis & Nikolakopoulos 2017). Among UAV the class of multi rotor-crafts can be further divided into tri-copters, quad-rotor, hexa-rotor and octa-copters. In the UAV community, the quadrotor is more popular than the others. It can be operated both indoors and outdoors and is a fast and agile platform for demanding maneuvers, e.g., hovering or moving along a target in close quarters.

An attitude solution aims to obtain the real-time accurate attitude angle of the aircraft. Accurate acquisition of attitude information is the premise for drones to achieve stable flight. The accuracy and speed of attitude calculation will directly affect the stability and reliability of the flight control algorithm. With the attitude calculation accuracy requiring, how the attitude data fusion refines is still a hot topic in current research. In the commonly used MEMS inertial navigation devices, gyroscopes will bring zero drift and integration error. Accelerometers are easily interfered by noise or vibration. Magnetometers are susceptible to interference from external magnetic fields (Kang et al. 2012).

To solve the problems and improve the measurement accuracy of the attitude angle, many algorithms based on the Kalman filtering principle were used. For instance, Feng et al. proposed an improved extended Kalman filter algorithm (BPNN-EKF) to improve the accuracy greatly (Feng et al. 2017). He et al. used a modified iterative EKF (IEKF) algorithm, which improved filtering precision (He et al. 2015). Wang et al. used Kalman filter algorithm to improve the accuracy of the attitude measurement of the aircraft (Wang &Yang 2013).

Pan et al. used the attitude estimation algorithm based on Unscented Kalman Filter (UKF), which achieved higher stability, solution accuracy and convergence speed than EKF (Pan et al. 2012). Tang et al. used the Cubature Kalman Filter (CKF) algorithm to estimate the performance better than UKF (Tang et al. 2012).

Although previous works improved the accuracy of attitude calculation, they are not suitable for low-cost embedded attitude measurement systems, because they often suffered from a large amount of mathematical computations. In this paper, an improved algorithm is introduced. It uses magnetometers and accelerometers to reduce attitude errors caused by gyroscopes deviation in MEMS sensor. In addition, to correct gyroscope bias, a proportional-integral (PI) controller is proposed into the control system. The experiment demonstrates that improved algorithm has merits in small computational complexity, good real-time performance and high precision.

2 POSTURE REPRESENTATION

2.1 Gesture of posture

Without considering the influence of the earth rotation upon coordinate system, we establish reference frame n and vector coordinate b system. In the geographic reference system, X_n $Y_n Z_n$ is reference frame. Carrier coordinate system is established on controlled object with inertial device carrying. Two coordinate systems have same initial orientation (Xu 2016). Reference frame is transformed into carrier coordinate system by rotation matrix (Zhang et al. 2017). Rotation matrix (C_n^b) is as in (1):

$$C_b^n = \begin{pmatrix} \cos\theta\cos\phi & \cos\theta\sin\phi & -\sin\theta \\ \sin\psi\sin\theta\cos\phi - \cos\psi\sin\phi & \sin\phi\sin\theta\sin\psi + \cos\phi\cos\psi & \sin\psi\cos\theta \\ \cos\phi\sin\theta\cos\psi + \sin\phi\sin\psi & \cos\psi\sin\theta\sin\phi - \sin\psi\cos\phi & \cos\psi\sin\theta \end{pmatrix} \quad (1)$$

2.2 Quaternion representation of attitude

A quaternion consisting of four elements is expressed as equation (2):

$$Q = q_0 + q_1 i + q_2 j + q_3 k \quad (2)$$

where $q_0 = \cos\frac{\theta}{2}, q_1 = l\sin\frac{\theta}{2}, q_2 = m\sin\frac{\theta}{2}, q_3 = n\sin\frac{\theta}{2}$.

Then equation (1) can be expressed as a quaternion:

$$C_b^n = \begin{bmatrix} 1 - 2(q_2^2 + q_3^2) & 2(q_1q_2 - q_0q_3) & 2(q_1q_3 + q_0q_2) \\ 2(q_1q_3 + q_0q_3) & 1 - 2(q_1^2 + q_3^2) & 2(q_2q_3 - q_0q_1) \\ 2(q_1q_3 - q_0q_2) & 2(q_2q_3 + q_0q_1) & 1 - 2(q_1^2 + q_2^2) \end{bmatrix} \quad (3)$$

Update of quaternion can obtain by solving quaternion differential equation by first-order Runge-Kutta method:

$$Q(t + \Delta t) = Q(t) + \frac{\Delta t}{2} \begin{pmatrix} -w_x q_1 - w_y q_2 - w_z q_3 \\ w_x q_0 - w_y q_3 + w_z q_2 \\ -w_x q_2 + w_y q_1 + w_z q_0 \end{pmatrix} \quad (4)$$

The quaternion representation of attitude angle can be acquired by comparing equations (1) and (3), as in equation (5):

$$\begin{cases} \varphi = \arctan\dfrac{2(q_0 q_1 + q_2 q_3)}{q_0^2 - q_1^2 - q_2^2 + q_3^2},\varphi \in (-\pi,\pi) \\[3mm] \theta = -\arcsin 2(q_1 q_3 - q_0 q_2),\theta \in \left(-\dfrac{\pi}{2},\dfrac{\pi}{2}\right) \\[3mm] \psi = \arctan\dfrac{2(q_0 q_3 + q_1 q_2)}{q_0^2 + q_1^2 - q_2^2 - q_3^2},\psi \in (-\pi,\pi) \end{cases} \qquad (5)$$

3 OUR APPROACH

Among MEMS sensors, a gyroscope is able to generate high-precision attitude estimation data in a short period of time, whereas the data generated by accelerometers and magnetometers is comparably poor. It is worth noticing that the gyroscope suffers from integral errors due to drift over time, leading the degradation to the attitude accuracy gradually. By contrast, the measurement errors of accelerometers and magnetometers do not increase (He & Chen 2016). In other words, with respect to the dynamic characteristics the gyroscope outperforms accelerometers and magnetometers. Thus, complementary filtering algorithm (Wang et al. 2016) was proposed to fuse data of the three types of MEMS sensors, in order to improve the final accuracy of attitude calculation and dynamic characteristic of control systems.

However, since the stop band attenuation of a low-pass filter is slow, the noise becomes large. Besides, if the filter's parameters are fixed, it is difficult to achieve optimal estimation values. In order to address these issues, we propose a new algorithm where the PI controller is introduced to reduce errors from MEMS sensors, i.e., gyroscopes, accelerometers, and magnetometers. The algorithm is improved based on the classical method called complementary filtering (Wang et al. 2016). Dynamic compensation coefficients are K_p and K_i. Weights of accelerometers and magnetometers are reduced, when noise interference size increases. In this case, it mainly relies on the gyroscope to perform attitude determination, improving the fusion precision of the attitude data. The implementation flowchart of the improved algorithm is displayed by Figure 1.

The procedure of our algorithm is as follows:

1. Unitize accelerometer and magnetometer measured value. In the carrier coordinate system, a measured value of accelerometers is recorded as $a_b = [a_{bx}\ a_{by}\ a_{bz}]^T$. After unitization, it is denoted by $\bar{a}_b$. The measured value of magnetometer is represented as $m_b = [m_{bx}\ m_{by}\ m_{bz}]$. After unitization, it is expressed as $\bar{m}_b$.
2. Convert reference vector under gravity field and geomagnetic field into the carrier coordinate system through attitude transformation matrix. In the navigation coordinate, a standard acceleration of gravity is defined as $g_n = [0\ 0\ 1]^T$, which can be described as $g_b = [g_{bx}\ g_{by}\ g_{bz}]$ after being transformed by C_n^b matrix. Magnetometers measure magnitude and direction of the geomagnetic field. Set reference frame x-axis to coincide with north direction. Reference vector is described as $b_n = [b_{nx}\ 0\ b_{nz}]^T$.

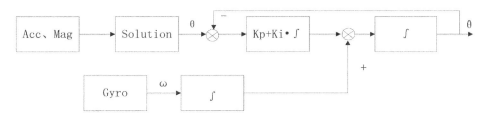

Figure 1. The implementation flowchart of the improved algorithm.

Magnetometers data of *n*-series derived from *b* lineage. In other words, converted magnetic field vector expression of $t_n = [t_{nx}\ t_{ny}\ t_{nz}]^T$ is obtained by $t_n = C_b^n \cdot \bar{m}_b$. On plane *XOY* (n-series), a vector size of the magnetometers must be the same, and here are the equations: $b_{nx} = \sqrt{t_{nx}^2 + t_{ny}^2}, b_{nz} = t_{nz}$.

Reference vector b_n can be converted to w_b by the following formula:

$$w_b = C_n^b \cdot b_n \qquad (6)$$

3. Calculate error vector. There exist errors between $\bar{a}_b$ and g_b. Make vector cross product $(e_a = \bar{a}_b \times g_b)$ between $\bar{a}_b$ and g_b, which can compensate and correct data produced by gyroscopes. Simultaneously, error vectors of magnetometers are $e_m = m_b \times w_b$.
4. Use error vector to correct gyroscope error. $e = e_a + e_m$ is error vector, which compensates angular velocity of gyroscope. By means of the PI controller, gyroscope error is revised, which gives the formula of PI controller $(\lambda = K_p \cdot e + K_i \cdot \int_0^t edt)$. In quaternion renewal equation, angular rate expression is $w_t = w + \lambda$.
5. Attitude quaternion is calculated by taking the corrected gyro value into equation (4).
6. On the basis of equation (5), standardization quaternion is transformed into attitude angle.

4 EXPERIMENT

4.1 *Experimental environment*

To verify feasibility and effectiveness of our algorithm, we established a test platform and used a small four-axis aircraft to be controlled. The main control chip adopts STM32F411 that has a rich peripheral interface. The attitude chip uses the nine-axis MPU9250, which owns DMP (Digital Motion Processor) and supports MPL. MPU9250 is connected to the main control chip by software IIC. Real-time attitude and solved attitude angle information are received by computer through serial communication. Measuring range of gyroscope output is ±2 000°/s, and cutoff frequency of low-pass filters is 50 Hz. In PI controller, parameters are $k_p = 10$ and $k_i = 0.008$.

4.2 *Experimental comparison and analysis*

In static environments, attitude angles calculated by Kalman filtering algorithm (KL) (Wang et al. 2016), first-order complementary filtering algorithm (FF) (Nan et al. 2017), second-order complementary filtering algorithm (SF) (Nan et al. 2017), DMP solution (He & Chen 2016) and improved complementary filtering algorithm (OURES) are calculated, test results shown in Figure 2. Attitude angles respectively calculated by accelerometer (AC) and gyroscope (GY) are shown in Figure 3.

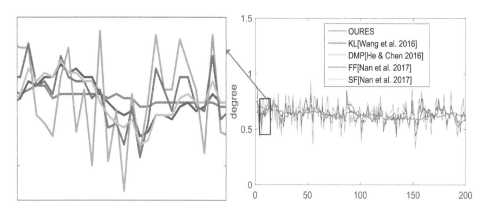

Figure 2.   Comparison of experimental results of OURES, KL, DMP, FF, SF.

It can be seen from Figure 3 that angles directly calculated by accelerometers own noise interference. Experimental results prove that angles calculated by gyroscopes drift nearly 1° within one minute. According to Figure 2, it reveals that calculation of first-order and second-order complementary filtering algorithm filters out some spikes and still exists large fluctuation, having a much better effect. Improved complementary filtering algorithm shows very good results, with relative static standard error of 0.01°.

To verify effectiveness of algorithm in dynamic environment, several dynamic calculations are performed on the four-rotor aircraft hardware platform. Test results of different algorithm are shown in Figure 4. Attitude angles respectively calculated by accelerometers and gyroscopes are indicated in Figure 5.

Figure 4 reflects that both Kalman filter algorithm and improved complementary filter algorithm can effectively filter out some interfere of high frequency signal. Compared with improved complementary filtering, second-order complementary filtering, first-order complementary filtering and Kalman filtering have some delay. Since second-order complementary filtering calculation is relatively large, real-time tracking performance is worse than first-order filtering effect. It is smoother than first-order complementary filtering curve, curve relatively lagging. Improved algorithm that can effectively filter out some signal interference is smoother than traditional algorithm filtering curve, because proportional-integral (PI) controller is introduced. Improved complementary filter which mirrors a better solution than traditional complementary filter can improve stability of system attitude solution. Table 1 compares standard deviation of time consumption and angle estimation.

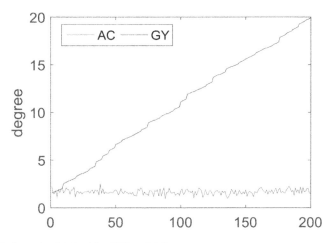

Figure 3.   Attitude angle obtained by MG and AC.

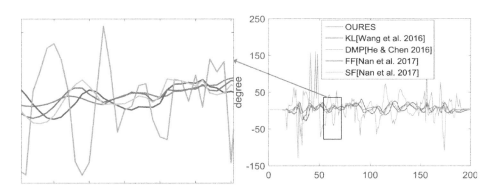

Figure 4.   Comparison of experimental results of OURES, KL, DMP, FF, SF.

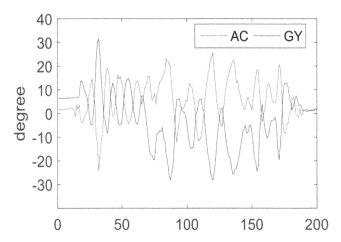

Figure 5. Attitude angle obtained by MG and AC.

Table 1. A comparison of time operation and standard deviation of angle estimation.

Algorithm	OP/us	Roll SD/(°)
OURS	5742	5.61
SF (Nan et al. 2017)	5172	8.83
FF (Nan et al. 2017)	4364	11.62
DMP (He & Chen 2016)	5012	38.77
KL (Wang et al. 2016)	5938	10.32

5 CONCLUSION

Accurate acquisition of attitude data is the premise for drones to achieve smooth and stable flight. This paper studies processing method of aircraft attitude information fusion, and a proportional-integral (PI) controller is proposed into the control system. Motivated by traditional complementary filter fusion, an improved complementary filtering algorithm is presented to fuse multi-sensor data. A hardware test platform is set up to test and compare attitude information acquired from different methods in static and dynamic environments. The comparison results show that the present improved filter algorithm can avoid that gyroscopes solve attitude independently. The algorithm effectively compensates output error of attitude angle. Also, it can inhibit drift and noise of attitude angle in static conditions and track change of attitude angle quickly in dynamic condition. Thus, the precision of attitude angle estimation is improved and stability of the system attitude calculation is enhanced.

ACKNOWLEDGMENT

We thank all the anonymous reviewers who help to significantly improve this manuscript. Besides, this study was financially supported by Zhejiang Provincial Natural Science Foundation of China (No. LGG18F030013, No. LGN18E080003).

REFERENCES

Ammour, N., Alhichri, H., Bazi, Y., Benjdira, B., Alajlan, N. & Zuair, M. 2017. Deep Learning Approach for Car Detection in UAV Imagery. Remote Sensing 9(4): 1–15.

Feng, S.J., Xu, Z.Y. & Shi, M.Q. 2017. Research on Attitude Algorithm Based on Improved Extended Kalman Filter. Computer Science 44(9): 227–229.

He, H.P., Pei, M.L. & Yang, W.K. 2015. Research on attitude measurement method of quadrotor UAV based on IEKF. Computer Simulation 32(4): 56–60.

He, S. & Chen, X.W. 2016. Mpu9250-based UAV attitude information collection and processing. Journal of Fujian University of Technology 14(6): 587–592.

Kanellakis, C., & Nikolakopoulos, G. 2017. Survey on Computer Vision for UAVs: Current Developments and Trends. Journal of Intelligent and Robotic Systems 87(1): 141–168.

Kang, T., Fang, J. & Wei, W. 2012. Quaternion-Optimization-Based In-Flight Alignment Approach for Airborne POS. IEEE Transactions on Instrumentation and Measurement 61(11): 2916–2923.

Nan, Y.R., Wan, D.J. & Pan, S. 2017. Design of second-order complementary filter attitude solver. Journal of Zhejiang University of Technology 45(4): 416–420.

Pan, Y., Song, P., Li, K. & Wang, Y. 2012. Attitude Estimation of Miniature Unmanned Helicopter using Unscented Kalman Filter. Mechanical and Electrical Engineering, 1548–1551.

Tang, X., Wei, J. & Chen, K. 2012. Square-root adaptive cubature Kalman filter with application to spacecraft attitude estimation. International Conference on Information Fusion. IEEE, 1406–1412.

Wang, S.H. & Yang, Y. 2013. Quadrotor aircraft attitude estimation and control based on Kalman filter. Control and Decision 30(9): 1109–1105.

Wang, H.W., Chen, M. & Zhang, K. 2016. Four-rotor aircraft attitude estimation based on quaternion and Kalman filtering. Microcomputer and Application 35(14): 71–73.

Xu, Y.C. 2016. Design and simulation of attitude calculation algorithm for quadcopter. Science and Technology Vision (23): 17–18.

Zhang, D., Jiao, Y.M. & Liu, Y.Q. 2017. Fusion attitude and complementary method for complementary filtering and Kalman filtering. Sensors and Microsystems 36(3): 62–65.

Mechanical engineering

Automatic Control, Mechatronics and Industrial Engineering – He & Qing (Eds)
© 2019 Taylor & Francis Group, London, ISBN 978-1-138-60427-8

Dynamic simulation of four-ring roller coaster based on ADAMS software

K.R. Bian, S.P. Zhou & J. Li
East China University of Science and Technology, Shanghai, China

W. Zhao
Hebei Zhi Pao Amusement Equipment Manufacturing Co. Ltd., Hebei, China

ABSTRACT: For specialized equipment such as roller coasters, design and safety analysis are very important. In this paper, the dynamic performance of a four-ring (four-loop) roller coaster is investigated. The velocity, acceleration and force curves of various elements are simulated using ADAMS software. The acceleration of the roller coaster is experimentally measured by using an accelerometer, and the results align closely with the simulation (the maximum error is 4.5%). From the results, we found that the value of dynamic response of the first car is larger than the others, and it should be considered to represent a higher level of risk. The changes of dynamic responses of the roller coaster under different loads and different wind speeds are then analyzed. We found that the peak acceleration of the roller coaster decreases with the increase of wind speed, while the force on the connecting fork increases with the increase of load of the roller coaster.

Keywords: four-ring roller coaster, dynamic simulation, acceleration, force

1   INTRODUCTION

The roller coaster is an important feature of amusement parks, and with the progression towards more exciting, higher-speed rides, stricter requirements for the design of components and analysis of their safety have been proposed. However, the traditional roller coaster design method (design–manufacture–test–improve–test) has the disadvantages of long cycle times and high costs, and it is difficult to guarantee safety (Zheng, 2002).

Researchers inside and outside China have carried out simulation research on roller coasters from different perspectives and achieved results. Within China, Yin and co-workers (Yin et al., 2012) used spline curves to build the required track for a roller coaster simulation, establishing a reference for roller coaster dynamics modeling. Chen and Guan (2010) used ADAMS dynamics simulation software (MSC Corporation, Newport Beach, CA, USA) to establish a roller coaster dynamics model and carry out simulation analysis. The simulation validity was verified by comparing the speed data obtained with measured data. Outside China, Teng and co-workers (Teng et al., 2010) established a roller coaster dynamics model and carried out a dynamics analysis: the error between the analysis result and the actual measurement was only 6.4%, which establishes a reference for follow-up safety analysis.

The research outlined above has provided a basis for the dynamic simulation of roller coasters. In order to further study the dynamic parameters of roller coasters, this paper took a four-ring (four-loop) roller coaster as its research object, carried out dynamic simulation using ADAMS, and obtained the roller coaster velocity, acceleration and force curves of its components. On the basis of the model validation analysis, the dynamic parameters of the roller coaster under different wind speeds and different load conditions were then compared and analyzed. The conclusions can provide a reference for the design and safety analysis of roller coasters.

2 ESTABLISHMENT OF THE ROLLER COASTER DYNAMICS MODEL

The four-ring roller coaster principally consists of the car body and the track. There are four cars in total, which are equipped with five wheel-groups. Individual cars are connected by connecting forks, of which there are three. The roller coaster track is 471 m long and contains a lifting section, downhill section, vertical ring section, floating ring section and spiral section. The track is composed of track tube, supporting tube, sleeper, and column.

2.1 *The modeling of the roller coaster*

Without affecting the dynamic simulation, the car body is simplified and the key connectors between the car bodies are retained by using SolidWorks modeling software (Dassault Systèmes, Vélizy-Villacoublay, France). The center curve of the supporting pipe can be obtained from the three-dimensional coordinates of the supporting pipe, and then the center curve of the left and right tracks can be obtained by coordinate transformation according to the inclination angle of the track and the relative positions of the left and right branch tracks (Liang et al., 2006).

2.2 *Adding constraints and loading*

The constraints of the roller coaster dynamic simulation include fixed-pair, rotating-pair and point-line constraints. The contact relationship between real wheels and the track is very complex, and the modeling is difficult to calculate accurately. The point-line constraint is therefore used to simplify the wheel–rail relationship. The roller coaster will bear the gravity, friction, wind resistance, and traction when climbing. The force of gravity is applied to the roller coaster retrospectively, and then solved.

2.3 *Model verification*

In order to verify the accuracy of the model, the acceleration at six representative points on the track (see Figure 1) is measured with an accelerometer and compared with the simulated acceleration. The comparison results are shown in Table 1.

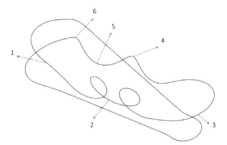

Figure 1. Acceleration contrast points.

Table 1. Comparison of acceleration values.

Point	Simulation (m/s^2)	Test (m/s^2)	Error (%)
1	−4.2	−4.1	2.4
2	13.8	13.2	4.5
3	−2.6	−2.7	−4.0
4	−9.4	−9.8	4.1
5	18.1	18.9	−4.2
6	−11.7	−11.3	−3.5

The comparison error for the six points is within 5%, and the maximum error is only 4.5%. The model and method are therefore credible.

3 ANALYSIS AND DISCUSSION OF SIMULATION RESULTS

Velocity, acceleration and force on connectors are important dynamic parameters of roller coasters for running safety. Vehicle load and wind speed are also important factors. Therefore, the vertical acceleration and the forces on the different connecting parts are compared under the conditions of full load and 15 m/s wind speed. The different running conditions are then also studied with reverse wind speeds (−5 m/s, −10 m/s and −15 m/s) and different loads (full load, half-load, and no load).

3.1 *Velocity analysis*

Under natural conditions, the wind may come from any direction, and the windward area of the roller coaster is constantly changing during operation. Only the most dangerous situation was therefore evaluated, that is, the wind direction always opposite to the direction of travel of the roller coaster and the largest windward area. The simulation results for different headwinds are shown in Figure 2.

As shown in Figure 2, the completion time of the simulation increases with the increase of wind speed, while the maximum peak velocity decreases with the increase of wind speed. The maximum difference in peak velocity is 2.5 m/s.

3.2 *Acceleration analysis*

Figure 3 shows that there is a difference in the vertical acceleration of different cars; the peak acceleration of the first car is −34.9 m/s^2, the acceleration of the rear car is −30.3 m/s^2, and the accelerations of the two middle cars are 29.6 and 27.1 m/s^2. Rides in the first and last cars are therefore more exciting than in the middle cars.

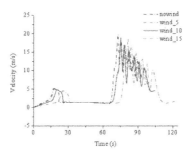

Figure 2. Different wind-force velocity curves.

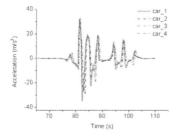

Figure 3. Vertical acceleration–time curves for different cars.

209

The simulation results under different wind speeds and loads are shown in Figure 4. From Figure 4a, it can be seen that the acceleration of the first vehicle under different wind speeds has obvious lag, and the peak acceleration decreases with the increase of the wind speed. The maximum acceleration difference is 3.7 m/s². From Figure 4b, it can be seen that the lag of acceleration for different cars is small under different loads, and the peak value and trend of change are not different. The load of the roller coaster has little effect on acceleration.

3.3 *Force analysis*

There are five wheel-groups and three connecting forks in the four-ring roller coaster, and the number of parts increases from the first car to the last car. The wheel–rail contact force of different wheel-groups and different connecting-fork forces is shown in Figure 5.

Figure 5a shows that the force on the front wheel group of the first car is the largest, with a peak value of 72,317 N. This is more than six times that on the other wheels, wherein the forces are similar to one another. Figure 5b shows that the maximum force on the connecting fork between the first car and the second car is 35,925 N, and the forces on the other connecting forks are similar. Therefore, the front wheel of the first car, and the connecting fork between the first and second cars should be taken as the key points in the safety inspection analysis.

The wheel–rail contact force and connecting-fork force under different load conditions are shown in Figure 6. As can be seen in Figure 6a, the wheel–rail contact force under different loads has a lag, and the lag increases gradually with the increase of load. Figure 6b shows that the force on the connecting fork under different loads increases with the increase of load, and the difference of peak force between full load and no load is 11,481 N. Therefore, it can be seen that the roller coaster is subject to higher forces (and is potentially more dangerous) under the conditions of higher loads and wind speeds, and these conditions should be the primary object of safety analysis.

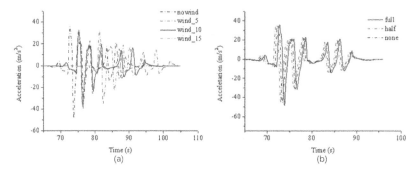

Figure 4. Vertical acceleration under: (a) different wind speeds; (b) different loads.

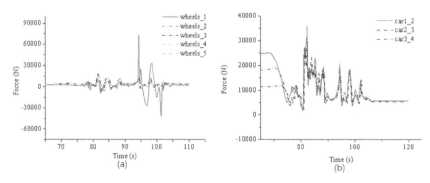

Figure 5. Forces on different cars and components: (a) different wheel–rail contact force curves; (b) different connecting-fork force curves.

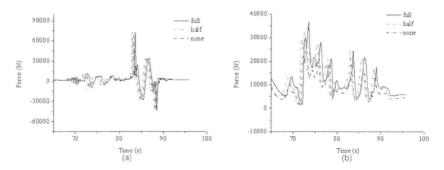

Figure 6.   Component forces under different load conditions: (a) wheel–rail force; (b) connecting-fork force.

4   CONCLUSIONS

A dynamic simulation model of a roller coaster is first established by using **ADAMS** software, and the speed, acceleration and force of connecting parts and time-varying curves of the wheel–rail contact force are obtained. The maximum error between the simulated acceleration and the measured acceleration is only 4.5%, which shows that the simulation result is credible.

Secondly, the maximum peak velocity and peak acceleration decrease with the increase of wind speed. Through comparing and analyzing the acceleration, fork force and wheel–rail contact force of different roller coaster cars, it is found that the acceleration peak value of the first car is largest; the wheel–rail contact force on the front wheel of the first car is 72,317 N, which is more than six times that of the other wheel-groups, and the fork force decreases gradually from the front to the rear of the train. The first car, under full load conditions, should therefore be the focus of safety analyses and checks.

ACKNOWLEDGMENT

The authors acknowledge financial support from the National Key Research and Development Program of China under Grant No. 2017YFC0805704.

REFERENCES

Chen, L. & Guan, W. (2010). Modeling and simulation of roller coaster. *Chinese Journal of Engineering Machinery*, 400–403.
Liang, Z.-H., Shen, Y., Ding, K.-G., Chen, G. & Huang, W. (2006). Dynamic simulation modeling method of roller coaster. *Journal of System Simulation*, *2006*(S1), 280–282.
Teng, R., Wei, G., Gao, S. & Chen, L. (2010). Dynamic modeling and simulation of roller coaster. In *Proceedings of 2010 International Conference on Computer Application and System Modeling (ICCASM 2010), Taiyuan, China* (vol. 5, pp. V5–340–V5–342). Piscataway, NJ: Institute of Electrical and Electronics Engineers. doi:10.1109/ICCASM.2010.5620012.
Yin, M., Yang, J.L. & Yang, R.G. (2012). Three-dimensional modeling and simulation of roller coaster track based on SolidWorks. *Lifting and Transportation Machinery*, 65–67.
Zheng, J.R. (2002). *ADAMS introduction and improvement of virtual prototyping technology*. Beijing: Machinery Industry Press.

Automatic Control, Mechatronics and Industrial Engineering – He & Qing (Eds)
© 2019 Taylor & Francis Group, London, ISBN 978-1-138-60427-8

Structural response of welded plates to implementation of different weld passes for a specific weld throat

A.H. Eslampanah

R&D Sector, Machine Sazi Arak Company, Arak, Iran

ABSTRACT: Finite Element welding simulation is a useful method in which temperature distribution, distortion and residual stresses can be predicted. A three-dimensional model with moving heat source has represented the most accurate results with the capacity of monitoring effects of different parameters like welding speed. This paper evaluates how making a specific throat of a fillet welded joint by one or two passes distorts the joint. Thus, an FE model using a moving heat source has been employed. One pass and two passes were simulated to somehow that in the two passes model, each weld line produced half of the weld area. Since other parameters including voltage and current welding have been assumed to be constant, weld speed in the two passes simulation became double that of the one pass model. Interesting results showed that this two pass welding leads to more unfavorable results.

Keywords: Finite element method, welding, residual stress, distortion

1 INTRODUCTION

Arc welding relies on the intense local heating of a joint where a certain amount of base metal is melted and fused with additional metal from the welding electrode. This intense local heating causes severe thermal gradients in the welded components while the non-uniform expansion and contraction of the weld and surrounding base material produces residual stress and distortion. A thermal elastic-plastic finite element method can be effectively used to predict welding residual stress and distortion for welded joints made by plates. Many scientists have worked on describing welding deformation and residual stress of fillet welds using appropriate models and its simulation (Barsoum & Barsoum, 2009 and Eslampanah et al., 2011). Now, the double ellipsoidal model (Goldak & Akhlaghi, 2005) is widely accepted and applied in simulations. One common distortion due to the welding process is buckling and this often happens in weldments. When thin plates are welded, compressive stresses occur in areas away from the weld and can cause buckling distortion. Where buckling distortion occurs, the magnitude of distortion tends to be very large. Therefore, Finite Element Method (FEM) could also be a useful tool to predict welding buckling distortion in thin-walled aluminum T-joints (Asle Zaeem et al., 2007). Aalami-aleagha and Eslampanah (2012) investigated residual stress and deformation on an unsymmetrical T-joint using both numerical and experimental methods. In that study, a surface heat flux simulates the transferred heat energy from arc combined with a volumetric heat flux which simulates the thermal energy of molten metal droplets. In this investigation, according to the nature of Shielded Metal Arc Welding (SMAW) and experimental observations, an ellipsoidal shape in which dissimilar dimensions at the front and rear was considered for both volumetric and surface heat flux. Then the mechanical constraint effect on residual stress and deformation was investigated.

Since evaluation of welding residual stress and distortion is a time-consuming analysis, it is very important to find a way to reduce this time. Eliminating one or more parameters can effectively reduce the time. In the case of elimination of one dimension only, several investigations were performed. For a butt weld pipe, the difference between 2-D and 3-D models was evaluated (Kyoungsoo et al., 2013). The results from the 2-D analysis have shown a similar stress distribution to that of the cross section of the 3-D analysis. Comparing the results between 2-D and 3-D analysis demonstrated an approximate 10% difference in axial stress, and 5% difference in circumferential stress. Although 2-D models save analyzing time, they decrease the accuracy of results. In this 2-D model, it is not possible to investigate welding speed and mechanical constraints. Eslampanah et al. (2015) presented a new simplified heat source model in which the parameter of time was eliminated during welding. Since this model can be 3-D, there is the possibility to evaluate constraint effects on welded specimens (Eslampanah et al., 2015).

In this study, a thermal elastic-plastic finite element method is used to estimate inherent deformations and residual stresses for a T-fillet welded model. The aim is to investigate how using more weld passes influences residual stress and distortion of the specimen. Hence, to perform a final weld throat two different models were assumed: one which made the throat in one weld pass; and the other in two passes (in which each pass produced a half of the weld area). In order to reduce heat input for the second model, the weld speed was doubled and the other parameters remained constant. Since weld speed is a crucial parameter here, a 3-D model was employed. To validate the results, data from literature have been used (Aalami-aleagha et al., 2012).

2   MODEL DESCRIPTION

The geometric dimensions of the samples are shown in Figure 1a. The web and the flange were tack welded first. The chemical composition and mechanical properties of the samples are that of general structural steel, Swedish type SS 2132 (Barsoum & Lundback, 2009).

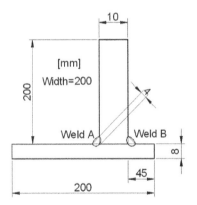

Figure 1a Specimen dimensions.

Figure 1a.   Specimen dimensions.

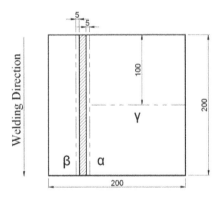

Figure 1b Path locations.

Figure 1b.   Path locations.

Table 1.   Welding parameters for one weld pass for each side and increased weld pass model.

	1 pass condition					2 pass condition			
	Time (s)	Current (A)	Volt (V)	Speed (mm/s)		Time (s)	Current (A)	Volt (V)	Speed (mm/s)
Weld A	59	150	21	2.7	Weld A1	29.5	150	21	5.4
					Weld A2	29.5	150	21	5.4
First Cooling	62	–	–	–		62	–	–	–
Weld B	59	150	21	2.7	Weld B1	29.5	150	21	5.4
					Weld B2	29.5	150	21	5.4
Final cooling	420	–	–	–		420	–	–	–

To evaluate the results, three paths have been considered according to Figure 1b. Shielded metal arc welding was used with electrode E6013 with a 3.2 mm diameter.

Table 1 shows the welding variables for the model that creates a 4 mm weld throat by one pass and represents the sequences for the model that makes a 4 mm throat by 2 weld passes.

3 FINITE ELEMENT SIMULATION

3.1 *Thermo-mechanical analysis*

The analysis procedure was based on uncoupled formulations for the calculation of welding residual stress, deformation and also temperature distribution.

A computational thermal elastic-plastic finite element procedure was developed to calculate welding deformations of fillet welded joints. The temperature-dependent thermal-physical properties such as specific heat, conductivity, density and temperature-dependent thermal-mechanical properties, such as Young's modulus, Poisson's ratio, thermal expansion coefficient and yield strength were used for thermal and mechanical analysis, respectively (Aalami-aleagha & Eslampanah, 2012). The mechanical properties of the weld metal are considered to be the same as the parent metal.

In this study, the thermo-mechanical behavior of the welded joint is simulated using uncoupled formulations: first, the temperature distribution is calculated using heat transfer analysis and then the results used for mechanical analysis subsequently. An 8-node linear transfer brick element (DC3D8) was employed for the 3-D thermal analysis and a 3-D 8-node bi-quadratic stress/displacement quadrilateral with reduced integration (C3D8R) was used in the structural analysis. A relatively dense mesh was adopted close to the weld line in order to represent the temperature and strain gradients accurately—the further away the weld line, the longer the size of the elements. The complete model contains approximately 11,750 elements. The smallest element size is $2 \times 2 \times 1.7$ mm approximately, based on mesh density analysis. As can be seen in Table 1, this simulation consists of four time steps, with each time step automatically divided into subsequences by the ABAQUS software. Boundary conditions were carried out according to Figure 2. Fixtures limited the angular deformation of the flange and web plates.

The model adopted the technique of element birth and death to simulate weld filler material. The elements in the weld beads which were born in the latter stages of the process have been subjected to change in the stiffness value (Aalami-aleagha & Eslampanah, 2012).

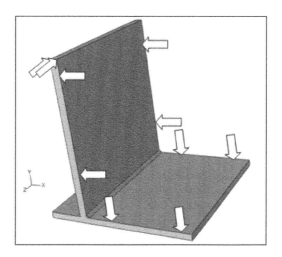

Figure 2. Constraints.

3.2 Heat transfer analysis

In order to consider the moving heat source in SMAW, a combination of surface and volumetric heat source was assumed (Barsoum, 2008). The rear and front of the electrode used in the welding process have dissimilar thermal situations due to the absorption of thermal energy by the molten metal in the front and ejection of energy from the rear due to the occurrence of the solidification process and latent heat. The surface heat flux for the front and rear of the welding zone has been modeled by the following equations (Aalami-aleagha & Eslampanah, 2012).

$$q_f(x',z',t) = \frac{3f_f Q_{Arc}}{ac_f p} e^{-3x'^2/a^2} e^{-3(z'-vt)^2/c_f^2} \tag{1}$$

$$q_r(x',z',t) = \frac{3f_r Q_{Arc}}{ac_r p} e^{-3x'^2/a^2} e^{-3(z'-vt)^2/c_r^2} \tag{2}$$

where x' and z' denote the local coordinates which are rotated 45° due to the working angle of the T-fillet weld in respect to the x and z axes.

The power of the welding arc (Q_{Arc}) is given in Equation 3 where Q_{Wire}, volumetric part of heat source, is attributed to the power of the molten metal droplets. In this equation η, I and U represent efficiency, arc current and arc voltage respectively:

$$Q_{Arc} = \eta\, IU - Q_{Wire} \tag{3}$$

In this analysis heat flux from the molten metal droplets is assumed to be 70% of the total heat, and the surface heat flux 30% of the total heat flux. The arc efficiency, η, is assumed to be 0.7. Using FORTRAN programming, the heat source was modeled by the ABAQUS user subroutine DFLUX for heat flux. The initial temperature was assumed to be 25°c. The user defined subroutine FILM was developed to simulate the combined convection and radiation (Eslampanah et al., 2015):

$$h = \begin{cases} 0.0668T & W/m^2 \quad 0 < T < 500°C \\ 0.231T - 82.1 & W/m^2 \quad T > 500°C \end{cases} \tag{4}$$

3.3 Mechanical model

The second stage of the analysis involves the use of temperature histories predicted by the thermal model as an input for a mechanical model. Due to the temperature gradient, the thermal strain is calculated by using the temperature-dependent coefficient of thermal expansion. Mechanical boundary conditions are applied in two different conditions, in the first to prevent rigid body motion, and in the other one to simulate fixtures, see Figure 2.

4 RESULTS AND DISCUSSION

Residual stresses and deflection have been obtained from the analysis. Figure 3 shows the vertical displacement of the flange along γ path located at the middle of weld line. Experiment results (Aalami-aleagha & Eslampanah, 2012) are related to one pass welding condition at each side. Obviously, the simulation is in good agreement with the experiment results, so the numerical results are considered acceptable. Here, the effect of applying two passes instead of one pass on deflection is observed. The curves show that displacement increases when the specimen is heated by a welding heat source with double the speed.

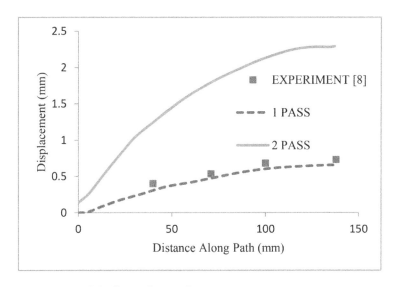

Figure 3. Displacement of the flange along path γ.

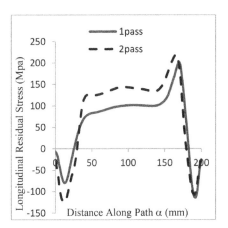

Figure 4a. Longitudinal residual stress along path α.

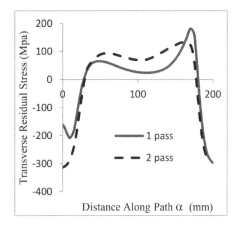

Figure 4b. Transverse residual stress along path α.

Surprisingly the displacement magnitude rises considerably (2.3 times greater at the edge) when each weld line is heated twice. Moreover, increasing the welding speed results in higher temperature gradients in the plates.

Figure 4 compares residual stresses between the two models along the alpha (α) path as defined in Figure 1. The graphs show that at the end of the path, after the peaks, both conditions resulted in the same magnitude and trends while, at the start, compressive residual stresses for the two pass condition reached considerably higher values. At the middle of the path, tensile residual stresses for the two pass model have worse results also, with a longitudinal residual stress and transverse residual stress of almost 40 MPa and 35 MPa higher than the one pass model, respectively. On the other hand, welding by two passes could mitigate maximum transverse residual stress from 180 MPa to 148 MPa as per Figure 4b, meanwhile it caused the peak longitudinal stress to raise around 13 MPa.

Figure 5 shows that applying two passes would exacerbate the situation in path β. A maximum stress was recorded of 135 MPa for the two pass model here.

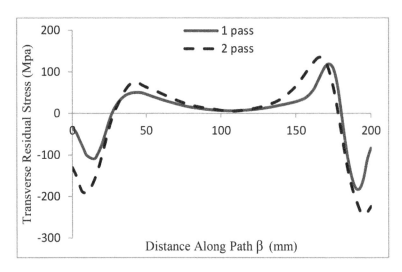

Figure 5. Transverse residual stress along path β.

5 CONCLUSION

Residual stress and distortion in fillet welded joints were evaluated for two different conditions for a weld throat of 4 mm: in one, the throat was produced by one pass and, in the other, in two passes. The speed in two pass model was considered twice that of one pass. Generally, the results demonstrated that the residual stresses and distortion of the two pass model were higher than the one pass model. Nevertheless, two pass conditions caused a considerably reduced maximum tensile transverse residual stress.

REFERENCES

Aalami-aleagha, M.E. & Eslampanah, A.H. (2012). Mechanical constraint effect on residual stress and distortion in T-fillet welds by three-dimensional finite element analysis. *Proceedings of the Institution of Mechanical Engineers, Part B:Journal of Engineering Manufacture, 227(2)*, 315–323.

Asle Zaeem, M., Nami, M.R. & Kadivar, M.H. (2007). Prediction of welding buckling distortion in a thin wall aluminum T joint. *Computational Materials Science, 38*(4), 588–594.

Barsoum, Z. (2008). Residual stress analysis and fatigue of multi-pass welded tubular structures. *Engineering Failure Analysis, 15*(7), 863–874.

Barsoum Z. & Barsoum, I. (2009). Residual stress effects on fatigue life of welded structures using LEFM. *Engineering Failure Analysis, 16*(1), 449–467.

Barsoum, Z. & Lundback, A. (2009). Simplified FE welding simulation of fillet welds 3D effects on the formation residual stresses. *Engineering Failure Analysis, 16*(7), 2281–2289.

Eslampanah A.H., Aalami-Aleagha M.E. & Feli S. (2011). Distortion and thermal analysis in T-fillet welded joint using FEM. *Proceedings of the 14th International Conference on Advances in Materials & Processing Technologies*, Istanbul, Turkey, July 13–16.

Eslampanah A.H., Aalami-aleagha M.E., Feli S. & Ghaderi M.R. (2015). 3-D numerical evaluation of residual stress and deformation due welding process using simplified heat source models. *Journal of Mechanical Science and Technology, 29*(1), 341–348.

Goldak, J.A. and Akhlaghi, M. (2005). *Computational Welding Mechanics*. New York, NY: Springer.

Kyoungsoo L., Maanwon K. & Sungho L. (2013). Three-dimensional finite element analysis for estimation of the weld residual stress in the dissimilar butt weld piping. *Journal of Mechanical Science and Technology 27*(1), 57–62.

Automatic Control, Mechatronics and Industrial Engineering – He & Qing (Eds)

Design optimization of metal inert gas welding parameters for recuperating an ultimate tensile strength of AISI SS304 using L9 Taguchi orthogonal array

S. Sharma
CSIR-Central Leather Research Institute, Regional Centre for Extension and Development, Leather Complex, Jalandhar, India

G. Singh
Department of Mechanical Engineering, Chandigarh University, Gharuan, Mohali, Punjab, India

N. Jayarambabu
Centre for Nanoscience and Technology, IST, JNTUH, India

G. Sahni
HOD & DGM (Design, Drawing & Development) Leader Valves Ltd., Jalandhar, Punjab, India
C Eng (Institution of Engineers India), QMS Lead Auditor, India

ABSTRACT: Metal Inert Gas (MIG) welding is one of the most broadly used welding processes in mass production, as well as in small scale industries. The main reason for this is that MIG welding can weld both ferrous and non-ferrous metals. The aim of the present study is to experimentally investigate the ultimate tensile strength during the MIG welding of AISI 304. The input parameters used for the experimental work were wire speed, gas flow rate and welding time. Mean effect plot and S/N ratio graphs have been used to optimize the welding parameters of MIG on AISI 304 stainless steel using the Taguchi method and ANOVA. It has been observed that wire speed has the largest effect on the tensile strength of AISI 304 steel MIG weldments. The optimum welding conditions for greater strength were wire speed (m/min), gas flow rate (liters/min) and welding time (sec).

Keywords: AISI 304, MIG, Ultimate tensile strength, Taguchi method & ANOVA

1   INTRODUCTION

Welding is a fabrication technology that is used to join materials, usually metals or thermoplastics, by causing fusion, which usually results by melting the base metal. Welding is considered to be the most economical and efficient way to join metals permanently. It is the only way of

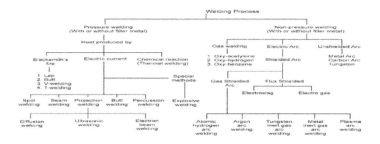

Figure 1.   Different types of welding techniques (Hooda et al., 2012; Kanwal & Jadoun, 2015).

joining two or more pieces of metal in order to make them act as a single piece. Welding ranks highly among industrial processes and involves more sciences and variables than those involved in any other industrial process. There are many ways to make a weld and many different types of welding. Some processes require sparks, while others do not even require extra heat. Welding can be done anywhere in all sorts of environments, outdoors or indoors, underwater and even in outer space. At present, welding is required in all kinds of fields, including agriculture, construction and vehicle manufacturing, and even on oil drilling rigs. Welding can be divided into various categories. Figure 1 shows the different types of welding techniques.

1.1 MIG welding

Metal Inert Gas (MIG) welding is a process in which a consumable wire electrode is fed into an arc and welds pool at a steady but adjustable rate, while a continuous envelope of inert gas flows out around the wire and shields the weld from contamination by the atmosphere. The MIG welding process has several advantages, which accounts for its popularity and increased use in the welding industries. The most common application of MIG welding is for automotive repair and it can be performed on any variety of vehicles. MIG welding can also be easily incorporated into robotics and can even be used to reinforce the surface of a worn out railway track. MIG welding will help to contribute to the creation of automobiles, the building of bridges and, even more importantly, is a more efficient way to weld.

1.1.2 Working principle

As shown in Figure 2, the electrode used in this process is in the form of a coil and it is continuously fed toward the work during the process. At the same time, inert gas (e.g. argon, helium) is passed around the electrode from the same torch. Inert gas, usually argon, helium or a suitable mixture of these, is used to prevent the atmosphere from contacting the molten metal and HAZ. When gas is supplied it gets ionized and an arc is initiated in between the electrode and the work piece. Heat is therefore produced. The electrode melts due to the heat and molten filler metal falls on to the heated joint. An arc may be produced between a continuously fed wire and the work. Continuous welding with a coiled wire helps the high metal depositions rate and high welding speed. The filler wire is generally connected to the positive polarity of the DC source forming one of the electrodes. The work piece is connected to the negative polarity. The power source could be a constant voltage DC power source, with electrode positive, and it yields a stable arc and smooth metal transfer with least spatter for the entire current range.

AISI 304 stainless steel has been selected as the work material to be welded by MIG welding. Two of the desirable mechanical properties of AISI 304 are high strength and high ductility. The chemical composition and properties of AISI 304 are shown in Tables 1 and 2.

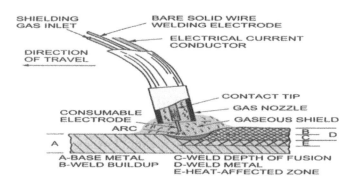

Figure 2. MIG welding process (Hooda et al., 2012; Kanwal & Jadoun, 2015).

Table 1. Chemical composition of AISI 304.

Elements	Percentage by weight (Maximum)
Carbon (C)	0.08
Manganese (Mn)	2.0
Silicon (Si)	0.75
Phosphorus (P)	0.045
Sulfur (S)	0.03
Chromium (Cr)	18–20
Nickel (Ni)	10.5
Nitrogen (N)	0.1

Table 2. Properties of AISI 304.

Property	Value
Ultimate tensile strength	505 MPa
Yield tensile strength	215 MPa
Brinell hardness	123
Modulus of elasticity	193–200 GPa
Poisson's ratio	0.29
Shear modulus	86 GPa
Density	8.0 g/cm^3
Thermal conductivity at 100°C	16.2 W/mK

2 LITERATURE SURVEY

The maximum yielding stress of two pieces of AISI 1040 medium carbon steel welded together with the help of metal inert gas welding. The parameters used to carry out this research were welding voltage, current, wire speed and gas flow rate. Response Surface Methodology (RSM) was applied to these welding parameters in order to find the optimized MIG welding parameters for maximum yielding stress (Hooda et al., 2012). Experiments were also performed on aluminum alloys of grades 6061 and 5083, which were welded with metal inert gas welding to find the hardness of the MIG welding joint using the Taguchi method. The parameters involved in this study were welding speed, welding current and welding voltage (Kanwal & Jadoun, 2015). Parametric optimization was determined when two dissimilar pieces of metal stainless steel (SS 304) and low carbon steel were welded together using MIG welding. The parameters on which the analysis was done were current, voltage and travel speed. ANOVA and the Taguchi technique were applied for the optimization using MINITAB 13 computer software (Chauhan & Jadoun, 2014). A research study was carried out to determine the optimized value of the MIG welding parameters needed to attain the maximum tensile strength of aluminum alloy AM- 40 (EN AW 5083). Welding current and welding voltage were the parameters used to judge this study (Saxena et al., 2015). The effect of MIG welding on AISI 1030's ultimate tensile strength was observed. The major parameters used in this study were welding speed and welding current. ANOVA and the Taguchi technique were practiced in order to generate an orthogonal array, so as to optimize the characteristics, such as signal to noise ratio (Patil & Waghmare, 2013). A research study has been presented on cold reduced low carbon steel IS 513 GR 'D' welded together using metal inert gas. Welding current, wire elongation and welding voltage were used as the welding parameters for the observation. The signal to noise ratio, ANOVA and orthogonal array of L_9 were used to optimize the input parameters and Taguchi methods were employed on weld width and weld height (Verma & Singh, 2014). The effect on the Ultimate Tensile Strength (UTS) of the ST-37 low alloy steel material was demonstrated, and the parameters being worked upon were current (A), voltage (V), gas flow rate and speed. The experiment was done by using a L_9 orthogonal array to find out the UTS. Confirmatory experiments were also performed to find out the optimal range sets of current, voltage speed and gas flow rate (Utkarsh et al., 2014). The effect of three process parameters on MIG welding was investigated, and the parameters chosen for the study were current, voltage and gas flow rate. Their effect on the tensile strength of welded joints on C20 carbon steel grade was studied. Taguchi's L_9 orthogonal array method was used for the optimization of these parameters. From the results we could conclude that voltage affected both the mean and variation of tensile strength, whereas the current had a significant effect on mean variation only (Ibrahim et al., 2012).

3 EXPERIMENTAL DETAILS: MATERIAL AND PROCESS

Work metal, AISI 304 stainless steel was welded together using AISI 308L filler material in MIG welding apparatus. This was used due to its comparable properties with AISI 304. An AISI 308L spool is shown in Figure 3.

Experiments were carried out based on Taguchi's L_9 and the ANOVA method. Three factors, i.e. wire speed, gas flow rate and welding time, were selected in order to conduct the experiment, with three levels each. The selected parameters and their levels are shown in Table 3.

During this study, tensile strength was measured after conducting a total of nine experiments. A total number of 18 pieces of AISI 304 were cut down from raw material. These pieces were then welded in pairs, as per the L_9 orthogonal array suggested by Taguchi's design. The welded joints are shown in Figure 4.

Figure 3. Welding spool.

Table 3. Electrical parameters and their levels to be used in MIG welding.

S.NO.	Input parameters	Levels corresponding to the selected parameters		
		1	2	3
1.	Wire speed (m/min.)	6	8	10
2.	Gas flow rate (liter /min.)	8	12	16
3.	Welding time (sec.)	5	7	9

Figure 4. Welded joints.

The tensile strength of the welded joints was tested on a universal testing machine. The tensile specimen used in this investigation is a standardized sample cross section, as shown in Figure 5.

Figure 5.   Specimens for testing in U.T.M.

Table 4.   Results for ultimate tensile strength.

Wire speed	Gas flow rate	Welding time	Tensile strength	S/N ratio
6	8	5	429.6	53.1709
6	12	7	375	52.6613
6	16	9	456.48	51.4806
8	8	7	463.88	53.1884
8	12	9	430.55	53.3281
8	16	5	510.18	52.6805
10	8	9	495.37	54.1545
10	12	5	503.7	53.8986
10	16	7	429.6	54.0434

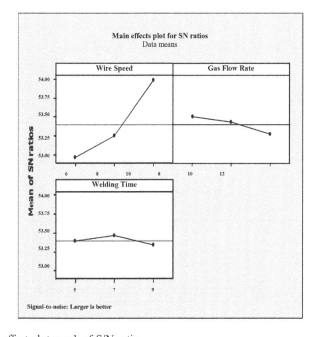

Figure 6.   Main effect plot graph of S/N ratios.

It could be noted that wire speed has the largest effect on the strength of AISI 304 steel at weld bead in MIG welding. The welding speed has the smallest outcome on the tensile strength. From the above main effect plot, it may be observed that the optimum conditions for hardness of weld parts are, A3, B1 and C2, i.e. wire speed (10 m/min), gas flow rate (8 liters/min) and welding time (7 sec).

The interaction plot, shown in Figure 7, which is the combination of all three parameters interacting with each other at different levels, gives the idea that wire speed, gas flow rate and welding time all had considerable effects on the weld bead hardness of AISI 304 stainless steel.

Table 5. ANOVA for S/N ratio of hardness.

Source	Sum of squares	Degree of freedoms (D.O.F.)	Adj. mean square	% Contribution
Wire speed	5018.13	2	2509.07	91.04
Gas flow rate	217.5	2	108.76	3.95
Welding time	64.80	2	32.40	1.18
Residual error	211.17	2	105.58	3.83
Total	5511.62	8	–	100

Table 6. Ranking of parameters for hardness.

Level	Wire speed	Gas flow rate	Welding time
1	52.95	53.50	53.39
2	53.24	53.42	53.46
3	53.98	53.27	53.33
Delta	1.03	0.23	0.13
Rank	1	2	3

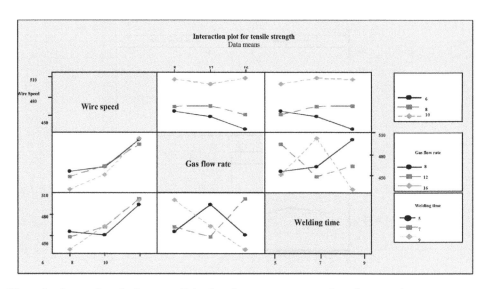

Figure 7. Interaction plot between all the three input parameters and tensile strength.

4 CONCLUSIONS

It can be concluded from the experimental investigation that, for the optimum ultimate tensile strength of the weld zone during the metal inert gas welding of AISI 304 stainless steel, wire speed and gas flow rate could have a significant contribution. The AISI 304 steel optimum welding conditions for higher tensile strength were, welding speed (10 m/min), gas flow rate (8 liters/min) and welding time (7 sec).

REFERENCES

Chauhan, V. & Jadoun, R.S. (2014). Parametric optimization of MIG welding for stainless steel (SS-304) and low carbon steel using Taguchi designmethod. *International Journal of Advanced Technology & Engineering Research*, 6(2), 2662–2666.

Hooda, A., Dhingra, A. & Sharma, S. (2012). Optimization of MIG welding process parameters topredict maximum yield strength in AISI 1040. *International Journal of Mechanical Engineering and Robotics Research*, 1(3), 203–213.

Ibrahim, A.I., Mohamat, S.A., Amir, A. & Ghalib, A. (2012). The effect of Gas Metal Arc Welding (GMAW) processes on different welding parameters. *Procedia Engineering*, 41, 1502–1506.

Kanwal, V. & Jadoun, R.S. (2015). Optimization of MIG welding parameters for hardness of aluminium alloys using Taguchi method. *SSRG International Journal of Mechanical Engineering*, 2(6), 53–56.

Park, H.J., Kim, D.C., Kang, M.J. & Rhee, S. (2008). Optimization of the wire feed rate during pulse MIG welding of Al sheets. *Journal of Achievements in Materials and Manufacturing Engineering*, 27(1), 55–58.

Patil, S.R. & Waghmare, C.A. (2013). Optimization of MIG welding parameters for improving strength of welded joints. *International Journal of Advanced Engineering Research and Studies*, 2(4), 14–16.

Saxena, V., Parvez, M. & Srivastava, S. (2015). Optimization of MIG welding parameters on tensile strength of aluminum alloy by Taguchi approach. *International Journal of Engineering Sciences &Research Technology*, 4(6), 451–457.

Utkarsh, S., Neel, P., Mahajan, M.T., Jignesh, P. & Prajapati, R.B. (2014). Experimental investigation of MIG welding for ST-37 using design of experiment. *International Journal of Advance Research in Science and Engineering*, 4(5), 1–4.

Verma, S. & Singh, R. (2014). Optimization of process parameters of metal inert gas welding by Taguchi method on CRC steel IS 513 GR D. *International Journal of Advance Research in Science and Engineering*, 3(9), 187–197.

Automatic Control, Mechatronics and Industrial Engineering – He & Qing (Eds)
© 2019 Taylor & Francis Group, London, ISBN 978-1-138-60427-8

The research status and prospect of cryogenic treatment on aluminum alloys

F.W. Jin & X.Y. Ji
School of Mechanical Engineering, Research Institute of Bijie Circular Economy,
Guizhou University of Engineering Science, Bijie, China

ABSTRACT: The deep cryogenic treatment of 2XXX series, 3XXX series, 6XXX series and 7XXX series of deformed aluminum alloy, and their castings, are analyzed and studied, together with its cryogenic treatment process and mechanism. The agreed procedure and the applicable mechanism of the deep cryogenic treatment of aluminum alloys ware discussed. Future research on cryogenic assembly and machining, as well as the theoretical results on the mechanism of existing cryogenic treatments is needed.

Keywords: Aluminum alloys, deep cryogenic treatment, properties, mechanism

1 INTRODUCTION

Research on the cryogenic treatment of materials have become a hot topic in recent years (Chetan & Reo, 2017; Sanjeev et al., 2017), in which the studies on cryogenic treatment of aluminum alloy are more common. At the end of the century, Tang and Huang (1998) was the first to report the results of cyclic treatment (i.e. cryogenic treatment) simultaneously improving the strength and plasticity of aluminum alloy LD10. Subsequently, Jin et al. (2000) reported that cryogenic treatment can improve the volume stability of an aluminum alloy (ZL108) piston, as well as its strength, elongation and hardness, and observed that the point-like eutectic (Si) in the matrix structure was more uniform. Chen and Li (2000a) and Chen and Li (2000b) reported the results of the cryogenic treatment of aluminum and aluminum alloys, in which they concluded that the most deformed aluminum alloys had an increase in tensile strength, but decreased plasticity. This was obtained by the cryogenic treatment of deformed aluminum alloys 1–8 series, and proposed the concept of grain rotation. Since then, research on the cryogenic treatment of aluminum alloys has attracted much attention. Research has mainly involved the improvement of microstructure and performance, the enhancement of dimensional stability and wear resistance, the elimination of residual stress, and the study of a corresponding mechanism. In particular, the influence of cryogenic treatment on the microstructure and properties of materials has mostly been studied. Aluminum alloys are divided into two categories: deformation and casting. Deformation aluminum alloys contain nine series and casting aluminum alloys are subdivided into six categories. Adding the state of every kind of aluminum alloy, the material of aluminum alloy can be said to be various. Most of the research on cryogenic treatment of aluminum alloys focuses on improving the structure, performance, process optimization and mechanism analysis of different materials. Based on the aluminum alloy material, the scope of the material of aluminum alloy refer to cryogenic treatment was summarized. As the deformation and cast aluminum alloy to the main object, the corresponding research achievements was summarized, especially, the results and corresponding mechanism were analyzed. By summarizing the common law of cryogenic treatment of aluminum alloys, this paper can provide a future reference for topic and material selection of cryogenic treatment research on aluminum alloys.

2 STUDY ON CRYOGENIC TREATMENT OF DEFORMED ALUMINUM ALLOYS

Deformable aluminum alloy is widely used in transportation, aerospace and other fields. At present, 2XXX, 3XXX, 6XXX and 7XXX series deformable aluminum alloys are most widely researched and used for cryogenic treatment. The effect and mechanism of cryogenic treatment on them are described below.

2.1 Cryogenic treatment on 2XXX series deformed aluminum alloys

The 2014 aluminum alloy has high strength and hardness, good plasticity and toughness as well as good forging, heat resistance and weldability, so it has been well applied in the aerospace field. With the rapid development of high-speed railway in recent years, 2014 aluminum alloy is often used in various parts of this industry. However, the service life of the parts is reduced because of its low corrosion resistance, which limits the range of the engineering application to varying degrees. To improve these shortcomings, Xu et al. (2014) studied 12 kinds of composite processes including cryogenic treatment; they discovered two important processes. One of them was that a solid solution treatment for 80 mins at 505°C with water cooling, plus a cryogenic treatment holding for 20h, followed by an aging treatment for 12h at 160°C. The second was a solid solution treatment for 80 mins at 505°C, water cooling plus an aging treatment for 12h at 120°C, followed by a regression treatment for 15 mins at 200°C, water cooling and an aging treatment for 12h at 160°C. The first process dealt with cryogenic treatment after water cooling, and when compared to the second process without cryogenic treatment, the hardness and corrosion resistance of the aluminum alloy was significantly improved. In addition to high strength and good heat resistance, 2024 aluminum alloy also has the advantages of low density. Therefore, it is often used for the structural parts of aircraft, such as covering, skeleton, rib beam and partition frame. The study by Wang et al. (2012) showed that 2024 aluminum alloy after cryogenic treatment not only reduced the residual stress of quenching and promoted the second phase precipitation, but also significantly improved fatigue and corrosion resistance performance.

2.2 Cryogenic treatment on 3XXX series deformed aluminum alloy

The cryogenic treatment of 3XXX series deformed aluminum alloy mainly focused on 3104 aluminum alloy. It is usually used for making aluminum cans, light bulb covers, roof panels and colored aluminum sheets because of its high strength, good formability and corrosion resistance. Wen and Zhang (2016) cryogenically treated the 3104 aluminum alloy sample before homogenization, and by this, the lamellar Al_6Mn phase appeared in the structure (which had the effect of inhibiting grain coarsening). Finally, the structure became more uniform and refine. Research by Fan et al. (2015) indicated that the strength and plasticity of 3104 aluminum alloy was significantly improved by cryogenic treatment, and it was believed that the best composite process of the alloy was a homogeneous treatment for 10h after cryogenic treatment for 6h. Zhang et al. (2014a) performed a cryogenic treatment on recycled 3104 aluminum alloy before homogenization, and this obtained the smallest Al_6Mn phase size; approximately 3.25 ± 1.18 μm, which is a quantitative result of cryogenic treatment affecting the precipitated phase size. Some studies (Zhang et al., 2014b) also showed that the tensile strength, yield strength and elongation of 3014 aluminum alloy was increased by 29%, 41% and 11% respectively, after the aluminum alloy was homogenized after cryogenic treatment, and the strength and toughness were also improved.

2.3 Cryogenic treatment of 6XXX series deformed aluminum alloys

The cryogenic treatment of 6XXX deformation aluminum alloys mainly involves 6005, 6061, 6005A and 6082 aluminum alloy. Among them, more uniform and fine precipitated strengthening phases were generated at the grain boundary in the 6005 aluminum alloy when its samples were subjected to cryogenic treatment after water quenching (Wan et al., 2013), which improved the alloy's hardness and strength, but reduced its plasticity and toughness. However, the needle-like β phases (Mg$_2$Si) were precipitated uniformly and diffusely in the intracrystalline, and the thick stick shaped β(Mg$_2$Si) phases were precipitated discontinuously in the grain boundary in 6061 aluminum alloy subjected to cryogenic treatment (Wang et al., 2016). It was also found that tensile strength, yield strength and elongation were improved by the treatment. Some researchers also discovered (Ni et al., 2018) that the elongation of 6061 aluminum alloy can be significantly improved by cryogenic treatment, but the abrasion resistance of the alloy decreased. After cryogenic treatment on the cold metal transition welding joint of the 6061 aluminum alloy (Liu et al., 2017), the tensile strength, elongation and microhardness of the joint were improved, which was consistent with literature (Wang et al., 2016). However, it is obvious that 6005 and 6061 aluminum alloys are contradictory in terms of the effect of cryogenic treatment on plasticity and toughness. More accurate understanding requires further research and exploration. The welding joints of 6065A and 6082 aluminum alloys were usually treated cryogenically because this gave good weld performance. The 6065A aluminum alloy welding joint that underwent three cycles of cryogenic treatment (Ji et al., 2016) appeared to have a strengthened phase β(Mg$_2$Si), and a series of fine precipitated phases, such as dispersion distribution Si and α-AlFeSi, which caused the joint to significantly increase in tensile strength and slightly decreased elongation. When the welding joint of 6082 aluminum alloy was treated at 540°C for a 50 mins solid solution water cooling and −196°C for a 20 h cryogenic treatment, the strength and plasticity of the joint had improved (Ge et al., 2015). It can be seen that both 6065A and 6082 aluminum alloy welding joints can dramatically improve their strength after cryogenic treatment, but whether the plasticity is improved is closely related to the specific cryogenic process.

2.4 *Cryogenic treatment on 7XXX deformed aluminum alloys*

The deformable aluminum alloy of 7XXX series is mainly divided into two types: one is a welded component material (Al-Zn-Mg series), the other is a high-strength alloy material (Al-Zn-Mg-Cu series). The application of the 7XXX aluminum alloys in the aerospace industry is very extensive. The crystalline grains of Al-Zn0.078-Mg0.018-Cu0.015 alloy was refined after cryogenic treatment (Li et al., 2015); this improved its mechanical properties. After aging treatment, the cryogenic treatment on the alloy induced a second phase precipitation discontinuously distributed along the grain boundary at the Guinier Preston zone in the aluminum matrix, which promoted the performance of the alloy. After solid solution treatment, 7A04 aluminum alloy was performed by cryogenic treatment, then a large number of precipitated phases were produced in and out of the grains whose distributions were uniform (Chen et al., 2012). Through such process, the hardness and tensile strength increased significantly, but plasticity and toughness improved slightly. After cryogenic treatment, the strength and corrosion resistance of the extruded 7A04 aluminum alloy, that underwent the solid solution aging treatment, was improved in different degrees (Xu and Bu, 2015), and the alloy's corrosion degree was related to the cryogenic time. After cryogenic treatment, the hardness, elastic modulus, hard elastic ratio (ratio of hardness and elastic modulus) and elastic recovery coefficient of LC9 (new brand 7A09) aluminum alloy were observably increased, but the friction factor was reduced, and its anti-indentation deformation ability was markedly improved (Liu et al., 2013). Cryogenic treatment can effectively enhance the tensile strength and microhardness of the 7A52 aluminum alloy welding joint (Zhang et al., 2014c), which was more obvious in the weld area.

3 CRYOGENIC TREATMENT OF CAST ALUMINUM ALLOY

Cast aluminum alloy has been widely used in the civil, aviation, marine and vehicle industry due to its excellent casting performance, corrosion resistance and mechanical properties. The research results and applications of cryogenic treatment have attracted more and more attention. The residual stress of the ZL205A transition ring decreased by nearly 20%, and its tensile strength and yield strength were promoted when it was treated by a cooling and heating cycle (i.e. cryogenic treatment) (Bi et al., 2016). The tensile strength of A319 alloy at room temperature improved slightly after cryogenic treatment, while the elongation increased significantly (Yang et al., 2014). However, if the alloy was treated with a solid solution plus an aging treatment before cryogenic treatment, both its tensile strength and elongation were markedly increased (Feng et al., 2014). Due to its advantages of low density, light weight, good thermal conductivity, low thermal expansion coefficient, excellent casting performance, volume stability, good abrasion resistance and corrosion resistance, hypereutectic high-silicon aluminum alloy was commonly used as the piston material for automobile engines. The hardness, tensile strength and wear resistance of the high-silicon aluminum alloy that was solidified rapidly, can be improved by cryogenic treatment. The improving effect was related to the order of cryogenic cooling and the number of cryogenic applications (Shi et al., 2012). The research results showed that solid solution treatment plus cryogenic treatment (for 24 h and on 2 occasions), together with an artificial aging treatment, was the best. The hardness, wear resistance and tensile strength of the hot-deposited and hot-extruded Al-Si alloy were significantly raised through deep cryogenic treatment (Wang et al., 2012), whose optimal treatment process was solid solution treatment plus cryogenic treatment and an aging treatment, which was consistent with the results in Shi et al. (2012). Cryogenic treatment can promote the precipitation of the second phase in hypereutectic Al-Si alloy (Tian et al., 2017) and refine the grain. Meanwhile, it can significantly increase the hardness, tensile strength and elongation of the alloy. The effects of improving performance were related to the cryogenic time.

4 MECHANISM OF CRYOGENIC TREATMENT ON ALUMINUM ALLOY

Tang and Huang (1998) proposed that circulation processing (i.e. cryogenic treatment) produces a large number of helical dislocations, dislocation loops and even dislocation cells. At the same time, the original grains are divided into many sub-crystals, which mutually interact and entangle, then markedly enhance structure stability; this also improves the performance of the mechanics of the materials. Meanwhile the residual stress of the alloy is reduced and the distribution of the stress becomes relatively uniform. The formation mechanism of dislocation and the precipitation law of the second phase has not been studied thoroughly by Tang and Huang (1998). Chen and Li (2000a) proposed the volume contraction and grain rotation effects in cryogenic treatment. The volume contraction effect is based on the law of thermal expansion and cold contraction of solids, which is actually the common law of all cryogenic materials. Volume shrinkage can lead to the following results: (1) the partial defects such as vacancies and micropores within the materials can be closed by the contraction; (2) the strain causes the lattice to contract, resulting in the atomic spacing to decrease, and the dislocation slip resistance to increase; (3) the contraction causes considerable internal stress in the material and induces a large number of dislocations; and (4) as a result, the internal energy is increased and sedimentary phases are precipitated. All these changes are conducive to improving the strength of materials. The grain rotation effect was proposed by Chen and Li (2000a) to solve the problem that the strength of aluminum alloy cannot always be improved by the process of cryogenic treatment. The so-called grain rotation refers to the fact that a large number of dislocations and sub-crystals are generated in the material due to volume contraction during the cryogenic treatment process. Recovery recrystallization is produced during the cryogenic recovery process, which causes the grains of aluminum alloy to rotate and form recrystallization texture in a selective orientation. When the grain orientation is conducive to preventing dislocation slip, the strength of the material can be promoted. It can be seen that the rotational effect of grain proposed is actually based on volume contraction. In addition, as mentioned by Chen and Li (2000b), the problem of grain rotation caused by cryogenic treatment is relatively complicated, and its preferred orientation depends not only on the alloy composition of the material, but also on the manufacturing process of the material. Some aluminum alloys have no improvement in their performance after cryogenic treatment, and there are other interpretation mechanisms besides the grain rotational effect (Jin et al., 2005). Based on the research of Tang and Huang (1998) and Chen and Li (2000a), the mechanism of the cryogenic treatment of aluminum alloys has been studied thoroughly by the authors of this paper. Volume shrinkage was taken as volume strain, and a quantitative relationship between stress and temperature was established by using Hooke's law (Jin et al., 2005). The quantitative relationship can be used to estimate the amount of pressure added to aluminum alloys during cryogenic treatment. The formation mechanism of dislocation and the precipitation mechanism of the strengthening phase of cryogenic treatments on aluminum alloys can be described qualitatively and quantitatively by using the defect theory of crystal. Aluminum alloys acquire compressive stress in the process of cryogenic treatment, which induces a large number of dislocations. At the same time, aluminum alloys under low temperatures also receive a lot of a saturation point defects (e.g. vacancy); under the effect of penetration, the dislocations proliferate. The dislocations on the one hand, through their own intertwining and pinning, improve the toughness of the alloy, but on the other hand, they interact with the solute atoms to result in sediment precipitation; thus, the dispersion is strengthened and improves the performance of the alloy.

5 CONCLUSIONS AND PROSPECTS

1. Both deformed and cast aluminum alloys can improve mechanical properties, fine structure and enhance corrosion resistance through cryogenic treatment.
2. The process of aluminum alloy cryogenic treatment has been basically understood. That is, the process effect of solid solution treatment plus cryogenic treatment, plus aging treatment, is better. Specific cryogenic parameters have been adjusted due to different materials.
3. The mechanism of the cryogenic treatment on aluminum alloys can be explained by the viewpoints put forward by Guangping Tang, Ding Chen and Fangwei Jin.
4. Further research work needs to be carried out in the future to make breakthroughs in cryogenic assembly and cryogenic machining, as well as design experimental research to confirm the theoretical results of mechanisms on cryogenic treatment.

ACKNOWLEDGMENTS

The authors acknowledge Guizhou University of Engineering Science's high-level talent scientific research project (Grant: G2018008).

REFERENCES

Bi, H.J., Hou, Z.Y. & Kang, T.T. (2016). Effects of thermal-cold cycling treatment on residual stress and mechanical properties of ZL205A. *Hot Working Technology*, *45*(6), 145–147.
Chen, D. & Li, W.X. (2000a). Cryogenic treatment of Al and Al alloys. *The Chinese Journal of Nonferrous Metals*, *6*, 891–895.
Chen, D. & Li, W.X. (2000b). Grain preferred orientation of Al and Al alloys through cryogenic treatment. *Journal of Central South University of Technology*, *6*, 544–547.
Chen, W.S., Li, C.Y. & Liu, L. et al. (2012). The study of cryogenic process in 7A04 aluminum alloy. *Modern Machinery*, *6*, 80–82, 90.
Chetan, G.S. & Reo, P.V. (2017). Performance evaluation of deep cryogenic processed carbide inserts during dry turning of Nimonic 90 aerospace grade alloy. *Tribology international*, *115*, 397–408.
Fan, Y.Q., Song, X.L. & Zhang, W.D. et al. (2015). Influence of heat treatment on microstructure and mechanical properties of 3104 aluminum alloy. *Foundry Technology*, *36*(4), 894–895.
Feng, G.F., Zhang, W.D. & Bai, P.K. et al. (2014). Effects of compound heat treatment on the microstructure and tensile strength of A319 Alloy. *Foundry*, *63*(3), 268–270.

Ge, P., Wang, M.J. & Wang, J. et al. (2015). Effect of cryogenic treatment on properties of 6082 aluminum alloy welded joint. *Heat Treatment of Metals*, *40*(4), 61–64.

Ji, K., Ni, G. & Li, X.B. et al. (2016). Effect of cryogenic treatment on microstructures and mechanical properties of welded joint of 6005A aluminum alloy. *Heat Treatment of Metals*, *41*(12), 127–130.

Jin, F.W. (2005). Study on improving performance of Al-Si alloy by deep cryogenic treatment and its mechanism. *Journal of Yunnan Agricultural University*, *5*, 729–733.

Jin, F.W., Huang, Y.Z. & Zhang, Y.H. (2000). Effect of deep cryogenic treatment on volume stability of cast aluminum alloy pistons *New Technology & New Process*, *5*, 24.

Li, C.M., Cheng, N.P. & Chen, Z.Q. et al. (2015). Deep-cryogenic-treatment-induced phase transformation in the Al-Zn-Mg-Cu alloy. *International Journal of Minerals, Metallurgy and Materials, 22*(1), 68–77.

Liu, H., Zhang, L.Y. & Jiang, Y.H. (2013). Effects of cryogenic treatment on mechanical properties and friction coefficient of LC9 aluminum alloy. *Heat Treatment of Metals*, *38*(6), 86–88.

Liu, Q., Zhang, N.N. & Zhang, Y.X. (2017). Effect of cryogenic treatment on microstructure and mechanical properties of 6061 aluminum alloy CMT welding joint. *Hot Working Technology*, *46*(19), 244–246.

Ni, H.J., Gu, T. & Lyu, Y. et al. (2018). Effects of cryogenic treatment on microstructure and property of 6061 aluminum alloy. *Hot Working Technology*, *47*(6), 239–241.

Sanjeev, K., Ajay, B. & Rupinder, S., et al. (2017). Effect of cryogenically treated copper-tungsten electrode on tool wear rate during electro-discharge machining of Ti-5Al-2.5Sn alloy. *WEAR, 386–387*, 223–229.

Shi, H F., Zhao, Z.W. & Wang, Y.L. et al. (2012). Effect of cryogenic treatment order and frequency on rapidly solidified high-silicon aluminum alloy. *Ordnance Material Science and Engineering*, *35*(2), 47–49.

Tang, G.P. & Huang, W.R. (1998). Effect of cyclic treatment on mechanical properties and microstructure of Al-alloys. *Heat Treatment of Metals*, *6*, 36–38, 45.

Tian, Z.Q., Wei, Z.X. & Wei, W. et al. (2017). Microstructure and mechanical properties of Al-Si alloy by deep cryogenic treatment. *Heat Treatment of Metals, 42*(2), 54–58.

Wan, P.H., Han, Y.J. & Zhao, H. et al. (2013). Research on microstructure and properties of 6005 Al alloy after cryogenic process. *Hot Working Technology*, *42*(14), 160–162.

Wang, Y.J., Sun, W. & Li, P. et al. (2012). Effect of cryogenic treatment on microstructure and properties of 2024 aluminum alloy. *Light Alloy Fabrication Technology*, *40*(9), 56–59.

Wang, M.H.,Yi, Y.P. & Huang, S.Q. (2016). Effect of thermal-cold cycling treatment on microstructure and tensile properties of 6061 aluminum alloy. *Transactions of Materials and Heat Treatment*, *37*(10), 63–67.

Wen, C.Z. & Zhang, K. (2016). Microstructure changes of 3104 aluminum alloy in composite heat treatment process. *Foundry Technology*, *37*(10), 2089–2091.

Wang, Y.L., Shi, H.F. & Fu, L.P. et al. (2012). Effect of cryogenic treatment time on microstructure and properties of rapidly solidified hypereutectic Al-Si-Fe alloy. *Heat Treatment of Metals*, *37*(3), 109–111.

Xu, J.Y. & Bu, J.R. (2015). Study on effect of cryogenic treatment on microstructure and corrosion resistance of 7A04 Al alloy. *Hot Working Technology*, *44*(8), 212–214.

Xu, L.D., Li, Y.F. & Ge, Y. et al. (2014). Study on the structure and properties of 2014 aluminum alloy for heat treatment. *Silicon Valley*, *7*(11), 27–28.

Yang, J., Zhang, W.D. & Zheng, K.D. et al. (2014). Effect of deep cryogenic treatment time on microstructure and tensile strength of A319 alloy. *Transactions of Materials and Heat Treatment*, *35*(S1), 181–184.

Zhang, W.D., Bai, P.K. & Yang, J. (2014a). Effect of homogenization and cryogenic treatment on microstructure of recycled 3104 aluminum alloy. *Transactions of Materials and Heat Treatment*, *35*(6), 66–70.

Zhang, W.D., Bai, P.K. & Yang, J. et al. (2014b). Tensile behavior of 3104 aluminum alloy processed by homogenization and cryogenic treatment. *Transactions of Nonferrous Metals Society of China, 24*(8), 2453–2458.

Zhang, Y.Y., Sun, X.J. & Zhu, X.B. (2014c). Effect of cryogenic treatment properties of 7A52 aluminum alloy welded joint. *Electric Welding Machine, 44*(10), 170–173.

Automatic Control, Mechatronics and Industrial Engineering – He & Qing (Eds)
© 2019 Taylor & Francis Group, London, ISBN 978-1-138-60427-8

Static stability analysis of space camera loaded cylinder using carbon-fiber-reinforced polymer

Y. Li
Northwestern Polytechnical University, Xi'an, China
Xi'an Institute of Optics and Precision Mechanics, CAS, Xi'an, China

W.J. Ge
Northwestern Polytechnical University, Xi'an, China

B. Hu
Xi'an Institute of Optics and Precision Mechanics, CAS, Xi'an, China

ABSTRACT: We describe research carried out on a loaded cylinder for future generations of space-based optical cameras. This work requires the application of Carbon-Fiber-Reinforced Polymer (CFRP). In order to meet the requirements of light weight and stability, a higher-modulus carbon-fiber-composite material may be used. In this paper, an equivalent model of the loaded cylinder is established. The static deformation of the loaded cylinder using M40J/epoxy resin and M55J/Cyanate Ester (CE) composites is analyzed using the finite element method. The analysis results show that the static deformation stability is improved by 32.8% using the M55J/CE composite material. The loaded cylinder manufactured with M55J/CE was tested and the results proved that the equivalent model and its finite element analysis are reasonable and reliable, laying the foundation for subsequent engineering application.

Keywords: Carbon-fiber-reinforced polymer (CFRP), space optical camera, loaded cylinder

1 INTRODUCTION

Due to its high modulus, light weight, high strength and low coefficient of thermal expansion, high-modulus carbon fiber has the advantages of high stiffness, high structural dimensional stability, good fatigue resistance and good vibration resistance, and is an ideal reinforcement for structural spacecraft composites (Zhou, 2017a, 2017b; Romeo, 1995; Romeo & Martin, 2007).

In recent years, space-based optical cameras have developed in the direction of high resolution, long focal length, large aperture, large field of view, large volume, and lighter weight. To reduce the weight of space cameras, lightweight structural materials with excellent properties must be used. Carbon-fiber composite is one of the best choices (Koyanagi et al., 2010; Utsunomiya et al., 2012).

Because the supporting structure between the optical elements involves great precision and requires high rigidity, the linear expansion coefficient of the construction material must be extremely constrained, and the space application requires the density to be as low as possible. Therefore, high-modulus Carbon-Fiber-Reinforced Polymer (CFRP) is gradually gaining wider application (Zou et al., 2017; Chen et al., 1998, 2000).

The development of a high-stability loaded cylinder is one of the important factors that determine the imaging accuracy of the camera system. The loaded cylinder is the main loaded structure between the main and secondary mirror elements. It must be of low density, high rigidity and good stability. A CFRP loaded cylinder made of M40J/epoxy resin composite has been used in satellite cameras. However, the influence of gravity should also be considered when the loaded cylinder is used for detection on the ground. If the rigidity

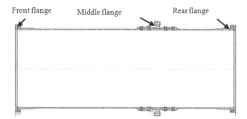

Figure 1.   Carbon-fiber-composite loaded cylinder.

of the loaded cylinder is insufficient, the optical axis of the main mirror and the secondary mirror will be tilted, which will seriously affect the imaging quality of the camera. Therefore, the structural design of a CFRP loaded cylinder must meet the requirements of the static load and space mechanics environmental test at the same time, and have sufficient margins of strength and stiffness.

This paper studies and compares two loaded cylinders of high-modulus carbon-fiber composites—M40J/epoxy resin and M55J/Cyanate Ester (CE) – to provide analysis methods and test data for subsequent application of M55J/CE composites.

The loaded cylinder of an existing coaxial camera (see Figure 1) is made of M40J/epoxy composites, mainly composed of carbon-fiber composites, but with a front flange, middle flange and rear flange made of titanium alloy. The middle flange is connected to the camera case, the front flange is connected to the main mirror part (which weighs 50 kg), and the rear flange is connected to the secondary mirror part (which weighs 13 kg). In ground-based detection, because of the effect of the earth's gravity, it was found that during the adjustment of the optical machine, even if the amount of tilt of the secondary mirror with respect to the primary mirror is only 0.1″, there is a coma on the image plane exceeding the requirements of the Modulation Transfer Function (MTF) of the optical system. An MTF that affects the entire optical system fails to meet the technical specifications.

2   EQUIVALENT MODEL ESTABLISHMENT

As illustrated in Figure 3, the loaded cylinder is simplified to a circular-section cantilever beam model. Using similar principles, it is necessary to determine the simulation load form and weight characteristics of a loaded cylinder sample.

The load-bearing deformation test results of a loaded cylinder of M40J/epoxy composite are measured when a 13 kg secondary mirror component load is mounted on the left end of the middle flange at 456.24 mm of the total length of the 694 mm loaded cylinder, and the right end of the middle flange is at 237.76 mm (see Figure 2). When the 50 kg main mirror part is mounted to simulate load, then the detected tilt of the primary and secondary optical axes is approximately 1″.

In order to simplify the model for follow-up engineering test verification, the equivalent conversion work needs to be performed.

According to the micro-deformation category, where the static loading deformation of the loaded cylinder is on a very small scale, the mechanical form of the 'double cantilever on both sides of the middle fixing branch' of the loaded cylinder of the M40J epoxy material can be equivalent to the theory of small deformation mechanics of the material. The mechanical form of the 'single cantilever on one side of the fixed side' is shown in Figure 2.

According to the linear equivalence principle, under small deflection the equivalent mass m_e is known from the torque equivalence relationship as:

$$m_e = \frac{13 \times 456.24 + 50 \times 237.76}{694} = 25.7 \,\text{kg} \tag{1}$$

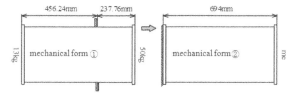

Figure 2. Loaded cylinder mechanical form equivalent diagram.

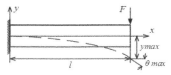

Figure 3. Single cantilever beam model.

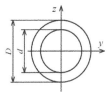

Figure 4. The moment of inertia of a circular-section beam.

In determining the weight of the simulated load of the loaded cylinder sample, the mechanical form ② of the analog loaded cylinder and the loaded cylinder sample can be simplified as a circular-section cantilever beam model. Thus:

$$\text{beam-end angle}: \theta_{max} = \frac{Fl^2}{2EI}; \text{maximum deflection}: y_{max} = \frac{Fl^3}{3EI}$$

where F is the beam-end force, l is the length of the cantilever beam, E is the elastic modulus of the material and I is the moment of inertia. The factors for the moment of inertia of a circular-section beam are shown in Figure 4, and it is calculated as follows:

$$I_y = I_z = \frac{\pi D^4}{64}(1 - \alpha^4)$$

Applying the principle of equivalence of beam-end angles, θ_{max}, and using the known parameters for the two types of loaded cylinder, which are:

$$F_1 = 25.7g, L_1 = 694, E_1 = 72, D_1 = 596, d_1 = 588;$$
$$F_2 = m_{e2}g, L_2 = 556.2, E_2 = 108, D_2 = 140, d_2 = 137$$

gives rise to the following equation:

$$\frac{25.7 \times 694^2}{72 \times 596^4 \times \left[1 - \left(\dfrac{588}{596}\right)^4\right]} = \frac{m_{e2} \times 556.2^2}{108 \times 140^4 \times \left[1 - \left(\dfrac{137}{140}\right)^4\right]} \tag{2}$$

This, when solved, indicates that $m_{e2} = 0.29$ kg. In addition, given that the cantilever end of the loaded cylinder sample has a flange with a mass of about 100 g, the simulation load of the loaded cylinder sample is determined to be 200 g.

3   FINITE ELEMENT SIMULATION ANALYSIS

In order to bring our engineering analysis closer to reality, the construction of the finite element analysis model mainly follows the principles of uniform dimensions, reasonable energy and load equivalence. The energy equivalent of the finite element model is embodied in the mass and stiffness equivalents. The meshing on the key force or heat transfer path should be more compact in order to accurately reflect the forces and thermal loads and ensure the equivalent stiffness of the structure; regardless of whether the load is static or dynamic, the point of application of the vector load, the size of the load, and the direction of the load should be handled. The magnitude and position of the scalar load must be exactly equivalent, and must be consistent with the actual project. Based on these principles, the finite element model is established as shown in Figure 5.

In terms of unit division, the loaded cylinder parts (including end flanges) in the finite element model all use hexahedral elements with high calculation accuracy. Any part that has

Figure 5.   The division of loaded cylinder model unit.

Table 1.   Properties of two typical high-modulus carbon-fiber composites.

Material	Young's modulus, E (GPa)	Poisson's ratio	Density, ρ (kg/m³)
M40J/epoxy resin	72	0.32	1,630
M55J/cyanate ester	108	0.28	1,650

Table 2.   Static deformation of loaded cylinder according to finite element analysis.

Material	Maximum deformation (μm)	Maximum deformation position
M40J/epoxy resin	2.92	Cantilever flange
M55J/cyanate ester	1.96	Cantilever flange

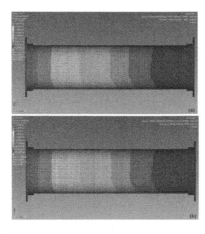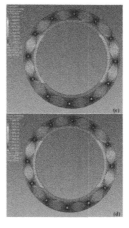

Figure 6.   Deformation clouds of the loaded cylinders: (a) M40J loaded cylinder (side view); (b) M55J loaded cylinder (side view); (c) M40J loaded cylinder (axial view); (d) M55J loaded cylinder (axial view).

236

Table 3. Primary and secondary optical axis tilt calculation results.

Material	Optical axis tilt ($''$)
M40J/epoxy resin	1.08287
M55J/cyanate ester	0.72723

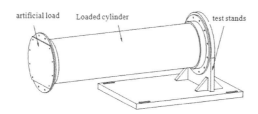

Figure 7. Experimental model of static loading deformation of loaded cylinder.

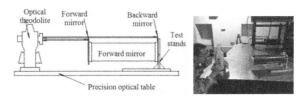

Figure 8. Loaded cylinder static test.

a connection with the loaded cylinder but is not critical (such as a test stand or simulated load) is replaced by a boundary constraint or a mass unit, and the connection relationship is a rigid connection of rbe2.

Some parameters of the M40J and M55J high-molded composites, based on actual test data, are shown in Table 1. The results of the calculation and analysis of the impact load tube under gravity are shown in Table 2. The deformation clouds of the load-bearing cylinders for the two CFRP materials are shown in Figure 6.

Based on the results of simulation analysis, in combination with the calculated formula for the tilt amount of the optical axis, the tilt values of the optical axis of the primary and secondary mirrors under the gravity load deformation of the two loaded cylinders are as shown in Table 3.

4 LOADED CYLINDER STATIC TEST VERIFICATION

The loaded cylinder was machined with M55J+ cyanate ester carbon-fiber composites, mounted on test stands (see Figure 7), and then mounted on precision optical tables with optical theodolites. A high-precision mirror (as shown in Figure 8) was used, aligned and zeroed using an optical theodolite, and the angle of the mirror recorded as a test reference.

The loaded cylinder sample was put on the left cantilever flange, because the load is thin and the shape is symmetrical, so that the deviation between the load center of gravity and the rotary axis of the loaded cylinder sample is less than ±1 mm. The angle of the front and rear mirrors is recorded again. The angular deviation of the two recordings can be converted into the tilting amount for the optical axis of the loaded cylinder.

In the test process, in order to reduce operational error, three groups of tests were conducted; in addition, in order to remove error introduced by the micro-deformation of the test

Table 4. Test data for loaded cylinder static-load deformation.

	Load status	Front mirror	Rear mirror	Optical axis tilt
1	No load	7′ 01.6″	11′ 68.5″	
	With load	7′ 03.1″	11′ 68.8″	0.7″
2	No load	7′ 01.7″	11′ 68.6″	
	With load	7′ 03.4″	11′ 69.0″	0.8″
3	No load	7′ 01.9″	11′ 68.8″	
	With load	7′ 03.6″	11′ 69.2″	0.8″
Optical axis tilt average				0.77″

support under load, the angular difference between the front and rear mirrors was measured before and after the load was added, as shown in Table 4.

As indicated in Table 3, the optical axis tilt calculated by the finite element analysis was 0.727″. The test results (average 0.77″) are basically consistent with this result, which verifies that the simplified model and the simulation results are credible.

5 CONCLUSION

The test results of the static load deformation of the loaded cylinder specimen are in good agreement with the finite element analysis results, effectively validating the reliability of the simplified model and the simulation analysis results. The results of finite element analysis show that the reliability of the space camera support cylinder has been greatly improved with the use of the M55J composite. For the same loaded cylinder model, M55J/cyanate ester improves the overall stiffness of the model compared to M40J/epoxy resin composites, reducing the tilt of the primary and secondary mirror axes from 1.08287″ to 0.72723″. The mechanical deformation stability increased by 32.8%.

REFERENCES

Chen, P.C., Bowers, C.W., Content, D.A., Marzouk, M. & Romeo, R. (2000). Advance in very lightweight composite mirror technology. *Optical Engineering*, *39*(9), 2320–2329.
Chen, P.C., Saha, T.T., Smith, A.M. & Romeo, R. (1998). Progress in very lightweight optics using graphite fiber composite materials. *Optical Engineering*, *37*(2), 666–676.
Koyanagi, J., Arao, Y., Terada, H., Utsunomiya, S., Takeda, S. & Kawada, H. (2010). Development of space telescope mirror made by light and thermally stable CFRP. In *Proceedings SPIE 7522, Fourth International Conference on Experimental Mechanics, Singapore* (pp. 1540–1549). doi:10.1117/12.851241.
Romeo, R.C. & Martin, R.N. (2007). Unique space telescope concepts using CFRP composite thin-shelled mirrors and structures. In *Proceedings SPIE 6687, UV/Optical/IR Space Telescopes: Innovative Technologies and Concepts III*. doi:10.1117/12.734648.
Romeo, R.C. (1995). CFRP composites for optics and structures in telescope applications. In *Proceedings SPIE 2543, Silicon Carbide Materials for Optics and Precision Structures* (pp. 154–161). doi:10.1117/12.225284.
Utsunomiya, S., Kamiya, T. & Shimizu, R. (2012). Development of CFRP mirrors for low-temperature application of satellite telescopes. In *Proceedings SPIE 8450, Modern Technologies in Space- and Ground-based Telescopes and Instrumentation II* (pp. 21–30). doi:10.1117/12.926133.
Zhou, H. (2017a). The review of US research history for high performance carbon fiber. *Synthetic Fiber in China*, *46*(2), 16–21.
Zhou, H. (2017b). The review of Japan research history for high performance carbon fiber. *Synthetic Fiber in China*, *46*(10), 19–25.
Zou, H., Li, W., Peng, G., et al. (2017). The development situation of high modulus carbon fiber and its applications in aerospace. *Synthetic Fiber in China*, *46*(6), 17–22.

Automatic Control, Mechatronics and Industrial Engineering – He & Qing (Eds)
© *2019 Taylor & Francis Group, London, ISBN 978-1-138-60427-8*

Experimental investigation of the influence of cutting parameters on surface roughness in machining Inconel 718

B. Sarkar, M.M. Reddy & S. Debnath
Curtin University, Miri, Malaysia

ABSTRACT: We investigated the influence of cutting parameters on surface roughness when machining Inconel 718 using a carbide (K5000-U) end mill. A series of experiments were carried out to find a suitable range of machining parameters. A design technique was used to reduce the number of experiments through differing combinations. Response surface methodology was used to develop a predictive model and an analysis of variance was employed to validate it. The model was also validated with additional experiments and showed good agreement. The results show that a combination of high cutting speed with a low feed rate and depth of cut achieved a good surface finish. It was observed that the depth of cut had a greater influence on surface quality than the feed rate or cutting speed.

Keywords: Nickel alloy, carbide end mill, surface roughness

1 INTRODUCTION

Nickel alloys are used extensively in the aerospace, automotive, petrochemical, marine, military and biomedical industries, as well as in the power plant and nuclear sectors (Pervaiz et al., 2014). Because these alloys maintain high strength and hardness at elevated temperatures, machining is difficult to perform. Properties such as low thermal conductivity, high strain hardening, high hardness at elevated temperature and high chemical reactivity are responsible for the poor machinability rating of nickel alloys (Jawaid et al., 2001). High cutting force and temperature develop at the tool–workpiece interface during machining. The high strength and hardness lead to rapid tool wear and poorly machined surfaces (Ezugwu & Bonney, 2004). Many researchers have suggested using coated tools where possible to minimize tool wear and improve surface quality (Dudzinski et al., 2004). Ali et al. (2017) studied wear behavior in high-speed machining of Inconel 718 with a solid SiAlON milling tool. The results show prominent diffusion wear and generation of high temperature at the cutting zone, which leads to the severe adhesion of workpieces to the flank and rake faces of the tools. Hao et al. (2017) used a SiAlON ceramic tool for milling and revealed that abrasive wear leads to poorly machined surfaces at low cutting speeds (50–200 m/min). Ucuna et al. (2015) performed dry milling with Diamond-Like Carbon (DLC)-coated and uncoated carbide tools. The result clearly shows that DLC-coated tools generated less Built-Up Edge (BUE) and burr formation than the uncoated ones. The literature suggests that tool geometry and cutting parameters have significant influences on the machinability of nickel-based superalloys (Pawade et al., 2008; Nalbant et al., 2007; Amini et al., 2014). Cantero et al. (2013) analyzed tool wear mechanisms in the turning of Inconel 718 in wet and dry cutting conditions using multilayer TiAl/TiAlN-coated carbide tools with different geometrical parameters. They reported that at a high velocity of 70 m/min, increasing the side cutting-edge angle reduced chipping wear, and avoided BUE and notching. At a lower cutting speed of 50 m/min, the tool life increased significantly. Pawade et al. (2007) observed a low degree of work hardening for a 20° chamfered edge insert with a cutting speed of 300 mm/min, a feed rate of 0.15 mm/rev and a depth of cut of 0.50 mm. Cutting parameters also have a significant influence on machinability and a good combination of cutting parameters can significantly improve the machinability of nickel-based superalloys. Zhaopeng et al. (2011) found BUE at a low cutting speed of 20 m/min, which was exacerbated by an increase in cutting speed. Hang et al. (2013) also observed BUE formation at a low cutting speed of 25 m/min with TiAlN-coated carbide tools. In an experimental study, Liao et al. (2008) suggested that a cutting speed in the range of 90–110 m/min is appropriate for slot milling, and 55–135 m/min is suitable for side milling of superalloys. Turning with TiCN/Al$_2$O$_3$-coated carbide tool inserts produces excellent surface quality at a cutting speed of 60 m/min (Devillez et al., 2011). Cutting parameters are important factors that influence the machined surface quality when cutting nickel-based alloys. In this research, the machining of Inconel 718 with a carbide tool was performed in end milling to investigate the effect of cutting parameters on surface roughness.

2 EXPERIMENTAL DETAILS

This experimental study was conducted in dry cutting conditions using a V30 Horizontal CNC machine (Leadwell CNC Corporation, Taiwan). The rectangular workpiece material used was Inconel 718 with dimensions of 202 mm × 102 mm × 12 mm. Detailed information on the chemical composition of the Inconel 718 alloy is provided in Table 1. Four-flute carbide end mills with a helix angle of 55° and a diameter of 8 mm were used for machining. A surface roughness tester (Mitutoyo SJ-210) was used to measure the machined surface at three different positions. The average surface roughness (Ra) value was recorded.

In the present study, an experimental design with a Central Composite Design (CCD) technique was used to minimize the number of experiments. For CCD, the number of experiments required can be determined from Equation 1:

$$n = 2^k + 2k + C \tag{1}$$

where n represents the number of experiments and k is the total number of parameters considered.

The experimental results are used to develop a predictive model using a Response Surface Methodology (RSM). The proposed relationship between the output response and input variables can be represented by Equation 2:

$$R = b_0 + b_1 \times A + b_2 \times B + b_3 \times C + b_{11} \times A^2 + b_{22} \times B^2 + b_{33} \times C^2 + b_{12} \times AB + b_{23} \times BC + b_{13} \times AC \tag{2}$$

where R is the response, and A, B and C are the different machining factors. The regression coefficient $b_i = 1, 2, 3,\ldots$ is computed by the regression method using the experimental results. The significance of factors and their interactions are computed using statistical analysis. The optimum response model and setting of parameters are generated according to the response model in Equation 2. A total of 20 runs or sets of experiments were conducted with different cutting parameters on the basis of CCD. The surface finish was recorded at three locations for each run. Table 2 shows the range of cutting parameters used in the

Table 1. Chemical composition of Inconel 718.

Element	Ni	Cr	Mo	Ti	Al	Nb	V	Cu	Fe	Co	C
% weight	53.50	18.60	2.95	0.97	0.59	5.15	0.024	0.140	17.30	0.19	0.58

Table 2. Machining parameters.

Cutting parameter	Notation	Unit	Range
Spindle speed	V_c	rpm	200 to 1100
Feed rate	f	mm/rev	30 to 120
Depth of cut	D_c	mm	0.05 to 0.5

experiments. The adequacy of the quadratic model was evaluated using an Analysis Of Variance (ANOVA). A design expert was used to establish the design matrix and to analyze the experimental data.

3 EXPERIMENTAL RESULTS AND DISCUSSION

Table 3 shows the surface roughness value for all 20 combinations of experimental runs. It was observed that the temperature at the cutting zone increased due to the increase of the cutting parameters during machining. The heat developed at the cutting zone may lead to poor surface quality and rapid tool wear.

The experimental results suggested a second-order predictive model for surface roughness. The mathematical model for the surface roughness is given in Equation 3:

$$R_a = 0.60 - 0.052A + 0.048B + 0.066C + 0.016AB - 6.869E^{-003}AC + 0.034BC \\ - 0.015A^2 - 7.432E^{-003}B^2 - 9.747E^{-003}C2 \tag{3}$$

The developed model was verified using ANOVA. The model's F-value is 3.78, which implies that the model is significant, and the values of 'Prob > F', at less than 0.0500, indicate that the model terms are significant. On the basis of the ANOVA result, the depth of cut has a more significant influence on surface roughness than the cutting speed or feed rate.

Figure 1a shows the perturbation plot for the cutting parameters. The plot indicates that surface roughness increases with increasing feed rate and depth of cut, while it decreases with an increase in cutting speed. The interaction effect of different cutting parameters on surface roughness is shown in Figures 1b, 1c and 1d. Figure 1b indicates that the surface roughness increases with an increase in feed rate and decreases with increasing speed. The influences of cutting speed and depth of cut on surface roughness are shown in Figure 1c: the graph shows that the surface roughness increases with an increase of depth of cut and decreases with increasing cutting speed. The effects of feed rate and depth of cut on surface roughness are

Table 3. Cutting parameters of experiments and resultant surface roughness in dry conditions.

No.	Run order	Spindle speed (V_c) rpm	Feed rate (f) mm/rev	Depth of cut (D_c) mm	Surface roughness (Ra) μm
1	16	650	75	0.275	0.36
2	10	1100	75	0.275	0.31
3	5	200	30	0.5	0.46
4	1	200	30	0.05	0.28
5	6	1100	30	0.5	0.16
6	19	650	75	0.275	0.36
7	7	200	120	0.5	0.64
8	8	1100	120	0.5	0.48
9	15	650	75	0.275	0.36
10	2	1100	30	0.05	0.15
11	12	650	120	0.275	0.37
12	3	200	120	0.05	0.39
13	20	650	75	0.275	0.36
14	13	650	75	0.05	0.23
15	9	200	75	0.275	0.31
16	14	650	75	0.5	0.44
17	11	650	30	0.275	0.30
18	18	650	75	0.275	0.36
19	17	650	75	0.275	0.36
20	4	1100	120	0.05	0.19

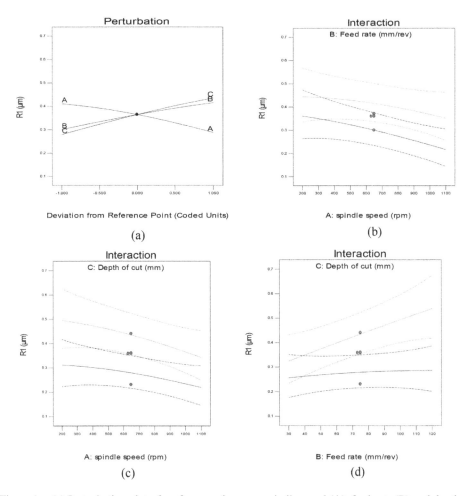

Figure 1. (a) Perturbation plot of surface roughness vs spindle speed (A), feed rate (B) and depth of cut (C); (b) interaction plot of surface roughness with spindle speed and feed rate; (c) interaction plot of surface roughness with spindle speed and depth of cut; (d) interaction plot of surface roughness with feed rate and depth of cut.

presented in Figure 1d, which indicates that the surface roughness increases with an increase of feed rate or depth of cut.

Figures 2a and 2b show surface roughness vs spindle speed and feed rate at a depth of cut of 0.275 mm. They show a good surface finish with increasing spindle speed, but a poorer finish with increasing feed rate. Figures 2c and 2d plot surface roughness vs spindle speed and depth of cut at a feed rate of 75 mm/rev. Surface roughness increases with an increase in depth of cut and decreases with an increase in cutting speed. Figures 2e and 2f show surface roughness vs feed rate and depth of cut at a spindle speed of 650 rpm: surface roughness increases with increasing feed rate and depth of cut. From these diagrams, it may be observed that poor surface roughness was generated due to an increase in feed rate and depth of cut, while surface roughness was improved with an increase of cutting speed.

Depth of cut is the most influential parameter on surface roughness, followed by cutting speed and feed rate. The lowest surface roughness was observed at a spindle speed of 1100 rpm; in other words, a cutting speed of 28 m/min, a depth of cut of 0.05 mm and a feed rate of 30 mm/rev.

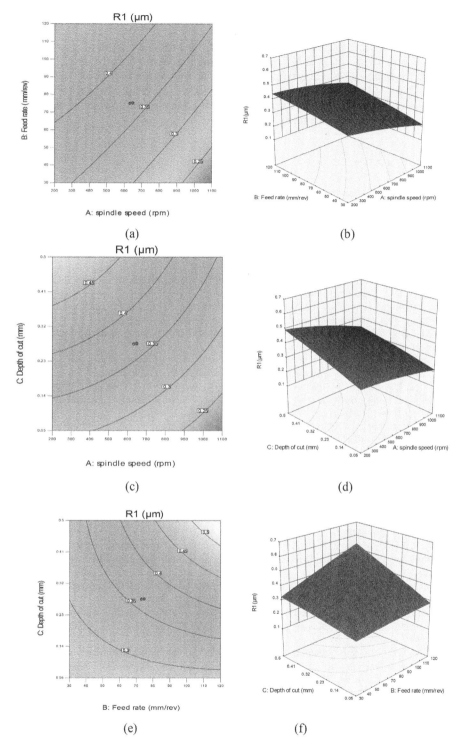

Figure 2.   (a) Contour plot of surface roughness vs spindle speed and feed rate; (b) 3D surface plot of surface roughness vs spindle speed and feed rate; (c) contour plot of surface roughness vs spindle speed and depth of cut; (d) 3D surface plot of surface roughness vs spindle speed and depth of cut; (e) contour plot of surface roughness vs feed rate and depth of cut; (f) 3D surface plot of surface roughness vs feed rate and depth of cut.

5 CONCLUSIONS

The effect of machining parameters (cutting speed, feed rate and depth of cut) on surface roughness was experimentally investigated and a mathematical predictive model developed using RSM for Inconel 718. The validity of the model was verified by ANOVA. The experimental validation shows close agreement with the model. The results show that a good surface finish is obtained at a higher cutting speed, low feed rate and low depth of cut in the selected range of cutting parameters.

ACKNOWLEDGMENT

The authors wish to thank the Ministry of Higher Education (Malaysia, FRGS grant) for providing financial support to conduct the experimental work.

REFERENCES

Ali, C., Alagac, M.S., Turan, S., Kara, A. & Kara, F. (2017). Wear behaviour of solid SiAlON milling tools during high speed milling of Inconel 718. *Wear*, 378–379, 58–67.

Amini, S., Fatemi, M.H. & Atef, R. (2014). High speed turning of Inconel 718 using ceramic and carbide cutting tools. *Arabian Journal of Science and Engineering*, *39*, 2323–2330.

Cantero, J.L., Dıaz-Alvarez, J., Miguelez, M.H. & Marın, N.C. (2013). Analysis of tool wear patterns in finishing turning of Inconel 718. *Wear*, *297*, 885–894.

Devillez, A., Le, G.C., Dominiak, S. & Dudzinski, D. (2011). Dry machining of Inconel 718, workpiece surface integrity. *Journal of Materials Processing Technology*, *211*, 1590–1598.

Dudzinski, D., Devillez, A., Moufki, A., Larrouquère, D., Zerrouki, V. & Vigneau, J. (2004). A review of developments towards dry and high speed machining of Inconel 718 alloy. *International Journal of Machine Tools and Manufacture*, *44*, 439–456.

Ezugwu, E.O. & Bonney, J. (2004). Effect of high-pressure coolant supply when machining nickel-base, Inconel 718, alloy with coated carbide tools. *Journal of Materials Processing Technology*, 153–154, 1045–1050.

Hang, Y., Zhao, P.H., Zheng, M.L., Sun, F.L. & Yang, S.C. (2013). Study of surface quality in machining nickel-based alloy Inconel 718. *International Journal of Advanced Manufacturing Technology*, *69*, 2659–2667.

Hao, Z.P., Hang, Y.F., Qiong, J.L., Ji, F.F. & Liu, X. (2017). New observations on wear mechanism of self-reinforced SiAlON ceramic tool in milling of Inconel 718. *Archive of Civil and Mechanical Engineering*, *17*, 467–474.

Jawaid, A., Koksal, S. & Sharif, S. (2001). Cutting performance and wear characteristics of PVD coated and carbide tools in face milling Inconel 718 aerospace alloy. *Journal of Materials Processing Technology*, *116*, 2–9.

Liao, Y.S., Lin, H.M. & Wang, J.H. (2008). Behaviors of end milling Inconel 718 superalloy by cemented carbide tools. *Journal of Materials Processing Technology*, *201*, 460–465.

Nalbant, M., Altın, A. & Gökkaya, H. (2007). The effect of cutting speed and cutting tool geometry on machinability properties of nickel-base Inconel 718 super alloys. *Materials & Design*, *28*, 1334–1338.

Pawade, R.S., Joshi, S.S. & Brahmankar, P.K. (2008). Effect of machining parameters and cutting edge geometry on surface integrity of high speed turned Inconel 718. *International Journal of Machine Tools and Manufacture*, *48*, 15–28.

Pawade, R.S., Joshi, S.S., Brahmankar, P.K. & Rahman, M. (2007). An investigation of cutting forces and surface damage in high-speed turning of Inconel 718. *Journal of Materials Processing Technology*, 192–193, 139–146.

Pervaiz, S., Rashid, A., Deiab, I. & Nicolescu, M. (2014). Influence of tool materials on machinability of titanium and nickel-based alloys: A review. *Materials and Manufacturing Processes*, *29*, 219–252.

Ucuna, I., Aslantasb, K. & Bedirc, F. (2015). The performance of DLC-coated and uncoated ultra-fine carbide tools in micromilling of Inconel 718. *Precision Engineering*, *41*, 135–144.

Automatic Control, Mechatronics and Industrial Engineering – He & Qing (Eds)
© 2019 Taylor & Francis Group, London, ISBN 978-1-138-60427-8

Parametric optimization of mechanical properties during friction stir welding of AL-6063

G. Singh
Department of Mechanical Engineering, Chandigarh University, Gharuan, Mohali, Punjab, India

S. Sharma
CSIR-Central Leather Research Institute, Regional Center for Extension and Development, Jalandhar, Punjab, India

A. Singh
Department of Mechanical Engineering, Chandigarh University, Gharuan, Mohali, Punjab, India

ABSTRACT: The present investigation includes the study of the Friction Stir Welding (FSW) of aluminum alloys in order to determine the optimum parameters of welding, such as tool rotation speed, tool materials and the number of passes needed to meet the objective of maximizing the hardness of the weld zone. A large number of interactions exist at different levels between the various parameters selected, so in order to perform the minimum amount of experimentation with optimum results, a Design of Experiments (DOE) approach was employed. In order to attain the target and optimize the results, Taguchi's DOE approach and Analysis of Variance (ANOVA) were employed. After a literature survey, the appropriate orthogonal array L-9 was selected and the experimentation was performed at three levels. Following this, an experimentation analysis of the result was undertaken using Minitab 15 statistical software to find the effect of the variation and percentage contribution of each factor. The experimentation was carried out to maximize the hardness of the welded zone.

Keywords: Friction stir welding, Mechanical properties, Aluminum alloy, Hardness

1   INTRODUCTION

Friction Stir Welding (FSW) is a joining process in a solid state in which a non-consumable rotating tool is used to join two facing surfaces. Temperature is produced between the tool and the material due to the rubbing action between them and this creates a very soft region near the FSW tool. At that point the two bits of metal are mechanically intermixed at the site of the joint, at which point the mollified metal (because of the hoisted temperature) can be joined by utilizing the mechanical weight, which is connected by the apparatus, much like a joining clay or mixture (Bahrami et al., 2004). The strong state nature of FSW prompts a few focal points over combination welding strategies as issues related with cooling from the fluid stage are evaded. Deformities, for example, porosity, solute redistribution, cementing breaking and liquation splitting, do not emerge in FSW (Balasubramanian, 2008). By and large, FSW has been found to create a low convergence of deformities and is extremely tolerant of varieties in parameters and materials. FSW has numerous applications, such as for freezer panels, panels for deck and divider construction, Apple's cutting edge iMac, China's high speed railway carriages, NASA's Orion Spacecraft and so forth. During FSW a non-consumable rotating tool with a specially designed pin and shoulder is inserted into the abutting edges of the sheets or plates to be joined and traversed along the line of the joint (Biallas et al., 1999; Bradshaw et al., 1998; Bussu et al. 1999; Casalino et al., 2013). The length of the pin is slightly less than the weld depth required and the tool shoulder should be in intimate contact with the work piece surface. The pin is then moved against the work piece. The half-plate, where the direction of the tool rotation is the same as that of the welding, is called the advancing side, while the other side is designated as being the retreating side (Cavaliere et al., 2008; Cavaliere et al., 2009; Cederqvist et al. 2001; Colwell et al. 2013; Elangovan et al. 2007). The proceeding side is on the right, where the tool rotation direction is the same as the tool travel direction (reverse to the direction of metal flow) and the retreating side is on the left, where the tool rotation is opposite the tool travel direction (parallel to the direction of metal flow).

2 PROBLEM FORMULATION

The selection of parameters and their interactions is as per Taguchi's orthogonal arrays. The array has been designed in such a way that it can handle maximum numbers of factors in definite numbers of experimental runs, as compared to full factorial design. The columns are balanced and are of an orthogonal nature. This means that in each pair of columns, all of the factor combinations happen an equal number of times. The orthogonal designs allow the evaluation of the influence of each factor on the response independently of all other factors. When the degrees of freedom are known, then it is easy to select an orthogonal array.

3 EXPERIMENTAL DESIGN

The procedure used by the Taguchi method is that firstly the objective function is defined, then the selection of the parameters is considered, which will affect the system performance. Once these tasks are completed, the next assignment is to explain the number levels and the total degree of freedom. Then the experimental run, as per the defined levels of the parameters, is performed, and then the results are analyzed in order to find out the effects of the different parameters.

In the present study, the objective function is to maximize hardness and get a fine microstructure. In this research, the S/N ratio is used to measure the quality of the characteristics and the significant machining parameters through Analysis of Variance (ANOVA) and "F" test values. The experimental results are confirmed having hardness as Higher the Better values.

3.1.1 Base metal for welding
Base metal used: Aluminum 6063-T6 and base metal plate dimensions: $110 \times 40 \times 6$ mm (length × breadth × thickness).

Table 1. Taguchi L_9 experiments run.

Experiment No	RPM	Tool material	Type of pass
1	1400	HSS	SP
2	1400	HCHCr	DP
3	1400	EN 19	RE
4	1800	HSS	RE
5	1800	EN 19	SP
6	1800	EN 19	DP
7	1120	HSS	DP
8	1120	EN 19	RE
9	1120	EN 19	SP

Table 2. Chemical composition and mechanical properties of AA 6063-T6 (wt %).

Silicon carbide	Iron	Copper	Manganese	Chromium	Zinc	Titanium	Magnesium	Aluminum
2–10	0–0.35	0–0.10	0–0.10	0–0.10	0–0.10	0–0.10	0.45–0.90	Remainder

Table 3. Mechanical properties of 6070 AA.

Tensile strength	Yield strength	Elongation	Hardness
216 MPa	80 MPa	Up to 20%	43 VHN

Figure 1. Vertical milling machine used in process.

Table 4. Specifications of FSW machine.

Make	HMT
Spindle speed (rpm)	45–1800
Transverse speed (mm/min)	5–51.5

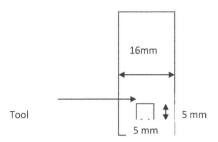

Figure 2. Tool geometry.

3.1.2 *Tool material*

The three tool materials that are used for welding the work piece are *Mild Steel*, *High-Speed Steel (HSS or HS)* and *High Carbon High Chromium.*

3.2 *Sample preparation*

Before going for welding on a vertical milling machine, the samples were prepared according to requirement on a power hacksaw, machined from the edges and then grinded to get them to fit in the vice. To get a good weld, the joint edges of the plates were matched properly (face to face). Properly matched edges are imperative for a good quality weld; failure to do this can lead to defects.

3.3 *Tool geometry for FSW*

The standard tool geometry of an FSW tool is shown in Figure 2. It is a single piece tool having three main parts. It is prepared by the turning of a 20 mm diameter mild steel rod. Both the shank diameter and the shoulder diameter will be 16 mm, and the pin diameter and pin length are 5 mm.

4 EXPERIMENTAL PROCEDURE

During the experimentation, the two AA 6070 plates with dimensions $110 \times 40 \times 6$ mm (L × b × t) respectively are placed on a vice in a manner that prevents the displacement of the plates during welding and fixed along the travel line of the welding tool. The tool is fixed firmly in the tool collect of respective dimension and rotated at the required rpm. The tool pin is plunged vertically into the joint line between the work pieces, while the tool is rotating. Due to the velocity difference between the rotating tool and the stationary work piece, heat is produced by frictional work and material deformation begins. To perform the welding, the rotating tool is traversed along the line, while the shoulder of the tool remains in intimate contact with the plate surface. The shoulder confirms the underlying material so that void formation and porosity behind the probe are prevented. As the heat is transferred into the surrounding material, the temperature rises and the material softens without reaching the melting point (hence known as solid state process). As the pin is moved in the direction of the welding, the leading face of the pin, assisted by a specified pin profile, forces the plasticized material to the back of the pin while applying a substantial forging force to consolidate the weld metal. When the weld distance is covered, the tool is pulled out of the work piece, leaving behind an exit hole as a footprint of the tool.

5 RESULTS AND DISCUSSION

5.1 *Hardness*

The hardness values for all of the specimens, as observed in Table 6, have been used to calculate the signal to noise ratio for hardness.

Table 5. Average hardness for FSW weld specimens.

Exp. no	Tool rotation speed	Tool material	Pass	Hardness 1	Hardness 2	Hardness 3	Average hardness
1	1400	HSS	Single	47.2	46.4	47.6	47.07
2	1400	HCHC	Double	42.9	43.0	43.2	43.03
3	1400	EN19	Reverse	45.2	45.6	41.1	44.00
4	1800	HSS	Reverse	46.3	44.2	45.1	45.20
5	1800	HCHC	Single	47.0	46.7	47.0	46.90
6	1800	EN19	Double	43.1	43.5	43.8	43.50
7	1120	HSS	Double	48.4	46.7	46.4	47.16
8	1120	HCHC	Reverse	46.5	46.4	47.1	46.67
9	1120	EN19	Single	47.8	48.1	48.0	47.96

Table 6. S/N ratio for hardness test.

RPM	Tool material	No. of passes	Hardness	S/N ratio
1400	HSS	SP	47.07	33.4549
1400	HCHCr	DP	43.03	32.6754
1400	EN 19	RE	44.00	32.8691
1800	HSS	RE	45.20	33.1028
1800	EN 19	SP	46.90	33.4235
1800	EN 19	DP	43.50	32.7698
1120	HSS	DP	47.16	33.4715
1120	EN 19	RE	46.67	33.3808
1120	EN 19	SP	47.96	33.6176

Table 7. Analysis of variance for S/N ratios for hardness test.

Source	DOF	Seq sum of squares	Adj sum of squares	Adj MS	F	% contribution
RPM	2	0.20014	0.09900	0.04950	3.39	20.53
Tool MATERIAL	2	0.40312	0.32336	0.16168	11.74	41.35
No. of PASSES	2	0.34233	0.34233	0.17116	11.09	35.11
Residual Error	2	0.02917	0.02917	0.01459		2.99
Total	8	0.97475				

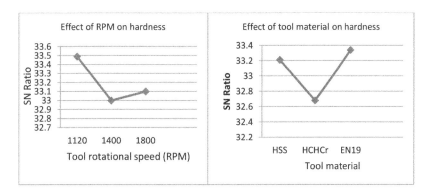

Figure 3. Effect of tool rotational speed and tool material on hardness.

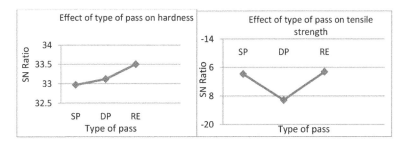

Figure 4. Effect of type of pass on hardness and tensile strength.

5.1.1 *Analysis of variance for Vickers hardness*

The results were analyzed using ANOVA in order to identify the significant factors affecting the Vickers hardness. The ANOVA for Vickers hardness at a 95% confidence interval is given. The variation data for each factor and their interactions were F-tested to find the significance of each. The principle of the F-test is that the larger the F value for a particular parameter, the greater the effect on the performance characteristic due to the change in that process parameter. The ANOVA Table shows that rotational speed (F 11.09 value), tool material (F 3.39 value) and no. of passes (F 11.74 value) are the factors that significantly affect the hardness. The type of pass has the highest contribution to hardness. The results of the main effects for S/N ratio to hardness are shown in Figures 3 and 4, which show the variation of hardness with the input parameters.

6   CONCLUSIONS

The present experimentation of friction stir welding was performed on AA6063, which is widely used in the automobile industry, aircraft industry and ship manufacturing. The following was concluded from the results:

- The process parameters that significantly affect hardness are tool rotational speed, tool material and type of pass affect.
- Tool material is a dominating parameter and effect the hardness as compared to other process parameters in a given range.
- The hardness of the weld joint decreases with the increase in tool **RPM**.
- The optimum condition for maximum hardness value is 1120 **RPM**, HCHCr tool material and reverse pass.

REFERENCES

Bahrami, M., Givi, M.K.B., Kamran, D. & Nader, P. (2014). On the role of pin geometry in microstructure and mechanical properties of AA7075/SiC nano-composite fabricated by friction stir welding technique. *Journal of Materials & Design, 53*, 519–527.

Balasubramanian, V. (2008). Relationship between base metal properties and friction stir welding process parameters. *Journal of Materials Science and Engineering, 480*, 397–403.

Biallas, G., Braun, R., Dalle Donne, C., Staniek, G. & Kaysser, W.A. (1999). Mechanical properties and corrosion behaviour of friction stir welded 2024-T3. *Proceedings of First International Symposium on Friction Stir Welding, Thousand Oaks, California, USA, 14–16 June, 1999*. TWI, Abington.

Bradshaw, L.M., Fishwick, D., Slater, T. & Pearce, N. (1998). Chronic bronchitis, work related respiratory symptoms, and pulmonary function in welders in New Zealand. *Occupational and Environmental Medicine, 55*(3), 150–154.

Bussu, G. & Irving, P.E. (1999). Static and fatigue performance of friction stir welded 2024-T3 aluminium joints. *Proceedings of First International Symposium on Friction Stir Welding, Thousand Oaks, California, USA, 14–16 June, 1999*. TWI, Abington.

Casalino, G., Campanelli, S. & Mortello, M. (2014). Influence of shoulder geometry and coating of the tool on the friction stir welding of aluminium alloy plates, *Procedia Engineering, 69*, 1541–1548.

Cavaliere, P. & Panella, F. (2008). Effect of tool position on the fatigue properties of dissimilar 2024–7075sheets joined by friction stir welding. *Journal of Materials Processing Technology, 206*, 249–255.

Cavaliere, P., Santis, A. De, Panella, F. & Squillace, A. (2009). Effect of anisotropy on fatigue properties of AA2198 Al–Li plates joined by friction stir welding engineering failure analysis. *Journal of Materials Science and Engineering, 16*, 1856–1865.

Cavaliere, P., Santis, A. De, Panella, F. & Squillace, A. (2009). Effect of welding parameters on mechanical and microstructural properties of dissimilar AA6082–AA2024 joints produced by friction stir welding. *Journal of Materials and Design, 30*, 609–616.

Cavaliere, P., Squillace, A. & Panella, F. (2008). Effect of welding parameters on mechanical and microstructural properties of AA6082 joints produced by friction stir welding. *Journal of Materials Processing Technology, 200*, 364–372.

Cederqvist, L. & Reynolds, A.P. (2001). Factors affecting the properties of friction stir welded aluminium lap joints. *Welding Research Supplement, 80*, 281–287.

Colwell, K.C. (2013). *Two metals enters one leaves–the miracle of friction stir welding.* Retrieved from https://www.caranddriver.com/features/two-metals-enter-one-metal-leaves-the-miracle-of-friction-stir-welding-tech-dept.

Elangovan, K. & Balasubramanian, V. (2007). Influences of pin profile and rotational speed of the tool on the formation of friction stir processing zone in AA2219 aluminium alloy. *Journal of Materials Science and Engineering, 459*, 7–18.

Fu, R., Sun, Z., Sun, R., Li, Y., Liu, H. & Liu, L. (2011). Improvement of weld temperature distribution and mechanical properties of 7050 aluminum alloy butt joints by submerged friction stir welding. *Journal of Material Design, 32*, 4825–4831.

Fujii, H., Cui, L., Maeda, M. & Nogi, K. (2006). Effect of tool shape on mechanical properties and microstructure of friction stir welded aluminum alloys. *Journal of Material Science Engineering: A, 419*(1–2), 25–31.

Hwang, Y.M., Fan, P.L. & Lin, C.H. (2010). Experimental study on friction stir welding of copper metals. *Journal of Materials Processing Technology, 210*, 1667–1672.

Jata, K.V., Sankaran, K.K. & Ruschau, J. (2000). Friction stir welding effects on microstructure and fatigue of aluminum alloy 7050-T7451. *Metallurgy and Material Science Journal, 31*, 2181–2192.

Automatic Control, Mechatronics and Industrial Engineering – He & Qing (Eds)
© 2019 Taylor & Francis Group, London, ISBN 978-1-138-60427-8

Mechanical properties of an AlSi10Mg sandwich structure with different origami patterns under compression

L.Z. Su
Xi'an Institute of Electronic Engineering, Xi'an, China
School of Aerospace Engineering, Tsinghua University, Beijing, China

Z. Li
Xi'an Institute of Electronic Engineering, Xi'an, China

Z.X. Ge
School of Automotive Engineering, Harbin Institute of Technology, Weihai, China

Y.P. Zhou & K.J. Wang
Xi'an Institute of Electronic Engineering, Xi'an, China

ABSTRACT: The mechanical properties of origami folding pattern core sandwich plates made of AlSi10Mg aluminum alloy have been investigated in this paper. The sandwich panel was fabricated using 3D printing technology and tested under a quasi-static out-of-plane compressive load. An explicit Finite Element Analysis (FEA) of Miura-ori models was developed and validated using the experimental results. Furthermore, parametric studies were performed on Miura-ori models that were subjected to quasi-static out-of-plane compression, through which the relationships between the mechanical properties and the geometric parameters were established. The results show that the compressive strength of the sandwich structure with a Miura-ori core decreased approximately linearly with the increase of the folding angle, and first increased and then decreased with the increase of the dihedral angle α. Finally, the mechanical properties of sandwich structures with a Miura-ori core and with a zigzag-based core were compared. It was shown that they have almost equal performance in terms of compressive force.

Keywords: sandwich structure, origami patterns, mechanical properties

1 INTRODUCTION

Sandwich structures are widely used in the aerospace and automotive industries due to their high stiffness/weight ratio and energy absorption capacity (Wadley et al., 2003). Typical sandwich structures consist of two thin, stiff and strong face sheets separated by a lightweight core. The mechanical behavior of sandwich structures depends on the material used for construction, the geometry of the face sheets and, especially, the core topology design. Rigid origami materials have attracted interest for the innovative design of mechanical metamaterials (Li & Wang, 2017). Investigations were carried out on sandwich structures that had an origami structure as their core, and they were found to improve shell bending stiffness and give higher strength/weight ratios (Kilchert et al., 2014). Compared to conventional foam and honeycomb-based structures, origami-based sandwich structures provide a good interior space for multiple functions, such as cross-flow heat exchange, shape morphing and high intensity dynamic load protection (Kooistra et al., 2004). Miura-ori is a classic origami folding pattern whose main constituents are parallelogram facets.

Miura-ori based fold cores have been regarded as a potential substitute for conventional honeycomb cores due to their excellent properties, such as a simple structure and continuously adjustable stiffness. As a result, many studies on Miura-ori inspired folded core sandwich structures are available in the literature. Heimbs et al. (2010) studied the mechanical behavior under compression, shear and impact loads of the folded cores made from prepreg sheets of carbon fiber/aramid fiber. Zhou et al. (2014) presented a parametric study on the mechanical properties of a variety of Miura-based folded core models that were virtually tested by quasi-static compression, shear and bending using the finite element method. Baranger et al. (2011) studied the out-of-plane compression performance of individual Miura-ori origami cores through experiments and calculations. In addition, Eidini & Paulino (2015) present a technique for creating zigzag-based mechanical metamaterials by combining origami folding techniques with kirigami. The studies preserve the remarkable properties of Miura-ori and expand on its design space. The zigzag-based folded material is appropriate for a wide range of applications, from mechanical metamaterials to deployable structures. However, few articles have studied the effect of origami core parameters on the mechanical properties of the AlSi10Mg sandwich structure and whether having an origami configuration as the folded core leads to better mechanical properties.

In this paper, the sandwich structures with Miura-ori based cores and zigzag-based cores using AlSi10Mg aluminum alloy by 3D printing technique. The detailed mechanical responses of the sandwich structures under a compressive load were studied by means of experiments and finite elements. The compressive load was measured and used to verify the accuracy of the numerical results. The influences of the folded core's topology parameters on the performance of the sandwich structures were investigated, based on the FEA simulations. Finally, the mechanical characteristics of the two structures, both the Miura-ori based core and the zigzag-based folded core, were compared and analyzed.

2 EXPERIMENTAL PROCEDURE AND METHODS

2.1 *The geometry of Miura-ori and zigzag-based cores*

The Miura-ori pattern can be constructed by the tessellation of multiple units. A unit consists of four identical parallelograms, as seen in Figure 1(a). The mountain and valley crease lines are marked as solid and dotted lines, respectively. For the sake of simplicity, in this paper, we set the length of the connected sides of the parallelogram to be equal and the length of the side is defined as a = b = 1. Its folding motion can be described by the folding angle φ and the dihedral angle *a*. From the geometric relations, it can be seen that the angles α, β and φ are

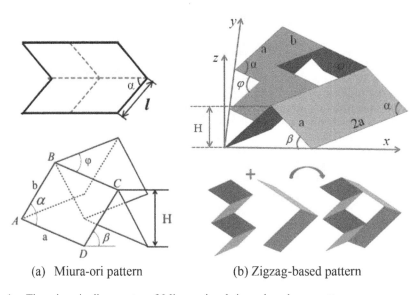

(a) Miura-ori pattern   (b) Zigzag-based pattern

Figure 1.   The origami cell geometry of Miura-ori and zigzag-based core patterns.

not completely independent. The rigid folding motion of the patterned sheet can be represented (Qiao, 2016) as follows:

$$\cos a = \cos\frac{\varphi}{2}\cos\beta \tag{1}$$

$$a = b = l = H / \sin\beta \tag{2}$$

when the height H is given as a constant 10 mm and the motion of the unit only depends on the angle α and angle φ.

Figure 1(b) shows the BCHn (Basic unit Cell with Hole) zigzag-based pattern. The top view is of a V-shaped fold, including two identical parallelogram facets connected along the ridges with length a. Its geometry can be defined by the facet parameters a, b, and α, and by the folding angle φ. The length of the side is defined as a = b = l, the same as the Miura-ori parameter definition. Two different scales of V-shapes, with the same angle φ, are connected along joining fold lines. The length b of the parallelogram facets in the left zigzag strip of V-shapes is half that of the strip on the right in the unit cell shown.

2.2 Sample preparation

In order to simplify the simulation calculation, the origami core unit cell was selected as the research object. The relative density ($\bar{\rho}$) of the folded core is calculated as:

$$\bar{\rho} = \frac{\rho_c}{\rho_m} = \frac{t_m S_m}{S_u H} \tag{3}$$

where, ρ_c, t_m, S_m and ρ_m are, respectively, the core density, the thickness of the folded core, total area and material density. S_u is the base area of the unit cell and H is the height of the unit cell as a constant 10 mm. The relative density $\bar{\rho}$ is equal to 0.2, and this is used for all of the models studied in this paper. The proposed sandwich core structures are fabricated using a 3D printer, as shown in Figure 2(a). Both the faces and the core are assumed to be made of AlSi10Mg aluminum alloy. The detailed material parameters are summarized in Table 1 (Kempen et al., 2012). The structural unit cell parameters of the prepared origami sandwich structure are shown in Table 2.

2.3 Measurement of mechanical properties

The compression test on the sandwich structures with an origami folding pattern core of AlSi10Mg aluminum alloy was performed on a universal testing machine INSTRON 3369.

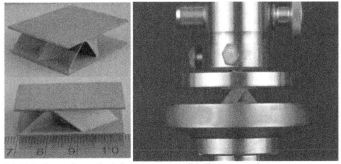

(a) Fabricated sandwich structure   (b) Compression test

Figure 2.   Fabrication and compression test of the origami folding pattern core sandwich plates.

Table 1. A slightly more complex table with a narrow caption.

Material	ρ_m (kg/m³)	E (GPa)	σ_{ys} (MPa)	σ_{uts} (MPa)	V	Alloying element (wt %)		
						Al	Si	Mg
AlSi10Mg	2690	69	185	285	0.33	Rest	9-11	0.2-0.45

Table 2. The geometric properties of fabricated models.

Model	$\alpha(°)$	$\varphi(°)$	H(mm)	t_m(mm)	Pattern
M1	50	45	10	0.292	Miura-ori
M2	50	60	10	0.3	Miura-ori
B1	50	60	10	0.34	BCHn

The testing device is shown in Figure 2(b). The lower plate is completely fixed, and the uniform compressive load is exerted through the upper plate, with an applied nominal displacement rate of 0.5 mm/min at the ambient temperature. The compressive stress-strain curve and the force-displacement curve are recorded. Finally, in order to explore the structural changes of the Al-based folded core during the compression process, a digital camera was used to record the deformation process under different strains. The experimental data showed good consistency with the simulation results from Figure 3(a).

2.4 *Characteristics of compressive responses*

The deformation behaviors in the compression are shown in Figure 3(b). It can be seen that the stress-strain curve of the Al-based origami sandwich structure is similar to that of general metal porous materials (Maskery et al., 2016), and it also has three distinct stages, i.e. elastic zone, plateau zone and densification zone. In the first stage, the folded cores exhibited linear-elasticity up to the yielding point where the peak load was reached. This was followed by strain hardening, where localized buckling instability occurred in the folded core panel corresponding to the load, which began to decline as the displacement increased. As the compressive load increased, the buckling panel would deform further and come into contact with the adjacent panels to form new supports. In the final stage, the folded core of the entire structure was completely compacted resulting in the stress being raised rapidly.

2.5 *FEA models*

In order to investigate the effects of geometric parameters on the mechanical properties under out-of-plane compression, a numerical analysis model was created through LS-DYNA. The FEA model was meshed with four-node shell elements according to the defined element size. Two rigid faces were applied to the upper and lower panels of the sandwich structure and a displacement load was applied to the upper rigid face. The lower rigid face was limited by displacement constraints. For quasi-static loading cases, the strain rate effect is not considered.

2.6 *Influence of unit cell parameters on the deformation of the Miura-ori core*

It can be seen from Figure 4 that when the height of the rod unit structure and the deployment angle are not changed, the compressive strength of the sandwich structure with the Miura-based folded core will decrease approximately linearly with the increase of the deployment angle. When the included angle is 30°, the compressive strength of the structure is 7.4 MPa, and when the included angle is 70°, the compressive strength of the structure is only 2.8 MPa. Figure 5 shows that when the height of the rod unit structure and the angle of the two faces are constant, the compressive strength of the sandwich structure with the Miura-based folded

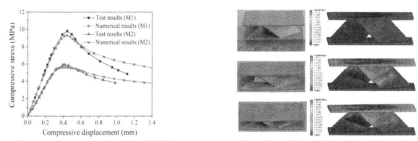

(a) Measured compressive stress/displacement curve   (b) Deformation of the Miura-ori structure

Figure 3.   Comparison of the experimental and numerical compressive stress-displacement curves for the sandwich structures with the origami folded core.

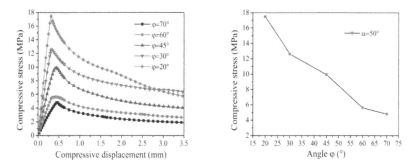

Figure 4.   Effect of folding angle on compression stress of sandwich structure ($\alpha = 50°$).

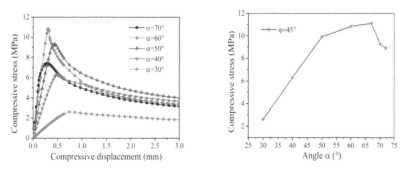

Figure 5.   Effect of folding angle on compression stress of sandwich structure ($\varphi = 45°$).

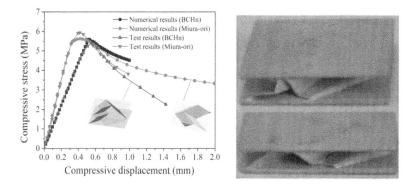

Figure 6.   Comparison of experimental and numerical compressive stress-displacement curves of two structures with Miura-ori and zigzag-based cores.

core will firstly increase and then decrease with the increase of the expansion angle α. The maximum value is reached at around 67°.

2.7 Comparison between Miura-ori and zigzag-based cores of equal areal density

In this section, the mechanical properties of the Miura-ori cores are compared to BCHn pattern cores with the same relative density, i.e. $0.2 \rho_m$. The detailed structural parameters can be seen in Table 2. Figure 6 shows the compression stress-strain curves of two structures with Miura-ori and BCHn-based cores. The yield points of the two structures are very similar, indicating that they have the same load bearing capacity. Therefore, when two sandwich plates are used as a structure in applications, the sandwich structure with the BCHn pattern core may have more advantages, because of the hole, for multiple functions, such as alignment.

3 CONCLUSIONS

In this paper, the sandwich plates with Miura-ori and zigzag-based cores of AlSi10Mg aluminum alloy were fabricated using 3D printing technology. Quasi-static out-of-plane compressive tests were conducted on the manufactured specimens. FEA models of the structural unit cell with a folded core were developed and validated with the experimental results. The effects of geometric parameters on the mechanical properties of the Miura-ori structure were investigated. The results showed that the compressive strength of the sandwich structure decreased approximately linearly with the increase of the folding angle, and firstly increased and then decreased with the increase of the dihedral angle α. Finally, the models with a Miura-ori core and with a zigzag-based core were compared. The deformation patterns were almost the same.

REFERENCES

Baranger, E., Guidault, P. & Cluzel, C. (2011). Numerical modeling of the geometrical defects of an origami-like sandwich core. *Composite Structures, 93*, 2504–2510.

Eidini, M. & Paulino, G.H. (2015). Unraveling metamaterial properties in zigzag-base folded sheets. *Science Advances, 1*, e1500224.

Heimbs, S., Cichosz, J., Klaus, M., Kilchert, S. & Johnson, A.F. (2010). Sandwich structures with textile-reinforced composite foldcores under impact loads. *Composite Structures, 92*, 1485–1497.

Kempen, K., Thijs, L. & Humbeeck, J.V. (2012). Mechanical properties of AlSi10Mg produced by selective laser melting. *Physics Procedia, 39*, 439–446.

Kilchert, S., Johnson, A.F. & Voggenreiter, H. (2014). Modelling the impact behaviour of sandwich structures with folded composite cores. *Composites Part A: Applied Science and Manufacturing, 57*, 16–26.

Kooistra, G.W., Deshpande, V.S. & Wadley, H. N.G. (2004). Compressive behavior of age hardenable tetrahedral lattice truss structures made from aluminium. *Acta Materialia, 52*, 4229.

Li, T.T. & Wang, L.F. (2017). Bending behavior of sandwich composite structures with tunable 3D-printed core materials. *Composite Structures, 175*, 46–57.

Maskery, I., Aboulkhair, N.T. & Aremu, A.O. (2016). A mechanical property evaluation of graded density Al-Si10-Mg lattice structures manufactured by selective laser melting. *Materials Science and Engineering-al A, 670*, 264–274.

Qiao, J.X. (2016). *The quasi-static and impact crushing of periodic cellular structures* (Dissertation). Tsinghua University, Beijing.

Wadley, H.N.G., Fleck, N.A. & Evans, A.G. (2003). Fabrication and structural performance of periodic cellular metal sandwich structures. *Composites Science and Technology, 63*, 2331–2343.

Zhou, X., Wang, H. & You, Z. (2014). Mechanical properties of Miura-based folded cores under quasi-static loads. *Thin-Walled Structures, 82*, 296–310.

Mechatronics

Automatic Control, Mechatronics and Industrial Engineering – He & Qing (Eds)
© 2019 Taylor & Francis Group, London, ISBN 978-1-138-60427-8

Research on the steering characteristics of the crawler robot based on the gyroscope

H.H. Liu, C.S. Ai, Y.Q. Bao, G.C. Ren & Y.C. Du
School of Mechanical Engineering, University of Jinan, Shandong, China

ABSTRACT: The crawler robot has the advantages of small grounding pressure, wide supporting area and better maneuvering characteristics; it is suitable for working in complex environments such as soft ground and muddy roads. But achieving accurate steering control of crawler robots is still a technical problem when the crawler is used as the robot moving chassis. According to the theory of steering dynamics, we analyzed the steering characteristics of the crawler robot and knew that the current output from the motor on both sides of the crawler affects the longitudinal speed, lateral speed and steering angle of the crawler robot. Therefore, we designed the detailed scheme of the crawler robot steering control system and achieved control of the current output from the motor on both sides of the crawler based on the dSPACE hardware-in-the-loop simulation control platform. Considering the gyroscope as the steering control angle detection component, it plays a key role in the accuracy and rapidity of the steering angle control in the robot steering control. So, we designed the gyroscope dynamic performance test system and the specific experiment for the gyroscope rotation dynamics. The experimental results show that the steady-state error of the gyroscope angle is 0.8 degree. The error range can meet the measurement requirements of the steering angle of the crawler robot, and precise control of the robot steering can be realized.

Keywords: gyroscope, crawler robot, steering control, angle error

1 INTRODUCTION

With the continuous development of mobile robot technology, the application range and functions of mobile robots have been greatly expanded and improved and have been widely used in the fields of industry, agriculture, national defense, medical care, rescue, and service. There are many types of mobile robots, and the movement methods of the robot can be divided into wheel type, leg type, led-wheel type, crawler type (Huang et al., 2015) and so on. Among them, the crawler robot is different from the general wheeled or leg robots. It has the advantages of small grounding pressure, wide supporting area and better maneuvering characteristics (Chen & Chen, 2007) and it is suitable for working in complex environments such as soft ground and muddy roads. But achieving accurate steering control of crawler robots is still a technical problem when the crawler was used as the robot moving chassis. Therefore, it is of great significance to study the steering control of crawler robots.

Unlike other mobile robots, crawler robots realize the steering function by using the sliding mechanism of the crawler and the ground. During the steering process, the large steering resistance torque is generated by the wide contact area between the crawler and the ground. And the imbalance between the steering resistance torque and the motor output torque causes a large non-systematic error, which affects the accuracy and flexibility of the steering of the robot (Gao, J. 2015). At present, there is much research on the steering characteristics of crawler robots. But most of this is in the theoretical research stage and there is relatively little literature on experimental research on the steering performance of crawler robots. Chen et al. (2007) established a vehicle motion control system model based on the control system simulation platform and did simulation analysis of the steering characteristics of the robot under different rotational speeds and different turning radii. The correctness of the model was verified. Rao et al. (2015) established the steering resistance torque model of the crawler robot at low speed and high speed steering and analyzed the steering parameters of the robot's structural parameters, steering radius, steering power and flexibility during large radius and small radius steering motions. The model is simulated and the variation law between the steering radius, steering power, slip angle and steering flexibility of the robot is obtained, but the steering performance is not verified by experiments. Rui et al. (2015) comprehensively tested the crawler and acquisition methods of the steering performance parameters of the crawler robot, and proposed a method of measuring the steering trajectory using the GPS-based steering performance test system to obtain the actual turning radius of the crawler vehicle. They combined the NI test system, storage speed torque meter and other devices to achieve the test for the steering performance parameters of the robot. The test result of the steering performance of the robot is very good, but the test cost is too high.

In summary, the main purpose of the current steering performance test of the crawler vehicle is still based on the steering performance of the vehicle. Usually, the steering performance parameters of one or several turning radii are mainly tested, and then the relevant parameter model is established. Finally, the model is verified by simulation experiments. However, the specific experimental verification is lacking. In order to realize the comprehensive and systematic research on the steering characteristics of the crawler robot, this paper first analyzes the steering characteristics of the crawler robot from the perspective of steering kinematics and dynamics, and shows the steering system control scheme of the crawler robot. Considering the gyroscope as the steering control angle detection component which plays a key role in the accuracy and rapidity of the steering angle control in the robot steering control, the gyroscope dynamic performance test system is designed, and a specific experiment is designed to analyze the gyroscope rotation dynamic characteristics. Finally, the experimental results are analyzed and studied.

2 STEERING DYNAMICS ANALYSIS OF THE CRAWLER ROBOT

Since the crawler is selected as the robot moving chassis, precise motion control is the key to the effectiveness of the steering control of the crawler robot. Therefore, before the control of the crawler robot, it is necessary to perform kinematics and dynamics analysis on the robot mobile platform, obtain its dynamic model, and then add it to the steering motion controller. In this paper, we use the force dynamic equilibrium method (Xu & Zhou, 2008) to determine the dynamic model of the crawler platform.

Assuming that the crawler robot is running on a horizontal plane, any moment of its motion can be regarded as the rotation around a certain point P. The point P is called the instantaneous center of rotation of the crawler robot, referred to as the instantaneous center. When the robot moves in a straight line, the instantaneous center P can be considered to be at infinity. Figure 1 depicts the motion diagram of the crawler robot moving platform. In

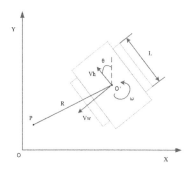

Figure 1. Motion schematic diagram of the crawler robot mobile platform.

260

Figure 1, L is the length of the crawler and ground contact section, O' is the geometric center of the robot platform, R is the instantaneous turning radius of point O', V_h and V_w are the longitudinal forward speed and the lateral sliding velocity of the robot platform, and ω is the rotational angular velocity of the robot platform. The generalized coordinate vector of the defined platform is $q = [x, y, \theta]^T$, where x, y represents the coordinate value of the platform geometric center point O' in the world coordinate system XOY, and θ represents the angle between the platform longitudinal axis and the world coordinate system Y axes. Let the mass of the robot platform be m, and the moment of inertia corresponding to O' is I. Refer to the previous research (Xu & Zhou, 2008) in which the dynamic equation of the crawler mobile platform is used. And in that equation, G is the gravity of the crawler platform at the O' point, u is the platform steering resistance coefficient, f is the running resistance coefficient, i_o is the current output by the outer crawler motor, i_i is the current output by the inner crawler motor, k is the conversion coefficient between the motor output torque, the transmission efficiency and the transmission ratio of the driving wheel are constant values, andρ is the middle convert of the variable $\rho = -m * (\dot{x} \sin\theta + \dot{y} \cos\theta) \dot{\theta}$.

3 THE SCHEME OF THE STEERING CONTROL SYSTEM

As shown in Figure 2, the steering control system of the crawler robot is a double closed-loop control system and consists of a tracked robot moving platform, a angle controller, two speed controllers, two motors, two incremental encoders, and an MPU6050 gyroscope.

In the steering control system of the crawler robot, the attitude angle control loop is used as the outer loop, and the motor speed control loop on both sides of the crawler is used as the inner loop, and the control system is realized based on a dSPACE hardware-in-the-loop simulation control platform. In the outer loop, the angle of deviation Δθ from the reference angle θ_d and the feedback angle θ_f is converted into the reference speed ω_d, wherein, the reference angle θ_d is given by the task decision layer or is set by itself, and the feedback angle θ_f is measured using the MPU6050 gyro sensor. In the inner loop, the reference speed ω_d is respectively used as the reference speed of the closed-loop control of the motor speed on both sides, and the speed deviation Δω1 from the reference speed ω_d and the feedback speed ωf1 which is measured by the incremental encoder1 controls the left motor through the speed controller1. In the same way, the speed deviation Δω2 from the reference speed ω_d and the feedback speed ωf2 which is measured by the incremental encoder2 controls the right motor through the speed controller2. This makes it possible to carry out differential control of the two-sided motor of the robot, thereby realizing steering control of the crawler robot. However, in order to achieve accurate and effective steering control of the crawler robot, accurate angle control must be achieved. To achieve precise angle control, it is necessary to realize the angle measurement. Therefore, it is necessary to analyze the dynamic characteristics of the attitude angle of the gyro sensor.

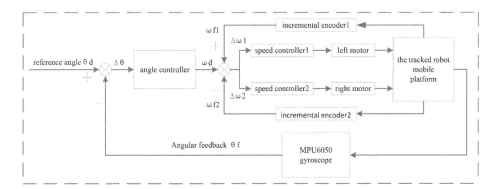

Figure 2.   Steering control system of the crawler robot.

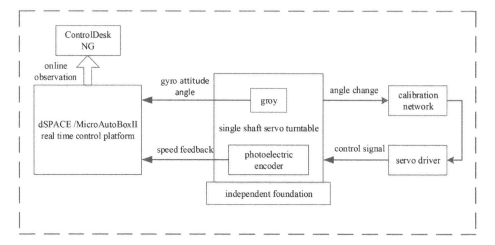

Figure 3.   Test system of gyro dynamic characteristics.

4   EXPERIMENTAL DESIGN FOR THE GYRO DYNAMIC ANALYSIS

4.1   *Test system of the gyro dynamic characteristics*

The test system is shown in Figure 3. The system uses a single shaft servo turntable placed on an independent foundation as the experimental platform, a dSPACE/MicroAutoBoxII controller as the control platform, and a ControlDesk NG as the data recording, online observation, and processing platform. When in use, the gyroscope is installed perpendicular to the plane of the turntable using a special fixture. The rotation speed signal of the turntable is connected to the servo motor on the turntable shaft through the calibration network and servo driver to form a servo control loop.

4.2   *Test method of the gyro dynamic characteristic*

The gyro angle dynamic characteristic test experiment takes the output angle of the single-axis servo motor as the reference angle, fixes the gyro sensor on the single-axis servo motor shaft, makes the gyroscope geometric center and the servo motor shaft coaxial, and makes the electricity for the controller, the gyro, the instrument and servo motor turntable. The experiment mainly tests the dynamic characteristics of the gyro sensor following the servo motor start acceleration phase, and the virtual operation software ControlDesk NG records the angle value of the gyro sensor output and the dynamic response time of the gyroscope following the turning of the servo motor. In order to fully investigate the angular error of the gyro sensor at small angles and large angles, the experiment is performed at various angles differing by 30 degrees between 0 and 360 degrees. Each angle is tested three times, and the test at different angles includes forward and reverse rotations.

5   EXPERIMENTAL RESULTS AND ANALYSIS

In the case of forward rotation, the angle measured mean is obtained from the measured gyro angle and the dynamic response time at the corresponding rotation angle, and the relative error of each angle is obtained according to the angle measured mean and the servo motor reference angle value. The calculation results are shown in Table 1. The reverse situation is the same as the forward rotation, and the relative error is also obtained from the angle measured mean. The calculation results are shown in Table 2.

Table 1. The measured mean value and relative error during forward rotation.

Reference angle (°)	30	60	90	120	150	180	210	240	270	300	330	360
Angle measured mean (°)	29.61	59.25	89.59	119.64	149.42	178.67	208.07	236.98	268.06	297.89	327.68	358.12
Angle relative error (°)	−1.30	−1.25	−0.46	−0.30	−0.39	−0.74	−0.92	−1.26	−0.72	−0.70	−0.70	−0.52
Response time mean (ms)	44	86	131	174	223	271	318	364	396	445	497	560

Table 2. The measured mean value and relative error during reversal rotation.

Reference angle (°)	30	60	90	120	150	180	210	240	270	300	330	360
Angle measured mean (°)	29.59	60.13	89.48	119.79	149.97	179.44	208.65	238.01	268.31	297.29	327.36	357.36
Angle relative error (°)	−1.37	0.22	−0.58	−0.18	−0.02	−0.31	−0.64	−0.83	−0.63	−0.90	−0.80	−0.73
Response time mean (ms)	47	90	136	179	221	278	314	369	392	458	501	563

The average of the relative errors in the forward and reverse directions was measured, and from this we get: in the case of forward rotation, the average relative error of the angle is 0.77%; in the case of reverse, the error is 0.56%.

6 CONCLUSION

In this paper, according to the theory of steering dynamics, the steering characteristics of the crawler robot were analyzed. Then the detailed scheme of the steering control system of the crawler robot is given and the dynamic performance test of the gyroscope was designed. The experimental results show that the steady-state error of the gyroscope angle is less than 0.8%. The results show that the steady-state error of the gyroscope is 0.8 degrees. The angle error range can meet the measurement requirements of the steering angle of the crawler robot.

REFERENCES

Huang, R.-Z., Li, B.-C. & Chen, G. (2015). Design and research for the leg-wheel walking robot. *Mechanical, 42*(08), 44–49.
Chen, S.-Y. & Chen, W.-J. (2007). Review of crawler mobile robot [J]. *Electromechanical Engineering, 12*, 109–112.
Gao J. (2015). The *navigation and control technology of tele-autonomous for small crawler mobile robot*. Beijing: Beijing Institute of Technology.
Chen, S.-Y., Sun, F.-C. & Zhang, C.-N. (2007). Collaborative simulation and control strategy of crawler vehicles electric drive system. *Computer Simulation, 4*, 247–251.
Rao, W., Wang, Z.-J. & Shi, J.-D. (2015). Research on steering characteristics of crawler mobile robot. *Journal of Central South University, 46*(07), 2474–2480.
Rui, Q., Wang, H.-Y. & Wang, X.-L. (2015). Research on the acquisition of steering performance parameters of armored vehicle based on experiments. *Journal of Mechanical Engineering, 51*(12), 127–136.
Xu, D. & Zhou, W. (2008). *Perception, orientation and control of indoor mobile service robots*. Beijing: Science Press.

Automatic Control, Mechatronics and Industrial Engineering – He & Qing (Eds)
© 2019 Taylor & Francis Group, London, ISBN 978-1-138-60427-8

Design and analysis of an autonomous cleaning robot for large scale solar PV farms

M. Sundaram
Department of EEE, PSG College of Technology, Tamil Nadu, India

S. Prabhakaran & T. Jishnu
Department of Robotics and Automation, PSG College of Technology, Tamil Nadu, India

S. Sharma
CSIR-Central Leather Research Institute, Regional Center for Extension and Development, Jalandhar, Punjab, India

ABSTRACT: In this era of depleting natural resources, renewable energy resources like solar power must be captured effectively to meet future power needs. As far India is considered, the installed electrical power capacity is 329.20 GW. Out of this, only 12.28 GW is supplied by solar power plants. One of the main reasons for the poor output from solar farms is the dust deposition on solar panels. Different observations of the effect of dust and other particles on solar panels states that the efficiency of solar farms will reduce drastically. To get the maximum output from solar panels, frequent cleaning is necessary. This work focuses on the design, development and analysis of portable solar panel cleaning equipment which can be installed in any kind of solar Photovoltaic (PV) farm.

Keywords: solar power, efficiency, design, analysis, solar panels

1 INTRODUCTION

The use of electricity is increasing every day. But generation is not on par with consumption. There are two main general categories of electricity generation—renewable and non-renewable energy sources. Solar PV systems, wind energy systems, hydro power plants, biomass and geothermal power plants come under renewable energy sources. But when considering the present scenario, hydro power plants must be considered under non-renewable energy sources since the availability of water is reducing day by day (El-Shobokshy & Hussein, 1993). Coal, oil, and natural gas, which come under the fossil fuel category, can be mentioned as non-renewable energy sources. In India, 59.2% of the total power generation is contributed by thermal power plants which mainly use coal. About 14.8% is contributed by hydro power plants and only 16% of total power generation is contributed by renewable energy sources which include solar PV, wind energy and biomass (Goossens et al., 1993). Depending more on non-renewable energy sources is no longer sustainable under current demand profiles. The importance of renewable energy sources is increasing each day. Out of the available renewable energy sources, solar PV is easily available everywhere with very little one-time investment. Dust accumulation is a significant problem for the solar power industry, so several different types of solar panel cleaning solutions are available across the world. Most of them are constrained to only one particular solar farm and the same system cannot be installed directly on other solar farms with different type of solar panels and ground structure (Shobokshy & Hussein 1993; Sulaiman et al., 2011; Garg, 1973). Eccopia is one of the famous companies which provides solar panel cleaning solutions. It uses a simple rail-like structure in which the cleaning head moves across a bank of panels of size 1600 mm × 990 mm (Eccopia E4). It's a dry cleaning method and it uses a microfiber cloth to clean the solar panels. Another model is the Serbot Swiss Innovations (SERBOT Swiss Innovations) Geeko Solar Farm in which a ground vehicle is used to carry water and other necessary supplies to clean solar panels. A large brush in size of 2 to 3 moves above the solar panels to clean them. Another model from the same brand is Geeko Solar which is used to clean rooftop solar panels. This uses a mini robot-like device which moves above the panels to perform cleaning. Clean Solar Solutions (Clean Solar Solutions) is a UK based solar panel cleaning company that offers automated cleaning solutions for solar panels. They have various solutions for cleaning solar panels such as spray systems, manual cleaning using hoses and water with specially designed brushes, and semi-automated cleaning for solar farms. This robot is similar to SERBOT Swiss innovations. The dimensions of the robot are equal to the width of the panel row. The dirt is cleaned by combining rotating brushes with water. In this work, a solar panel cleaning robot has been developed. It is capable of adapting for any kind of solar farms and easily programmable. The main advantage is that it can be transported to any location of solar farm with ease. This robot is capable of tracking the edges of the solar panels to move in a straight direction and it can turn once it reaches any edge of panels.

2 SOLAR PANEL CLEANING ROBOT AND STANDARD SOLAR PANEL DIMENSIONS

A standard solar panel with the most common dimensions has been selected for the testing and design of a solar panel cleaning robot. Table 1 details the standard solar panel dimensions.

Table 2 details the dimensions of the cleaning robot. Consider if the solar panels are kept vertical, from the dimensions it can be stated that the robot can clean more than half of a single solar panel on one path since the length of brush is more than half of the breadth of the solar panel. On the return path, it can clean the remaining area. There are a few other options when designing the brush for the cleaning robot. Designing a brush for the entire breadth of solar panel will be a tedious task and it cannot be fabricated so easily. If the length of brush is kept below that of half of the solar panel width, it requires more than three trips to complete the cleaning of a single panel. A brush with more than half the width of the solar panel will be a good option and that has been fabricated.

Table 1. Solar panel dimensions.

Parameter	Dimensions/details
Make	Solar Semiconductor
Model	SSI-3M6-245
Type	Ploy Crystalline
Length	1660 mm
Breadth	990 mm
Thickness	42 mm
No. of cells	60
Weight	20 Kgs
Peak Power	245 W
Solar panel inclination (South India)	10.5° to 11.5°

Table 2. Cleaning robot dimensions.

Parameter	Dimensions/details
Length	660 mm
Breadth	535 mm
Width	320 mm
Length of brush	550 mm
Type of brush	Cylindrical

Table 3. Brush speed and cleaning efficiency comparison.

Sl. No.	Brush speed (RPM)	Brush position (From panel)	Remarks
1	100	20 mm	Brush not touching panel. No effective cleaning.
2	100	15 mm	Brush touching panel. Cleaning not as effective as expected. Requires more brush pressure.
3	125	10 mm	Correct brush pressure, speed needs to be increased
4	150	10 mm	Correct brush pressure, speed needs to be increased
5	200	10 mm	Effective cleaning. Dust/dirt not spreading to nearby panels/areas.
6	250	10 mm	Dust flying onto nearby panels and deposited. Motor speed above 800 rpm is not suggested in normal conditions.

3 SELECTION OF OPTIMUM CLEANING MECHANISM

The major dust seen on solar panels are fine particles of soil, sand, dust carried by wind, and bird droppings. It is possible to avoid bird droppings since the chance of birds coming to solar farms is less. To clean the remaining dust particles, a simple scrubbing mechanism will do the work. A set of cylindrical brushes with fine nylon bristles can be selected to clean the solar panels. Hard bristles will damage the solar panels and their efficiency may reduce. An occasional cleaning using water is preferred so that it can wipe off all those hard dust particles which are stuck on the panels. The cleaning method selected here is finalized after many tests and trial runs. The type of brush is decided initially and the type of bristles used and their grade is decided after trial runs. This brush consists of six rows of bristles with a length of 50 mm each. Initially the brushes are driven by a Brushless DC (BLDC) motor with a gearbox assembly. Various speeds are set on the BLDC motor and the cleaning efficiency is analyzed. At higher speeds, the chance for flying dust is more. The comparison of various speeds and pressures applied by brush are discussed in Table 3. The pressure of the brush on the solar panel is varied using two linear actuators. Changing the height of the cleaning head will result in varying brush pressure on the solar panel.

From Table 3 it is evident that effective cleaning is achieved at 200 rpm when the distance between brush and solar panel is kept at 10 mm since it allows the brush bristles to bend a little and clean effectively.

4 BLOCK DIAGRAM OF SOLAR PANEL CLEANING ROBOT

A CAD model of the proposed solar panel cleaning robot has been developed in CAD modeling software. Aluminum sheets of the required thickness have been selected and are cut to the specific dimensions using a laser cutting technique. These pieces are then bent using conventional methods and assembled to form a desired robotic structure. The dimensions are same as that mentioned in Table 2. The working robot is controlled by an ATMEGA 2560 microcontroller. The entire system is powered by a 12 V DC source. Eight infrared sensors are installed at the four corners of the robot to detect the edges of the solar panel for tracking a path as well as preventing the robot from falling from the solar panels. Four DC geared motors are used in this robot; two for driving the cleaning brushes and two for the wheels for robot movement. The motors used to drive the robot are equipped with encoders so that precise movement of robot can be achieved. Feedbacks from the encoders is given to the microcontrollers to get a closed-loop system. The brush motors are programmed to run at a constant speed (200 rpm) in opposite directions. There are two motor drivers each capable of driving two motors at a time with bidirectional feature is used to drive four motors in this design. The maximum current the driver can withstand is 10 A. The driver unit also has overcurrent protection as well as blocked rotor detection features. The purpose of the linear actuators is to raise or lower the entire unit according to the brush conditions. If the robot is equipped with a new set of brushes with long bristles, the clearance required between the solar panel and the robot will need to be more. As the brushes deteriorate with use, the contact between the brush and the solar panel gets reduced. In such conditions, the entire robot can be lowered to make the brush touch the solar panel. This adjustment can be made manually using the HMI unit. The linear actuators also have a feedback mechanism which gives analog outputs from which the robot can detect the position of the stroke of the linear actuator. The feedback from the linear actuators is then given to the microcontroller for a close-loop operation. The HMI method used in this robot consists of a Nokia 5110 1.5″ 84 × 48 display with a backlight. The operator can communicate with the robot through this unit. Infrared sensors used to detect the edges of the solar panel and track a path to achieve proper cleaning. The selection of motors and linear actuators is done after a thorough study on the weight of robot, inclination to be kept and the time required by the robot to travel from bottom to top of the solar panel etc. The specifications of the DC geared motor with encoder are shown in Table 4. The linear actuators used in this robot are from the Actuonix brands L12 series. These linear actuators are driven by DC geared motors with a rack and pinion method to achieve linear movement. The specifications of the linear actuators are shown in Table 5.

Table 4. Wheel motor specifications.

Parameter	Value
Voltage	12 V
Torque	110 gf-cm
Rated motor speed	5950 RPM
Rated current	900 mA
No load speed	7300 rpm
Gear box specifications	
Gear ratio	100
Rated torque	660 Nm
Rated speed	60 RPM
Rated power	7 W

Table 5. Linear actuator specifications.

Parameter	Value
Voltage	12 V
Gear ratio	210:1
Torque	31 N @ 7 mm/sec
Stroke length	50 mm
Rated current (mA)	200 mA
Maximum speed (mm/sec)	6.5 mm/sec

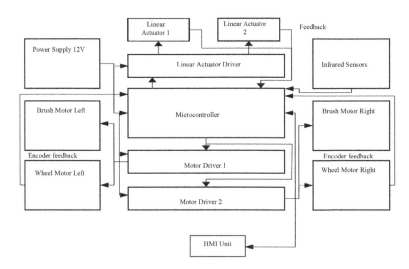

Figure 1. Block diagram representation of a solar panel cleaning robot.

5 WORKING AND PROGRAM FLOW

The solar panel cleaning robot is fully controlled by an **ATMEGA** 2560 microcontroller. This controller is programmed in a particular flow such that it will initially check the status of the brushes and arrange the height of the robot accordingly. Initially the operator will be asked whether to change the brush position or not. If the operator selects yes, the next two options will be displayed in the **HMI** unit. The operator can enter the distance to move the robot up or down manually or select any predefined values. After entering these distance values, the robot will ask from which corner the clearing should start – either left or right. In the next step, robot will start cleaning. The eight infrared sensors placed at the four corners of the robot track the edges of the solar panel so it moves straight. If the initial starting side is selected as left, the two **IR** sensors on the left of the robot will act as edge tracking sensors. In normal running conditions, the two edge tracking sensors will be outside the solar panel. If they come inside the panel, this will be detected and the corresponding motor speed will be varied to keep it on track. Consider a solar array with two panels kept vertical. The solar panel cleaning robot is kept on the right side of the panel. From there, it starts moving forward tracking the edges of solar panel with the help of the **IR** sensor. Once it reaches the end of second panel, the robot makes a – turn and comes down. Then it moves to the next set of panels in that array and the same steps are repeated. Once four of the sensors located at front and any of the sides come out, the robot will stop cleaning and stay at the same place. This occurs when the robot reaches the end of the last panel in the array. The operator has to manually take the robot and put it on the next array to start cleaning it. The major drawback mentioned in existing systems is rectified in this model by making small changes in the software program to match the solar farm layout. This model is also independent of the number or size of solar panels. The programming for the robot's microcontroller ATMEGA 2560 is done using Arduino programming software. The flow of the program is shown in Figure 2. Initial coding for the Nokia 5110 display has been done. The first part of the program directs the flow to the next section according to inputs from the operator. i.e., if the operator selects a brush changing option, the program will jump to the linear actuator section and, after executing that task, it will return to the main screen. Once the brush setting is complete, the program will shift to the tracking edges and the section activating the motors. The outputs

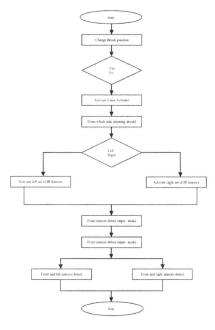

Figure 2.   Flowchart for the program.

from IR sensors are used to operate the wheel motors and the output from wheel motor encoders is used to obtain a constant speed for the operation. The brush motors are set to run constantly at 200 rpm without any changes. The trial run results are really amazing as there are no dust particles left on solar panels.

The edges of the panels are neatly cleaned which is a main location for dust to accumulate. For hard and stuck dust, the robot is made to run two times on the same route to ensure that they are neatly cleaned. This robot can clean one solar panel in 15 seconds which means that an array of 20 panels can be cleaned in less than seven minutes taken into consideration the time taken by the robot to turn at each end-point. The flowchart for the program is shown in Figure 2.

6 RESULTS AND OBSERVATIONS

The following details represent the after effect of dust deposition on the efficiency of solar panels. An experimental study of solar farms has been made to understand the exact power loss occurring because of dust. Solar farm details are provided in Tables 6 and 7.

These technical specifications of solar farm and solar panels come under standard test conditions. In real time application, their efficiency depends on the various factors which are mentioned in the previous section. Initially the test is done for one single panel and then determined for the entire farm. A single panel is entirely covered with dust to analyze the effect on its performance. The maximum and minimum efficiency of a single solar panel without dust is 13.36% and 9.25% respectively. The maximum and minimum efficiency of a single solar panel with dust is 5.64% and 2.33% respectively. So, if this is considered for the whole solar farm, the overall efficiency will be drastically reduced due to the effect of dust. The proposed plant's installed capacity is 1 MW. With the effect of dust, we can expect a loss in efficiency of between 30 and40% depending on several factors like the amount of dust, the type of dust, and its concentration. By utilizing the proposed solar panel cleaning robot, the power drop due to dust deposition can be completely overcome and thereby achieving the desired output from the solar panels.

Table 6. Solar farm details.

Solar farm size	1 Megawatt (MW)
Land utilized	4.8 acre
AC power rating of array	80% of DC power rating
Maximum power capacity of PV system	800 kW (AC)
Estimated annual power production	1.7 million kilowatt hours (kWh)

Table 7. Solar panel specifications.

No of panels installed	5,040
Length	1,660 mm
Breadth	990 mm
Thickness	42 mm
Panel combination	195 W and 200 W capacity types
Module efficiency	15.3% to 17.4%
Cell efficiency	17.8% to 20.1%
Maximum voltage output per panel	55.8 V
Maximum current per panel	3.59 A
Panel weight	20 kg

Array of four solar

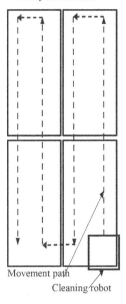

Movement path

Cleaning robot

Figure 3.   Robot movement path.

Figure 4   Test setup inside lab.   Figure 5.   Solar panels before and after cleaning.

7   CONCLUSIONS

The design and development of a solar panel cleaning robot has been done effectively. From the results of trial run, it is found that the cleaning is efficient and less time consuming.

The main advantage of this solar panel clearing method is that it is easily portable and adaptable to any kind of solar farms unlike the other permanent fixing methods mentioned in the studies. The cleanliness of the solar panel before and after cleaning has been deeply studied and it shows that the output from the solar panel after cleaning is 30–40% above panels covered with dust.

REFERENCES

El-Shobokshy, M.S., & Hussein, F.M. (1993). Degradation of photovoltaic cell performance due to dust deposition on to its surface. *Renewable Energy, 3,* 585–590.
Garg, H.P. (1973). Effect of dirt on transparent covers in flat-plate solar energy collectors. *Solar Energy, 15,* 299–302.
Goossens, D., Offer, Z.Y. & Zangvil, A. (1993). Wind tunnel experiments and field investigations of eolian dust deposition on photovoltaic solar collectors. *Solar Energy, 50*(1), 75–84.
Sulaiman, S.A., Hussain, H.H., Leh, N.S.H.N. & and Razali, M.S.I. (2011). Effects of dust on the performance of PV panels. *World Academy of Science, Engineering and Technology, 58,* 588–593.

271

Automatic Control, Mechatronics and Industrial Engineering – He & Qing (Eds)
© 2019 Taylor & Francis Group, London, ISBN 978-1-138-60427-8

Application of a factor graph in MEMS IMU self-alignment technology

H.F. Xing, H. Yang, C. Wang & M. Guo
Tsinghua University, Beijing, China

ABSTRACT: In this work, we describe a novel and feasible technology for the self-alignment of Micro Electro Mechanical System (MEMS) Inertial Measurement Units (IMUs) on swing bases. To the best of the authors' knowledge, this is the first report on the joint use of the Rotation Modulation Technique (RMT), Inertial Space Gravity Based Alignment (ISGBA) and factor graph for self-alignment in MEMS IMUs. RMT is adopted to suppress the impact of inertial sensors bias. Hence ISGBA allows small alignment errors when used with RMT under swing condition. As a result, a factor graph is used to further optimize the state estimation and solve for attitude angles through a message-passing algorithm. Analysis of the experimental data showed that the standard deviations of the pitch, roll, and heading solved using the proposed method were 0.011°, 0.008°, and 1.057°, respectively. These results validate the effectiveness of the proposed approach.

Keywords: MEMS IMU, swing base, self-alignment, rotation modulation technique, inertial space gravity based alignment, factor graph

1 INTRODUCTION

The performance of micro electro mechanical system (MEMS) inertial measurement units (IMUs) has greatly improved following advances in science and technology. MEMS IMUs have achieved wide commercial applications such as in mobile phones, unmanned aerial vehicles (UAVs) and robots owing to their small size, low cost, and light weight (Deppe et al., 2017). Therefore, MEMS IMU has always been a burning research topic.

Initial alignment involves calculating the transformation matrix from the body frame to the navigation frame, and needs to be performed prior to solving for inertial navigation (Liu et al., 2016). Due to the low precision of MEMS gyros, the initial alignment of MEMS IMUs generally require the heading information provided by other sensors such as magnetometers or satellite receivers (Meiling et al., 2017). The MEMS gyroscope north seeker performs its function through a rotation modulation technique (RMT) on a static base without relying on other sensors (Iozan et al., 2012). However, the error is significant under swing circumstances. The eastward gyro drift has the greatest influence on the heading accuracy, but RMT can effectively suppress the effect of the MEMS inertial sensors constant bias that makes it possible to achieve alignment on a swing base (Xing et al., 2018). A swing base is defined as a disturbing condition, such as a moored ship, or a vehicle affected by engine vibration.

This paper proposes the joint use of RMT, inertial space gravity based alignment (ISGBA) and a factor graph to achieve self-alignment in MEMS IMUs on a swing base. A factor graph is a typical probability graph model and a multivariate global function that is factored into the product of several local functions. The message-passing algorithm for factor graphs is the sum-product algorithm (SPA) (Kschischang et al., 2002). Due to the versatility of the factor graph message-passing algorithm, it has a wide range of applications in adaptive filtering, channel equalization, encoding and decoding, and signal processing (Loeliger, 2004).

Hauke Strasdat compares factor graphs with Kalman filter-based algorithms and concludes that factor graphs are more efficient in visual simultaneous localizations and mapping (SLAM) applications (Strasdat et al., 2012). Factor graphs can achieve the optimization of state in terms of navigation with multi-sensors (Chen et al., 2016). Some scholars developed real-time navigation algorithms and systems based on factor graphs, with a system framework that offers good adaptability and flexibility (Chiu et al., 2014, Chiu et al., 2013). Factor graphs can visualize the processing of complex data, shows good scalability and flexibility, and can greatly save computing resources (Kaess et al., 2008). At present, factor graphs have been successfully applied in many fields. The key behind its popularity is that factor graphs use natural and human-readable graphical language to express problems (Ta et al., 2014).

This paper introduces factor graphs in the self-alignment process in MEMS IMUs. It was used to further optimize attitude estimation based on the results obtained by applying the RMT and ISGBA methods.

2 TECHNICAL APPROACH

This section analyzes the self-alignment technology for MEMS IMUs under swing condition. The technology proposed in this paper involves the joint use of three techniques: RMT, ISGBA, and factor graphs. For better explanation of these techniques, definitions of coordinate frames are first provided.

In this paper, the e frame is the earth-fixed frame, the i frame represents inertial frame, the n frame is the navigation frame (East-North-Up coordinate frame), the b frame is the body frame, the s frame is the IMU frame that rotates synchronously with the IMU while the i_{b0} frame is the body inertial frame formed by consolidating the b frame in inertial space. Next, we will discuss the key techniques involved in this paper.

2.1 *Rotation modulation technique*

To simplify the analysis, the RMT principle is analyzed under static condition and it is assumed that the b and n frames coincide.

As shown in Figure 1, the counter clockwise direction is defined as the positive direction, whereas ω is the angular rate of rotation around the z-axis. The s frame is coincident with the b frame at initial time.

Since the RMT has similar effects on the accelerometer bias and the gyro drift, we only analyze the gyro drift here. After time t, the gyro drift in the n frame is as shown below:

$$\varepsilon^n = C_b^n C_s^b \varepsilon^s = \begin{bmatrix} \cos(\omega t) & -\sin(\omega t) & 0 \\ \sin(\omega t) & \cos(\omega t) & 0 \\ 0 & 0 & 1 \end{bmatrix} \begin{bmatrix} \varepsilon_x^s \\ \varepsilon_y^s \\ \varepsilon_z^s \end{bmatrix} = \begin{bmatrix} \varepsilon_x^s \cos(\omega t) - \varepsilon_y^s \sin(\omega t) \\ \varepsilon_x^s \sin(\omega t) - \varepsilon_y^s \cos(\omega t) \\ \varepsilon_z^s \end{bmatrix} \quad (1)$$

where $\varepsilon^s = \begin{bmatrix} \varepsilon_x^s & \varepsilon_y^s & \varepsilon_z^s \end{bmatrix}^T$ is the gyro drift represented in the s frame.

After rotating time T 360°, integration of the gyro drift yields the following formula:

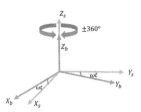

Figure 1.   Rotation modulation technique (RMT) illustration.

$$\int_0^T \varepsilon^n dt = \begin{bmatrix} 0 \\ 0 \\ T\varepsilon_z^s \end{bmatrix} \tag{2}$$

It can be seen that the bias of the inertial sensor is modulated into a periodic signal in the n frame, and integrates to zero in a complete rotation period. The constant bias of inertial sensors in the direction of the rotation axis cannot be modulated. This is the basic principle of the RMT. Further, to reduce the influence of the scale factor error, this paper uses the reciprocating rotation method shown in Figure 1.

2.2 *Inertial space gravity based alignment*

Given that the rotation speed of the earth is constant, the rotation angle of gravity g in the inertial space can be accurately calculated as long as the time is accurate. The direction change of g in the inertial space contains the northward information, so the IMU can be aligned according to the movement of g in the inertial space.

The purpose of ISGBA in rotary MEMS IMUs is to calculate the transformation matrix from the s frame to the n frame expressed by:

$$C_s^n = C_i^n C_{i_{b0}}^i C_s^{i_{b0}} \tag{3}$$

The matrix C_i^n can be calculated based on the earth's rotation rate ω_{ie}, the latitude L and the time interval Δt_k.

$$C_i^n = C_e^n C_i^e = \begin{bmatrix} -\sin\omega_{ie}\Delta t_k & \sin\omega_{ie}\Delta t_k & 0 \\ -\sin L\cos\omega_{ie}\Delta t_k & -\sin L\sin\omega_{ie}\Delta t_k & \cos L \\ \cos L\cos\omega_{ie}\Delta t_k & \cos L\sin\omega_{ie}\Delta t_k & \sin L \end{bmatrix} \tag{4}$$

The gravity in the i frame can be described as:

$$g^i = \begin{bmatrix} -g\cos L\cos\omega_{ie}\Delta t_k \\ -g\cos L\sin\omega_{ie}\Delta t_k \\ -g\sin L \end{bmatrix} \tag{5}$$

The velocity value at time t_k can be determined using (6).

$$V^i(t_k) = \int_{t_0}^{t_k} g^i dt \tag{6}$$

According to the principle of RMT, the impact of accelerometer bias can basically be ignored when the MEMS IMU rotates about the z-axis periodically, thus the velocity value at time t_k in the i_{b0} frame is:

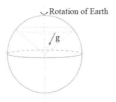

Figure 2.   The movement of g in the inertial space.

$$V^{i_{b0}} = \int_{t_0}^{t_k} C_s^{i_{b0}} f^s dt = -C_i^{i_{b0}} \int_{t_0}^{t_k} g^i dt \qquad (7)$$

where, f^s is the output of the accelerometers.

Next, $C_{i_{b0}}^i$ can be acquired using (8).

$$V^i = C_{i_{b0}}^i V^{i_{b0}} \qquad (8)$$

Then the solution for $C_{i_{b0}}^i$ can be determined by the double-vector attitude determination. We can then update $C_s^{i_{b0}}$ using (9), as follows:

$$\dot{C}_s^{i_{b0}} = C_s^{i_{b0}} [\omega_{i_{b0}s}^s \times] \qquad (9)$$

where, $\omega_{i_{b0}s}^s$ is the output of the gyros containing gyro drift. The introduction of the RMT drastically reduced the effect of gyro drift.

The solution of C_i^n can be obtained based on $C_i^n, C_{i_{b0}}^i$ and $C_s^{i_{b0}}$ from (4), (8), and (9). Then attitude angles including pitch, roll and heading are easily calculated based on the transformation matrix C_s^n. But because of the impact of random noise of the MEMS IMU, state optimization is always needed for better accuracy.

2.3 *Factor graph*

The factor graph is a graphical model composed of variable nodes, factor nodes, and lines. It is intuitive and easy to understand when expressing functions. The basic idea behind this algorithm is connecting the state variables and measurement information using a graph model. After the available measurement values are given, the maximum posterior probability (MAP) estimate of the joint probability distribution function (PDF) is calculated for all states.

Factor graphs are introduced into state optimization, and SPA, as explained in (Kschischang et al., 2002). The velocity error of the MEMS IMU under swing condition can be used as an observation. Therefore, the state model and the observation model can be obtained as follows:

$$\dot{x}(t) = F(t)x(t) + G(t)w(t) \qquad (10)$$

$$y(t) = H(t)x(t) + v(t) \qquad (11)$$

where $F(t) = \begin{bmatrix} -\omega_{ie}^n \times & 0_{3\times3} & -C_s^n & 0_{3\times3} \\ -g^n \times & 0_{3\times3} & 0_{3\times3} & -C_s^n \\ & 0_{6\times12} & & \end{bmatrix}$, $G(t) = \begin{bmatrix} -C_s^n & 0_{3\times3} \\ 0_{3\times3} & C_s^n \\ & 0_{6\times6} \end{bmatrix}$, $w(t)$ and $v(t)$ are system noise and observation noise, respectively.

The state vector is $x = \begin{bmatrix} \phi & \delta V & \varepsilon & \nabla \end{bmatrix}^T$, where ϕ, δV, ε and ∇ represent the misalignment angle vector, the velocity-error vector, the gyro drift vector and the acceleration bias vector, respectively. The observation matrix is $H = [0_{3\times3} \quad I_{3\times3} \quad 0_{3\times3} \quad 0_{3\times3}]^T$.

From the equations (10) and (11), it can be seen that the estimation of state, x is based on observations, y, and can be expressed as $f(x_1, x_2 \cdots x_n \mid y_1, y_2 \cdots y_n)$. According to Bayes formula, the conditional probability density function $f(x_1, x_2 \cdots x_n \mid y_1, y_2 \cdots y_n)$ can be factorized as

$$f(x_1, x_2 \cdots x_n \mid y_1, y_2 \cdots y_n) = \prod_{j=1}^{n} f(x_j \mid x_{j-1}) f(y_j \mid x_j) \qquad (12)$$

Figure 3 is the part of factor graph described in (12). The messages passed in the operation of the SPA are also shown in Figure 3.

According to the rules of SPA, the following formula can be obtained.

276

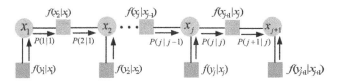

Figure 3.   Part of the factor graph.

(a)   (b)

Figure 4.   (a) The turntable experiment environment, (b) the experimental setup with Epsilon 20.

$$
\begin{aligned}
P_{j|j}(x_j) &= P_{j|j-1}(x_j)f(y_j \,|\, x_j) \\
&= N(x_j, \hat{m}_{j|j-1}, \sigma_{j|j-1}^2)N(y_j, Hx_j, \sigma_v^2) \\
&\propto N(x_j, \hat{m}_{j|j-1}, \sigma_{j|j-1}^2)N(x_j, y_j \,/\, H, \sigma_v^2 \,/\, H^2) \\
&\propto (x_j, \hat{m}_{j|j}, \sigma_{j|j}^2)
\end{aligned} \tag{13}
$$

where, $\hat{m}_{j|j}=\hat{m}_{j|j-1}+\sigma_{j|j-1}^2 H^T G(y_j - H\hat{m}_{j|j-1})$, $\sigma_{j|j}^2 = \sigma_{j|j-1}^2 - \sigma_{j|j-1}^2 H^T G H \sigma_{j|j-1}^2$, with $G \triangleq (\sigma_v^2 + H\sigma_{j|j-1}^2 H^T)^{-1}$ $\hat{m}_{j|j}$ denotes the minimum mean square error (MMSE) estimate of x_j based on the set of observations up to time j.

$$
\begin{aligned}
P_{j+1|j}(x_j) &= \int P_{j|j}(x_j)N(x_{j+1}, Fx_j, Q_w)dx_j \\
&\propto N(x_{j+1}, \hat{m}_{j+1|j}, \sigma_{j+1|j}^2)
\end{aligned} \tag{14}
$$

where, $\hat{m}_{j+1|j}=F\hat{m}_{j|j}$, $\sigma_{j+1|j}^2 = F\sigma_{j|j}^2 F^T + Q_w$.

The iterative solution to $\hat{m}_{j|j}$ this algorithm is the best estimate of the state vector x_j. For more details, refer to (Kschischang et al., 2002).

3   EXPERIMENTAL RESULTS

Turntable-based experiment on real data was carried out. As shown in Figure 4a, the MEMS IMU was installed on a three-axis turntable with the performance parameters listed in Table 1.

We set the parameters of the turntable control program so that the inner frame and the middle frame swung and the outer frame can be rotated reciprocally. Table 2 shows the parameters. Figure 5 displays the angular rate and acceleration of the MEMS IMU for one test. A total of 10 trials were performed to verify the reproducibility. As shown in Figure 4b, the actual attitude of the last stop position acquired with a high-end INS, Epsilon 20, was used as a reference. The roll, pitch and heading angle obtained using Epsilon20 were about 0°, 0° and 166° (anticlockwise).

The data were processed from the z-axis rotation until the end of rotation, as shown in Figure 5. The ISGBA algorithm was used to deal with the 10 sets of data obtained from the turntable experiment, with the results shown in Figure 6. The mean and standard deviation (SD) of the attitude angles at the end of alignment obtained using the ISGBA method is listed in Table 3.

Table 1. MEMS IMU parameters.

Parameter	Gyro	Accelerometer
Repetitiveness of bias	10°/h (1σ)	0.5 mg (1σ)
Random noise	$0.02°/\sqrt{h}$	$5\,\mu g/\sqrt{Hz}$
Measuring range	± 300°/s	± 20 g

Table 2. Parameters of the experiment.

Items	Inner frame	Middle frame	Outer frame
Amplitude (°)	6	10	–
Frequency (Hz)	0.125	0.1	–
Rotation rate (°/s)	–	–	20

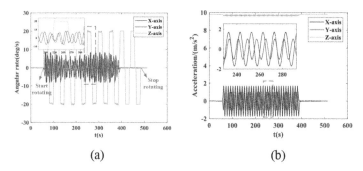

(a)   (b)

Figure 5. (a) Angular rate in the x-, y-, and z-axis, (b) acceleration in the x-, y-, and z-axis.

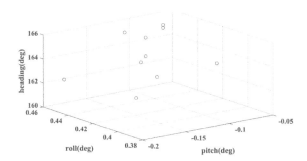

Figure 6. Calculated results of the ISGBA.

Table 3. Statistical results of the ISGBA method.

Parameters	Pitch (°)	Roll (°)	Heading (°)
Mean	−0.122	0.431	164.172
SD	0.031	0.019	1.497

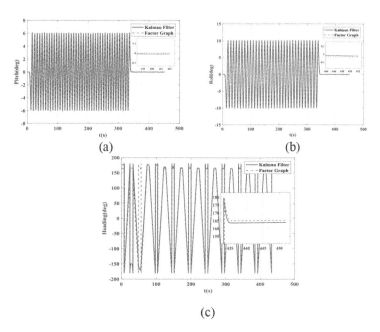

Figure 7. (a) The change of pitch angle solved using a Kalman filter and a factor graph, (b) the change of roll angle, (c) the change of heading angle.

Table 4. Statistical parameters of the estimated attitude angles.

Method	Pitch (°)		Roll (°)		Heading (°)	
	Mean	SD	Mean	SD	Mean	SD
Kalman filter	0.004	0.018	0.003	0.014	166.286	1.196
Factor graph	0.006	0.011	−0.004	0.008	165.856	1.057

As can be seen from the Table 3, the difference between the attitude angles calculated using the ISGBA algorithm and the real attitude angles obtained using Epsilon20 is small. These results are used for further optimization of the state estimation. A factor graph and a Kalman filter are used here for the optimal estimation of state. The effect of the two methods applied to one test is shown in Figure 7.

The mean and SD of the estimated attitude angles at the last time are shown in Table 4. It can be seen that the two optimal methods for the estimation of state can both improve the alignment accuracy. However, the factor graph offers slightly better result than Kalman filter, thereby verifying the effectiveness of the proposed approach.

4 CONCLUSION

This paper presented a self-aligned technology for MEMS IMUs under swing conditions. The RMT, ISGBA algorithm and the factor graph that make up the technology were illustrated in detail. The following conclusions can be drawn: (1) The RMT and ISGBA algorithms are critical in achieving self-alignment in MEMS IMUs and reduces the alignment error. (2) This paper verified that factor graphs can meet the requirements for the optimal estimation of state. The standard deviations of the pitch, roll, and heading obtained using the factor graph were 0.011°, 0.008°, and 1.057°, respectively. (3) Factor graphs are intuitive and flexible, and this is one of the reasons they have become a hot research topic in the field of navigation and information fusion in recent years.

The MEMS IMU self-alignment approach proposed in this paper does not require aiding sensors such as magnetometers; hence it improves the application range of MEMS IMUs. If MEMS inertial sensors with better performance are used, this method can achieve higher accuracy. Future work may focus on error analysis and design of a practical rotary MEMS IMU.

REFERENCES

Chen, W., Zeng, Q., Liu, J., Chen, L. & Wang, H. Research on the multi-sensor information fusion method based on factor graph. Position, Location and Navigation Symposium, 2016. 502–506.

Chiu, H. P., Williams, S., Dellaert, F., Samarasekera, S. & Kumar, R. Robust vision-aided navigation using Sliding-Window Factor graphs. IEEE International Conference on Robotics and Automation, 2013. 46–53.

Chiu, H. P., Zhou, X. S., Carlone, L. & Dellaert, F. Constrained optimal selection for multi-sensor robot navigation using plug-and-play factor graphs. IEEE International Conference on Robotics and Automation, 2014. 663–670.

Deppe, O., Dorner, G., König, S., Martin, T., Voigt, S. & Zimmermann, S. 2017. MEMS and FOG Technologies for Tactical and Navigation Grade Inertial Sensors—Recent Improvements and Comparison. *Sensors,* 17, 567.

Iozan, L. I., Kirkkojaakkola, M., Collin, J., Takala, J. & Rusu, C. 2012. Using a MEMS gyroscope to measure the Earth's rotation for gyrocompassing applications. *Measurement Science & Technology,* 23, 025005.

Kaess, M., Ranganathan, A. & Dellaert, F. 2008. iSAM: Incremental Smoothing and Mapping. *IEEE Transactions on Robotics,* 24, 1365–1378.

Kschischang, F. R., Frey, B. J. & Loeliger, H. A. 2002. Factor graphs and the sum-product algorithm. *IEEE Transactions on Information Theory,* 47, 498–519.

Liu, M., Gao, Y., Li, G., Guang, X. & Li, S. 2016. An improved alignment method for the Strapdown Inertial Navigation System (SINS). *Sensors,* 16, 621.

Loeliger, H. A. 2004. An introduction to factor graphs. *Signal Processing Magazine IEEE,* 21, 28–41.

Meiling, W., Guoqiang, F., Huachao, Y., Yafeng, L., Yi, Y. & Xuan, X. A loosely coupled MEMS-SINS/ GNSS integrated system for land vehicle navigation in urban areas. Vehicular Electronics and Safety (ICVES), 2017 IEEE International Conference on, 2017. IEEE, 103–108.

Strasdat, H., Montiel, J. M. M. & Davison, A. J. 2012. Visual SLAM: Why filter? ☆. *Image & Vision Computing,* 30, 65–77.

Ta, D.-N., Kobilarov, M. & Dellaert, F. A factor graph approach to estimation and model predictive control on unmanned aerial vehicles. Unmanned Aircraft Systems (ICUAS), 2014 International Conference on, 2014. IEEE, 181–188.

Xing, H., Chen, Z., Yang, H., Wang, C., Lin, Z. & Guo, M. 2018. Self-Alignment MEMS IMU Method Based on the Rotation Modulation Technique on a Swing Base. *Sensors,* 18, 1178.

Automatic Control, Mechatronics and Industrial Engineering – He & Qing (Eds)
© 2019 Taylor & Francis Group, London, ISBN 978-1-138-60427-8

Modeling and simulation of dynamic fire assignment for aircraft

M. Zhong, R.N. Yang, J. Wu & H. Zhang
Air Force Engineering University, Xi'an Shanxi, China

ABSTRACT: Dynamic Weapon-Target Assignment (DWTA) is crucially important under the condition when a limited number of weapons need to be employed immediately. We study the problem using the multi-objective optimization model consisting of Static Weapon-Target Assignment models (SWTA). The SWTA of the next wave is used as an input by the strike effect of the previous wave, and the model is updated. Aiming at the high speed of the algorithm, the Multi-Objective Evolutionary Decomposition Algorithm (MOEA/D) is improved, and the conjugate gradient method is mixed to search for optimization. The simulation results show that the improved algorithm improves the speed of the algorithm and completes the dynamic weapon-target assignment while retaining the advantages of the MOEA/D algorithm.

Keywords: weapon-target assignment, weapons science and technology, decomposition multi-objective optimization algorithm, conjugate gradient algorithm

1 INTRODUCTION

Weapon-Target Assignment (WTA) is a process in which weapons are reasonably allocated to specific targets, taking into account operational objectives, target characteristics, weapon performance and other factors (Zhang et al., 2015). WTA is a pivot in battlefield planning, and it is an important factor in determining combat effectiveness (Bayrak et al., 2013). It mainly includes Static Weapon-Target Assignment (SWTA) and Dynamic Weapon-Target Assignment (DWTA). Traditional SWTA does not take account of time. However, in actual combats, the battlefield situation changes rapidly. The DWTA is important for prompt decision-making in the fast changing battlefield situation.

DWTA is a multi-hit wave assignment model, which consists of several wavelet SWTA. The next wave decision-making adjusts and redistributes the target parameters according to the results of previous attacks. In recent years, the research on DWTA has been on the rise. The goal-based DWTA model (Xin et al., 2010; Xin et al., 2011) and the asset-based DWTA model (Hosein & Athans, 1990; Havens, 2002) have been proposed. The DWTA model proposed in this paper considers the target value, threat, weapon cost and other factors to make the model closer to the actual combat situation. Threats are associated with time, that is, high threat targets that have not been destroyed will be prioritized later on.

The WTA model is a completely Non-deterministic Polynomial (NP) problem, as it cannot be directly calculated, except through "guessing" to get the results. Researchers use multi-objective optimization algorithms and other heuristic algorithms to solve the problem. The Decomposition Evolutionary Algorithm (MOEA/D) searches by means of randomness. The algorithm is not easy to drop into local optimal solutions, but has good convergence and uniformity of Pareto optimal solution set. However, due to the real-time and complexity of DWTA, the algorithm has higher speed and convergence requirements. In this paper, an improved decomposition evolution algorithm based on hybrid conjugate gradient method is proposed. The algorithm not only retains the advantages of MOEA/D, but also has good convergence. The simulation results show that the proposed algorithm improves the computation speed and convergence over the original algorithm.

2   PROBLEM MODELING

2.1   *DWTA model analysis*

In actual combat, due to technical constraints, aircrafts cannot launch all the firepower units carried in one wave at a time. It is easy to destroy one target with all the firepower, but it will reduce the effectiveness of firepower distribution and result in the risk of undestroyed targets remaining. Real-time battlefield should divide attacks into multiple waves.

This paper proposes the use of the multi-wave weapon-target assignment model, (DWTA), which is based on the traditional WTA model.

$$DWTA = \sum_{i=1}^{s} SWTA(t) \tag{1}$$

The DWTA model consists of several SWTA processes. The input of the model is updated according to the attacking effect of the previous wave, the threat of the battlefield, the value of the target and other factors, and inputs into the SWTA model to generate a new wave of weapon-target assignment scheme. The model is shown in Figure 1.

2.2   *Target analysis of pre-strike*

Let's denote n as the number of pre-attack targets is n, and is (I = 1, 2, 3... N) The target value. The weight of each target value is determined by the Analytic Hierarchy Process (AHP), which constructs the judgment matrix of each index and obtains its corresponding eigenvectors. The relative weight of one index to another can be obtained after normalization. Assign 1 as unimportant, 3 as equal, 5 as important, 7 as extremely important, between each score take 2, 4, 6, the use of expert evaluation of the target value table shown in Table 1.

The target value evaluation table is converted into a matrix of $n \times n$, $\bar{v}_i$ is the 1/n power of the product of each row element in the matrix.

$$\bar{v}_i = \prod_{j=1}^{n} (v_{ij})^{1/n} \tag{2}$$

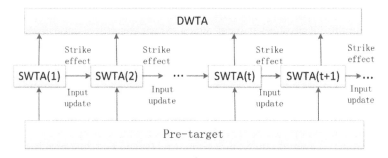

Figure 1.   Firepower distribution flow chart.

Table 1.   Target comparison evaluation form.

Bombing	Target 1	Target 2	...	Target n
Target 1	V_{11}	V_{12}	...	V_{1n}
Target 2	V_{21}	V_{22}	...	V_{2n}
......				
Target n	V_{n1}	V_{n2}	...	V_{nn}

282

The target V_i weight is:

$$v_i = \overline{v}_i \Big/ \sum\nolimits_{i=1}^{n} \overline{v}_i \tag{3}$$

Denote the target threat as $W_i(I = 1, 2, 3... N)$. If the target is not destroyed in this wave, it will still be threatened when the aircraft launches fire in the next wave, so the threat of the undestroyed target will increase with the times of attack. The threat value of the t-th wave at target i is:

$$w_i(t) = 2^{t-1} w_i \tag{4}$$

2.3 SWTA optimization model

Assume there is a total of m-types of weapons. The upper limit of the firepower of the aircraft is M, the number of pre-attack targets is n, P_{ij} $P_{ij}(t) \in [0,1]$ is the damage probability of the i-type weapon to the target j, indicating that in the t-th attack, the i-th fire unit probability of damage to target j is:

$$P_{ij}(t) = 1 - (1 - P_{ij})^{x_{ij}} \tag{5}$$

x_{ij} is the decision variable representing the number of i-type firepower unit attack target j in the t-wave order, which satisfies $x_{ij} \in [0, X_i(t)]$, where $X_i(t)$ is the i-type firepower unit available in t-waves. The probability of damage to target j for all firepower units at t-wave is:

$$P_j(t) = 1 - \prod_{i=1}^{m} \left((1 - P_{ij})^{x_{ij}} \right) \tag{6}$$

The function F1 (t) for target damage probability is constructed as:

$$F1(t) = \sum_{j=1}^{n} \left(v_j \left(1 - \prod_{i=1}^{m} (1 - P_{ij})^{x_{ij}} \right) \right). \tag{7}$$

The function for the target thread F2 (t) is:

$$F2(t) = \sum_{j=1}^{n} \sum_{i=1}^{m} \left(1 - \frac{w_j(t)(1 - P_j(t))}{c_i} \right) \tag{8}$$

F2 shows that the greater the probability that the target is destroyed, the smaller the target threat, and the greater the threat of undestroyed targets as the number of attack waves grows. $c_j \in N$ is the cost of using a firepower unit. Under the premise of destroying the target, the firepower unit used should be as small as possible.

The optimal fire distribution model for the t-th attack wave, SWTA (t) is set up as follows:

$$\begin{cases} \max(F1(t)); \\ \max(F2(t)); \\ x_{ij} \in [0, X_i(t)]; \\ P_{ij}(t) \in [0,1]; \end{cases} \tag{9}$$

2.4 DWTA optimization model

The DWTA model is based on SWTA model. With the change of attack wave number t, the number of each type of firepower unit, attack target state, target threat and other factors change accordingly.

After the t-wave attack ends, the target parameters are reset according to the battlefield feedback. If the target j is destroyed, the target value $v_j = 0$ and the target threat $w_j = 0$. The target parameters $v_j(t + 1)$ $w_j(t + 1)$ $P_j(t + 1)$ are updated to establish the t+1th-order firepower model SWTA (t+1).

3 HYBRID CONJUGATE GRADIENT METHOD FOR MOEA/D ALGORITHM

3.1 *MOEA/D algorithm*

Compared with other kinds of evolutionary algorithms, the core strategy of the MOEA/D algorithm is to decompose the multi-objective optimization problem into several single-objective optimization problems, and then perform computations on the decomposed single-objective problem (Tan, 2013). Compared with the MOGLS and NSGA-II evolutionary algorithms, their sub-problems can co-evolve without repetitive optimization sub-problems, which improves computational efficiency (Zhang et al., 2015; Ding, 2012). They also won the 2009 CEC contest because of their excellent algorithm performance (Zhang et al., 2009). However, due to its large coding length and large population size, the convergence speed of the algorithm is relatively slow, and there is still a certain gap between the speed requirements of real-time planning.

3.2 *Conjugate gradient method*

The conjugate gradient method (Jiang et al., 2007; Hong & Mo, 2009) is a deterministic algorithm based on steepest descent method (Wu et al., 2010) and the Newton method (Xue et al., 2008). Compared with Newton's method, it does not need to compute Hesse matrix and has good quadratic termination, but it is easy to fall into local optimum when solving optimization problems.

3.3 *Hybrid evolutionary algorithm*

3.3.1 *Analysis of hybrid evolutionary algorithm*
For solving the multi-wave firepower distribution model DWTA, not only the optimization of the algorithm solution is required, but also the real-time planning of each wave firepower distribution model is required according to the battlefield feedback. Also, the time window is short, and the convergence speed of the algorithm is also required.

According to the characteristics of the two algorithms, the two algorithms can be combined. Combining the advantages of good global convergence of the MOEA/D algorithm and fast convergence speed of the conjugate gradient method, a MOEA/D algorithm based on the conjugate gradient method is proposed. Previously, some researches have combined the conjugate gradient method with the MOEA/D algorithm. In their literature (Gen & Cheng, 1997), the conjugate gradient method is used to search all individuals in the population of the MOEA/D algorithm, but the computational complexity of the algorithm is too large and the speed advantage is lost. In order to improve the searching ability of the MOEA/D algorithm, the central individual of the population is used as the initial searching point, and the conjugate gradient method is used to search for new individuals, instead of the lowest fitness individuals in the original population. Let the population size be n and the population individual be $a_i \in R, (i = 1, 2, ... m)$, then the population center individual a_0 is:

$$a_0 = \frac{1}{m} \sum_{i=1}^{m} a_i \qquad (10)$$

3.3.2 *Algorithm flow*
The MOEA/D algorithm flow of the hybrid conjugate gradient method is as follows:

Step 1: Population initialization
Step 2: Individual neighbor selection

Step 3: Calculating the fitness value, selecting the individual a_0 of the population center
Step 4: Searching the initial search point a_0 using the conjugate gradient method, and after reaching the iteration precision or number of times, obtaining the search result set a_n^i
Step 5: Generate a new individual from a_n^i and calculate the fitness value, and replace the individual with the lowest fitness value in the population with the new individual
Step 6: Neighbors optimize each other to generate new individuals;
Step 7: The population is updated, and it is judged whether the loop is over. If it is not finished, go to step 2 and continue the loop until the loop ends;

4 SIMULATION RESULTS AND ANALYSIS

The model is simulated in the environment of CPU Core i7, main frequency 2.40 GHZ, RAM 8.0 GB, operating system win10 and MATLAB 2014 a.

4.1 *Setting of simulation parameters*

The number of algorithm population is 150, the neighbor size is 30, and the maximum iteration number is 400. Pre-hitting target is $M = 5$. The aircraft can put up to 8 firepower units in one wave and firepower unit type $M = 4$. The carrying quantity of each type of firepower unit is: $M_1 = 8$, $M_2 = 4$, $M_3 = 6$ and $M_4 = 6$, the values of weapons are: $c_1 = 0.5$, $c_2 = 1$, $c_3 = 0.7$, $c_4 = 0.7$. Target damage probability P_{ij} and target threat value w_j are shown in Table 2 with the target threat and damage probability table. The target value comparison evaluation table is shown in Table 3.

The weight of the target value is obtained according to the analytic hierarchy process. $v_1 = 0.08$, $v_2 = 0.19$, $v_3 = 0.19$, $v_4 = 0.08$, $v_5 = 0.46$.

4.2 *Simulation results analysis*

4.2.1 *Analysis of Pareto front-end diagram*
The MOEA/D algorithm and the hybrid conjugate gradient algorithm are used to plan the first wave, SWTA (1). In the simulation experiment using the traditional MOEA/D algorithm, the Pareto curve has been stabilized in the 200 s, and the convergence time is 22 seconds. Using a mixed MOEA/D algorithm to solve the target, the algorithm converges steadily after 80 iterations using 14 seconds. The Pareto front-end diagrams of the two methods are shown in Figure 2 and Figure 3.

Table 2.   Target threat and damage probability.

Target	Threat value w_j	M_1	M_2	M_3	M_4
T_1	1.5	0.65	0.9	0.8	0.6
T_2	5	0.5	0.6	0.8	0.5
T_3	3	0.75	0.9	0.7	0.8
T_4	1.5	0.8	0.5	0.7	0.4
T_5	10	0.4	0.85	0.5	0.7

Table 3.   Target value comparison evaluation form.

Target	T_1	T_2	T_3	T_4	T_5
T_1	1	1/3	1/3	1	1/5
T_2	3	1	1	3	1/3
T_3	3	1	1	3	1/3
T_4	1	1/3	1/3	1	1/5
T_5	5	3	3	5	1

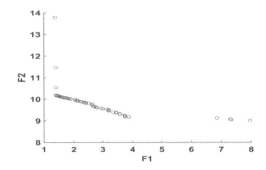
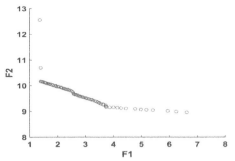

Figure 2. Pareto front-end diagram of traditional MOEA/D.

Figure 3. Pareto front end of mixed MOEA/D.

Table 4. SWTA (1) weapon-target assignment results.

Target	M_1	M_2	M_3	M_4
T_1	0	0	1	0
T_2	1	0	1	0
T_3	1	0	0	1
T_4	1	0	0	0
T_5	0	1	0	1

Table 5. SWTA (2) weapon-target assignment results.

Target	M_1	M_2	M_3	M_4
T_1	1	0	1	0
T_2	0	0	1	2
T_3	2	0	1	0

$P_1(2) = 0.93$, $P_2(2) = 0.95$, $P_4(2) = 0.98$.

Table 6. SWTA (3) weapon-target assignment results.

Target	M_1	M_2	M_3	M_4
T_1	2	1	1	0

$P_1(3) = 0.997$.

Figures 2 and 3 show that the hybrid conjugate gradient MOEA/D algorithm has a smoother Pareto curve than the traditional MOEA/D algorithm. There is no interruption in Figure 3, so the curve continuity is better. In summary, the improved MOEA/D algorithm has been improved in terms of convergence, speed and diversity of the solution distribution.

4.2.2 *Analysis of weapon-target assignment results*

The first wave of weapon-target assignment results is shown in Table 4.

The probability of target destruction is calculated on the basis of the first wave of weapon-target assignment. $P_1(1) = 0.8$, $P_2(1) = 0.9$, $P_3(1) = 0.95$, $P_4(1) = 0.8$, $P_5(1) = 0.95$. Assuming that the target with a destroying probability of more than 95% is destroyed, the target parameters are updated for SWTA (2) and SWTA (3). The results of weapon-target assignment are shown in Tables 5 and 6.

The model has carried on an effective firepower distribution to the target. It generates hits on high-value "key targets" undestroyed by the last wave after considering the target value and threat, and the cost of weapons.

5 CONCLUSION

In this paper, an improved decomposition multi-objective evolutionary algorithm is used to solve the dynamic firepower allocation problem with multi-wave strikes. In contrast with the traditional weapon-target assignment model, the proposed algorithm considers the target value, target threat, fire unit value and other elements all together, so is closer to the actual combat situation. It has an important practical significance.

REFERENCES

Bayrak, A.E. & Polat, F. (2013). Employment of an evolutionary heuristic to solve the target allocation problem. *Allocation Information Sciences, 222*, 675–695.

Ding, D.W. (2012). *Research on the application of decomposition based multi-objective evolutionary algorithm (MEA/D)* in Antenna Optimization Design [D]. Zhenjiang: Jiangsu University.

Gen, M., Cheng, R.W. (1997). Genetic algorithm and engineering design [M]. New York: John Wiley & Sons.

Havens, M.E. (2002). Dynamic allocation of fire and sensors [D].Monterey, CA, US: Naval Postgraduate School.

Hong, L. & Mo, L.L.A. (2009). New conjugate gradient method of global convergence. *Operations Research Transitions, 13*(1), 95–106.

Hosein, P.A. & Athans, M. (1990). *Some analytical results for the dynamic weapon-target allocation problem*. Cambridge, UK: MIT.

Jiang, J.S, HE, C.X. & Pan, S.H. (2007). Optimization computation method. Guangzhou; South China University of Technology Press, pp. 132–138.

Tan, Y.Y. (2013). *Several improved decomposition-based multi-objective evolutionary algorithms and their applications*. Xi''an: Xi'an University of Electronic Science and Technology.

Wu, F, Li, X.M. & Zhu, X.H. (2010). Some important improvement for the gradient method. *Journal of Guangxi University, 35*(4), 596–600.

Xin, B., Chen, J., Zhang, J., et al. (2010). Efficient decision makings for dynamic weapon-target assignment by virtual permutation and tabu search heuristics. *IEEE Transactions on Evolutionary Computation, 40*(6), 649–662.

Xin, B, Chen, J., Peng, Z.H., et al. (2011). An efficient rule-base constructive heuristic to solve dynamic weapon-target assignment problem. *IEEE Transactions on Evolutionary Computation, 41*(3), 598–606.

Xue, Y. (2008). *Principle and method of optimization*. Beijing: Beijing University of Technology Press.

Zhang, Q, Liu, W, Li, H. (2009). The performance of a new version of MOEA/D on CEC 09 Unconstrained MOP Test Instance. *IEEE Congress on Evolutionary Computation (CEC '09), Piscataway*, 203–208.

Zhang, Y., Yang, R.N., Zuo, J.L. & Jing, X.N. (2015). Improved decomposition-based evolutionary algorithm for multi-objective optimization model of dynamic weapon. *Acta Armamentarii, 36*(8), 1533–1540.

Zhang, Y., Yang, R.N., Zuo, J.L. et al. (2015). Enhancing MOEA/D with uniform population initialization, weight vector design and adjustment using uniform design. *Journal of Systems Engineering and Electronics, 26*(5), 1010–1022.

Pattern recognition

Microblog topic mining based on a combined TF-IDF and LDA topic model

B. Ge, W. Zheng, G.M. Yang, Y. Lu & H.J. Zheng
Anhui University of Science and Technology, Huainan, Anhui, China

ABSTRACT: Because keywords in a microblog topic are affected by time of release, influence and ways to emphasize special words (such as "#" and "【】"), a standalone Latent Dirichlet Allocation (LDA) topic model cannot accurately cluster microblog keywords into microblog topics. This paper proposes a mining method based on a combined Term Frequency-Inverse Document Frequency (TF-IDF) and LDA topic model, which can accurately mine microblog keywords and cluster them into microblog topics. This method can effectively overcome the data sparsity problem caused by the length restriction on the microblog. First, we collect microblogs containing keywords (such as "Internet") from the Internet, and perform a pretreatment on them. Then microblog weightings such as release time, influence and keywords are extracted using the TF-IDF algorithm. Finally, the set containing these weighted keywords is used as an input document to train an LDA topic model and achieve the topic mining of the microblog. After recall and precision rate evaluation, this method demonstrates better alignment with the topic of the microblog.

1 INTRODUCTION

Through market investigation, society and business are often aware of trending topics and the discussions around them. A microblog is a good channel for mining information, but the text information of a microblog contains extremely large and varied data. Within the text information stream of a microblog, the traditional data mining method based on word frequency is too simple and cannot intuitively and effectively explore a microblog's topic. In order to gather similar contents or microblogs together and accurately explore microblog topics, this paper proposes a microblog topic mining method based on a combined Term Frequency-Inverse Document Frequency (TF-IDF) and Latent Dirichlet Allocation (LDA) topic model, which involves the weighting of the microblog's time of release, influence and keywords in developing research and experiments.

Microblog topic mining can be divided into two parts: keyword extraction by TF-IDF, and clustering keywords into topics through the LDA topic model. The detailed process is as follows:

1. Collect the microblog texts containing the current keywords via the Internet and perform word segmentation on the microblog texts;
2. Use the TF-IDF algorithm and comprehensively consider the release time, influence and keywords of the microblog to extract a microblog keyword weighting and form a keyword document;
3. Train the LDA topic model with these keyword sets as input, and cluster the keywords into topics according to the probability distribution;
4. Finally, compare the above results with the topic mining results produced by the use of a standalone LDA topic model in order to draw conclusions.

2 RELATED WORK

2.1 *TF-IDF study*

The TF-IDF algorithm and its improved algorithm are widely applied and can be used to process a continuous data flow. This algorithm can realize the test of a real data flow and also capture the most frequent items (Erra et al., 2015). A new distance-based term weighting method is also based on TF-IDF. This method can prevent the term weighting from being compromised in the case of overly large news amounts. The experimental results show that this method can perform better than the standalone TF-IDF in terms of news classification and clustering (Chen, 2017). Combined with the news event evolution model, TF-IDF recalculated the weightings and completed the integration of the models. The results show that this method is better than the baseline (Wen et al., 2017).

2.2 *LDA topic model study*

Recent studies have shown that topic models such as LDA are widely applied. In order to prune irrelevant tweets, a novel method to improve topics learned from Twitter content without modifying the basic machinery of LDA was applied (Hajjem & Latiri, 2017). In order to improve the efficiency of the model, an Enriched LDA (ELDA) model was shown to improve the validity and accuracy of topic identification (Shams & Baraani-Dastjerdi, 2017). In addition, the LDA model can be used for text categorization. A news classification method in TF-IDF, Support Vector Machine (SVM) and LDA thematic model. The classification result precision is improved to some extent (Cao et al., 2017).

3 RESEARCH ON MINING MICROBLOG TOPICS BASED ON A COMBINED TF-IDF AND LDA TOPIC MODEL

The LDA topic model is a method to extract keywords based on word frequency and has been widely used in text mining. In this section, we describe how we combine a microblog's release time, influence and keywords into three aspects of a proposed mining method based on a combined TF-IDF and LDA topic model.

3.1 *Extracting microblog keyword weighting based on TF-IDF*

This paper extracts the keywords of microblog texts preprocessed by the TF-IDF model, which are combined with the microblog's weightings in terms of time of release, influence, and keywords.

3.1.1 *Microblog's release-time weighting*
The release-time weighting of the microblog is based on when the microblog was released and the index describing the importance of the keywords. The closer the microblog release time is to the point of topic mining, the greater the time weighting of the keywords. Thereby, all time-tagged microblogs are aggregated into documents, and the microblog keywords are extracted through TF-IDF. The time weighting corresponding to the microblog release is added to the keywords, and reflects the time value of the microblog. The microblog's release-time weighting is calculated using Equations 1 and 2:

$$P = \sum_{i}^{t} Ki \tag{1}$$

$$weight_{time}(word) = \alpha_0 + \beta \tag{2}$$

where K_i represents the full microblog in time period t. Based on the time of the microblog, the time weighting is added to the keywords; a_0 is the initial time weighting of the microblog keywords, β is the incremental weighting with time, and $weight_{time}(word)$ is the microblog's release-time weighting.

3.1.2 *Microblog's influence weighting*

The microblog's influence weighting is based on the number of fans and the number of likes on the blogposts that describe the importance of keywords. Therefore, in the extraction of keywords from a microblog with a high number of fans and high praise, the keywords acquire a greater weight of influence for the microblog. Thus, when weighted—setting a threshold on the number of fans and the number of likes of the microblog—keywords extracted from the microblog with more than 1,000 fans are given microblog 'fans weights', and keywords extracted from the microblog with more than 100 fans are given microblog 'like weights'. The sum of the microblog's fans weight and its like weight results in the microblog's influence weighting:

$$weight_{fans}(word) = \begin{cases} X_0 & N < 1000 \\ X_0 + (N - 1000) & N \geq 1000 \end{cases} \tag{3}$$

$$weight_{forword}(word) = \begin{cases} Y_0 & M < 100 \\ Y_0 + (M - 100) & M \geq 100 \end{cases} \tag{4}$$

$$weight_{effect}(word) = weight_{fans}(word) + weight_{forword}(word) \tag{5}$$

In Equation 3, $weight_{fans}(word)$ represents the number of microblog fans, X_0 represents the initial value of the microblog's fans weight, and N represents the number of fans. When the number of fans does not reach the threshold of 1,000, the microblog's fans weight takes the initial weight. When the number of fans is 1,000 or more, the microblog's fans weight increases with N. Equation 4 indicates the value of the microblog's like weight: Y_0 represents the initial value of the microblog's like weight, M represents the number of microblog likes, and when the number of likes does not reach the threshold of 100, microblog fans weight $weight_{forword}(word)$ with M increments. In Equation 5, $weight_{effect}(word)$ denotes the influence weighting of the microblog, which is the sum of the fans weight and the like weight.

3.1.3 *Microblog's keyword weighting*

The microblog's keyword weighting means the way to emphasize special words associated with the microblog, an indicator that describes the importance of keywords. The "#" and "【】" characters in the microblog are custom expressions by which the microblog emphasizes its content. The words enclosed by "#" or "【】" tend to indicate the topic of the document. Therefore, this paper gives higher weight to words enclosed by "#" and "【】" characters. In order to highlight the importance of the subject in the description of the microblog, it is necessary to highlight the importance of words in the description of the microblog's topic in "#" and "【】". But "#" and "【】" have been included in the Institute of Computing Technology, Chinese Lexical Analysis System (ICTCLAS) word segmentation tool in the table, and they must be removed from the participle table. The definition of the microblog's keyword weighting is shown in Equation 6:

$$weight_{new}(word) = 2a_0 \tag{6}$$

where $weight_{new}(word)$ is the microblog's keyword weighting, and a_0 is the initial weighting of the word.

In the process of extracting keywords, TF-IDF adds these three weightings to the keywords, namely, the microblog's weighting of release time, the microblog's weighting of influence, and the microblog's weighting of keywords. The sum of the three weightings is the keywords weighting, as shown in Equation 7:

$$weight_{word} = weight_{time}(word) + weight_{effect}(word) + weight_{new}(word) \tag{7}$$

where $weight_{word}$ is the keywords weighting, and keywords are extracted by TF-IDF in combination with the keywords weighting to form a keywords document. This keywords document is then input to the LDA topic model.

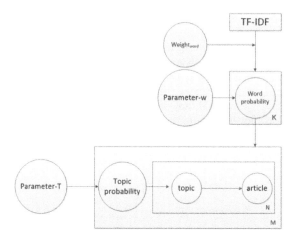

Figure 1. TF-IDF and LDA topic model.

3.2 *Keywords clustering based on the LDA topic model*

The keywords weighting is composed of three aspects of the microblog: release time, influence, and keywords. We combine the keywords weighting with the combined TF-IDF and LDA topic model to extract the microblog topic. We are extracting keywords from microblog texts by TF-IDF and giving keywords the appropriate keywords weighting. The LDA topic model can be used to model the texts' probability. The basic idea is that each text can represent the mixed distribution of a series of topics. It can be expressed as $P(z)$. Each topic represents the probability distribution of all words in the vocabulary and can be expressed as $P(w_i|z_i)$. Hence, the probability distribution of each word in the text can be expressed as follows:

$$P(w_i) = \sum_{j=1}^{T} P(w_i \mid z_i = j) P(z_i = j) \qquad (8)$$

The data set of keywords is used as the method of input training of the LDA model. K words are required to generate M documents containing N words. Word probability means the probability distribution of the words. Parameter-W and Parameter-T represent the parameters at word level of the material. Topic probability represents the variable at document level. Each document corresponds to one topic probability. In other words, the probability of generating the topic varies from one document to another. Sampling of topic probability should be conducted for each document. Topic and article are the variables at word level. The topic is generated by topic probability. The article is jointly generated by topic and Parameter-W. Each word corresponds to one topic. The model is illustrated in Figure 1.

4 RESEARCH AND ANALYSIS

Taking into account the huge number of users and amount of information on the Sina microblog, we selected only some of the user data for our research. We extracted more than 10,000 online microblog messages from 500 authenticated Sina Weibo users between 1 June 2017 and 31 December 2017.

4.1 *Comparison and analysis*

The keywords were extracted by D-algorithm and combined with the microblog's weightings of release time, influence and keywords to reduce the length of the keyword data in order to obtain new data text. We then clustered the keywords in the new data text using the LDA topic model, intercepting the first four keywords of each topic and comparing them with a standalone LDA topic model, as shown in Table 1.

In order to compare the accuracy of a standalone LDA topic model and the TF-IDF and LDA topic model that combines keyword weights in the mining of microblog topics, the measures used are precision P, recall R and F-measure. The scheme is shown in Table 2.

The specific definitions of P, R and F-measure are shown in Equations 9, 10 and 11:

$$P = \frac{a}{a+c} \tag{9}$$

$$R = \frac{b}{b+d} \tag{10}$$

$$F\text{-}measure = \frac{2*P*R}{P+R} \tag{11}$$

Figure 2 shows the results of a comparison of the LDA topic model and the combined TF-IDF and LDA topic model with keywords weighting in extracting the P, R and F-measure of the microblog's keywords.

Table 1. Keyword extraction of two topic mining strategies.

Topic mining method	Topic number	Keywords in the microblog topic
LDA topic model	1	Big data, Blocks, Artificial intelligence, Put away
	2	Artificial intelligence, One, Block, Put away
	3	Internet of things, Put away, Full text, 5 g
	4	Learning, Depth, Put away, Full text
Combined TF-IDF and LDA topic model	1	Artificial Intelligence, China, Innovation, Technology
	2	Depth, Learning, Many, Assembly
	3	Corporate, Global, Future, Development
	4	Blocks, Big Data, Internet of Things, AI

Table 2. Comparison results of keywords.

Category	LDA	TF-IDF + LDA
Keywords match correct	a	b
Keywords match error	c	d

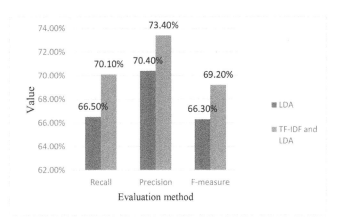

Figure 2. Comparison of P, R and F-measure of the LDA and the combined TF-IDF and LDA models.

From the experimental results comparing the two topic mining strategies, the TF-IDF and LDA topic model combined with keywords weightings produces higher precision, recall, and F-measure values than the standalone LDA topic model. This illustrates the value of the combination of TF-IDF and keywords weighting with the LDA topic model, which has the advantages of a stronger ability to mine information, a higher information accuracy, and effectively overcomes the data sparsity problem caused by the limited length of the microblog and can accurately mine the topics of the microblog.

5 CONCLUSION

With the rise of new media, represented by the microblog, it is difficult to extract the valuable information behind the data from the original massive, disordered and noisy plethora of microblog data. This paper presents a topic mining method based on a combined TF-IDF and LDA topic model of the microblog's specific topic; first, by collecting words containing keywords from the online microblog, and then performing a pretreatment on these microblogs. The weightings of the microblog's release time, influence and keywords are extracted using the TF-IDF algorithm. Finally, the set containing these weighted keywords is used as an input document to train and complete the topic mining of the microblog. Through this experiment, we found that the method of combining the microblog's release time, influence and keyword weightings using the TF-IDF and LDA topic model can ensure the accuracy of the current topic mining of microblogs from many angles, and make the semantics of each keyword of the topic more cohesive: the distinction between microblog topics is made more obvious. Precision, recall, and F-measure values have been improved. Therefore, in conclusion, the combination of the microblog's release time, influence and keyword weightings using the combined TF-IDF and LDA topic model strategy is much more aligned with a specific microblog topic than the standalone use of the LDA topic model.

REFERENCES

Cao, B., Liu, X., Liu, J. & Tang. (2017). Domain-aware Mashup service clustering based on LDA topic model from multiple data sources. *Information and Software Technology*, *90*, 40–54.

Chen, C.-H. (2017). Improved TFIDF in big news retrieval: An empirical study. *Pattern Recognition Letters*, *93*, 113–122.

Erra, U., Senatore, S., Minnella, F. & Caggianese, G. (2015). Approximate TF-IDF based on topic extraction from massive message stream using the GPU. *Information Sciences*, *292*, 143–161.

Hajjem, M. & Latiri, C. (2017). Combining IR and LDA topic modeling for filtering microblogs. *Procedia Computer Science*, *112*, 761–770.

Shams, M. & Baraani-Dastjerdi, A. (2017). Enriched LDA (ELDA): Combination of latent Dirichlet allocation with word co-occurrence analysis for aspect extraction. *Expert Systems with Applications*, *80*, 136–146.

Wen, A., Lin, W., Ma, Y., Xie, H. & Zhang, G. (2017). News event evolution model based on the reading willingness and modified TF-IDF formula. *Journal of High Speed Networks*, *23*(1), 33–47.

Automatic Control, Mechatronics and Industrial Engineering – He & Qing (Eds)
© 2019 Taylor & Francis Group, London, ISBN 978-1-138-60427-8

Application of collaborative representation in audio signal recognition

Q.F. Huang & X.J. Li
Zhonghuan Information College, Tianjin University of Technology, Tianjin, China

ABSTRACT: Feature extraction and classifier design are the main problems of acoustic signal recognition algorithms. In this paper, we extract the mel-frequency cepstral coefficients of the acoustic features of vehicles in complex scenes. Collaborative representation is introduced for the design of a classification scheme (Collaborative Representative Classification, CRC), which synthetically considers the relationship among samples. Experiments show that the proposed algorithm produces good performance in vehicle recognition for the case of a complex data set. Compared with other classification algorithms, the method improves the precision of recognition.

Keywords: collaborative representation, mel-frequency cepstral coefficients, vehicle recognition, complex scenes

1 INTRODUCTION

Recently, sparse coding or sparse representation (Zhang et al., 2011) has been widely studied to solve signal recognition, partially due to the progress of l_0-norm and l_1-norm minimization techniques. In pattern classification, sparse representation can also be used to classify a signal and detect a target.

Sparse representation codes a signal y over a dictionary φ, such that $y \approx \varphi \alpha$ where α is a sparse vector. The sparsity of α can be measured by l_0-norm, which counts the number of non-zeros in α. Because the combinatorial l_0-norm is NP-hard, the l_1-minimization, as the closest convey function to l_0-norm minimization, is widely employed in sparse coding: $\min_\alpha \|\alpha\|_1 \ s.t. \|y - \varphi\alpha\|_2 \le \varepsilon$, where ε is a small constant. Although l_1-minimization is much more efficient than l_0-minimization, it is still time-consuming, and hence many fast algorithms have been proposed to speed up the l_1-minimization process.

Most literature (e.g. Kopparapu & Laxminarayana, 2010; Boucheron et al., 2012) places too much emphasis on the role of l_1-norm sparsity in signal classification, while the role of collaborative representation, using the training samples from all classes to represent the query sample y, is largely ignored.

In this paper we propose a new classification scheme, namely Collaborative Representation Classification (CRC), which has significantly less complexity than Sparse Representation Classification (SRC), but leads to very competitive classification results. Section 2 briefly reviews the Mel-Frequency Cepstral Coefficient (MFCC) algorithms (Gang et al., 2010) that are used to extract the audio features. Section 3 briefly reviews SRC (Yang et al., 2011; Kua et al., 2011) and Section 4 presents the CRC scheme (McLaughlin et al., 2017; Larrain et al., 2017; Feng & Zhou, 2016). Section 5 presents the results of extensive experiments, and Section 6 concludes the paper.

2 MFCC ALGORITHMS

MFCC algorithms involve the following steps when used for parameter extraction:

1. Endpoint detection

   For an original audio signal, we need to detect the starting point and termination point, and then delete the silent segment. In this way, we can reduce the amount of calculation and improve the accuracy of feature extraction.

2. Pre-emphasis

   In order to obtain a flat signal spectrum, which can make spectrum analysis easier, we usually achieve it by the application of a digital pre-emphasis filter $H(z)$ after the signal has been sampled and quantized:

$$H(z) = 1 - a * z^{-1} (0.9 \leq a \leq 1) \tag{1}$$

3. Windowing

   Because the audio signal is a typical time-varying signal, we use a Hamming window to cut out N audio signal samples. Every sample continues for about 10 ms, so we get a series of approximately stable signals.

$$w(n) = 0.54 - 0.46\cos(2\pi n/(n-1)) \quad n = 2,3,...N-1 \tag{2}$$

4. Fast Fourier Transform (FFT) conversion and energy spectrum

   The audio signal is fast and unstable in the time domain, so we usually convert the signal to the frequency domain by using a FFT conversion to obtain the spectrum and energy spectrum of the signal.

5. MFCC parameter and differential cepstrum

   We use 24 triangular window filters for signal filtering, to simulate the masking effect of the human ear. We take the log of the output and then do a discrete cosine transformation. After this, we obtain the standard MFCC parameter; in order to better reflect the dynamic characteristic of the acoustic signals and improve recognition accuracy, we need the differential cepstrum:

$$d(n) = \frac{1}{\sqrt{\sum_{i=-k}^{k} i^2}} \sum_{i=-k}^{k} i \times c(n+i) \tag{3}$$

3 THE SRC SCHEME

Natural signals, such as sound, image or seismic data, can be stored in compressed form, in terms of their projection, given a suitable basis. When the basis is chosen properly, a large number of projection coefficients are either zero or small enough to be ignored. If a signal has only s non-zero coefficients, it is said to be s-sparse. If a large number of projection coefficients are small enough to be ignored, then the signal is said to be compressible. The signal acquisition model of Compressed Sensing (CS) is quite similar to a conventional sensing framework. If χ represents the signal to be sensed, then the sensing process may be represented as:

$$Y = \Phi \chi \tag{4}$$

where Φ is an m-by-n measurement matrix and Y is a measurement vector. The signal of interest χ can be expressed on a representational basis as:

$$\Psi x = \chi \tag{5}$$

where x is the s-sparse vector, representing projection coefficients of χ on Ψ. Measurement vector Y can now be rewritten in terms of x as:

$$Y = \theta x \tag{6}$$

where $\theta = \Phi\Psi$ is an $(m \times n)$ – dimensional, reconstruction matrix. In a signal reconstruction model, we need to solve $Y = \theta x$. However, in a signal classification model, instead of recovering the signal, we need to find the appropriate category for the signal; we call this Sparse Representation Classification (SRC). The SRC algorithm involves the following steps:

1. Normalize the columns of X to have unit l_2-norm.
2. Code y over X via l_1-minimization:

$$(\hat{\alpha}) = \arg\min_\alpha \| \alpha \|_1 \ s.t. \| y - X\alpha \|_2 < \varepsilon \tag{7}$$

where constant ε is to account for the dense small noise in y, or to balance the coding error of y and the sparsity of α.
3. Compute the residuals:

$$e_i(y) = \| y - X_i\hat{\alpha}_i \|_2 \tag{8}$$

where $\hat{\alpha}$ is the coding coefficient vector associated with class i.
4. Output the identity of y as:

$$identity(y) = \arg\min_i\{e_i\} \tag{9}$$

From the above, we can see that the SRC involves two key aspects: first, the coding vector of query sample y is required to be sparse; second, the coding of y is performed collaboratively over the dataset X, instead of each subset X_i.

4 COLLABORATIVE REPRESENTATION-BASED CLASSIFICATION (CRC)

Sparsity is the key insight in compressed sensing—most signals admit a decomposition over a reduced set of signals from the same class. Unfortunately, there is no known algebraic solution to such an l_1-regularized least-squares formulation. A combined l_1/l_2 regularization tends to robustify/group the coefficients of the solution while enforcing sparsity. The representation obtained, seen as a linear decomposition over a pool of samples, has a structural meaning in that the residuals and the solution coefficients reveal the importance of each sample (or group thereof) for the new input query. This information is used in the classification (or assignment) of the query sample to the most appropriate class of samples (the one with minimum residual error or largest coefficient impact).

The basic Ordinary Least Squares (OLS) problem aims at optimizing:

$$\hat{\beta}_{OLS} = \arg_\beta\min \| y - X\beta \|^2 \tag{10}$$

where X is the data matrix with $(m \times n)$-dimensional samples and β is the vector of coefficients from the representation of the query y. If $(X^TX)^{-1}$ exists, the algebraic solution is given by:

$$\hat{\beta}_{OLS} = (X^TX)^{-1}X^Ty \tag{11}$$

The Collaborative Representation with regularized least squares, here abbreviated as CR, solves:

$$\hat{\beta}_{CR} = \arg\min \| y - X\beta \|^2 + \lambda_{CR} \| \beta \|^2 \tag{12}$$

where λ_{CR} is a regulatory parameter. The algebraic solution becomes:

$$\hat{\beta}_{CR} = (X^TX + \lambda_{CR}I)^{-1}X^Ty \tag{13}$$

where I is the $m \times m$ identity matrix.

Usually, the information used for classification is the residual corresponding to each class c:

$$r_c(y) = \| y - X_c \hat{\beta}_c \| \tag{14}$$

where $\hat{\beta}_c$ and X_c are the coefficients and samples corresponding to class c from the full representation of y, defined by the coefficients $\hat{\beta}_c$ and the training samples X. The classification decision is taken using:

$$class(y) = \arg_c \min r_c(y) \tag{15}$$

For a Collaborative Representation Classifier with regularized least squares (CRC) we use (as above):

$$\hat{\beta}_{CR} = Py, \quad P = (X^T X + \lambda_{CR} I)^{-1} X^T \tag{16}$$

and the regularized residuals are taken as:

$$r_c(y) = \| y - X_c \hat{\beta}_c \| / \| \hat{\beta}_c \| \tag{17}$$

The CRC decision is taken as above. P does not depend on the query y and can be precomputed. This confers a large computational advantage on CRC over SRC, which runs a query-dependent optimization. The CRC algorithm involves the following steps:

1. Normalize the columns of X to have unit l_2-norm
2. Code y over X by $\hat{\rho} = Py$ where $P = (X^T X + \lambda \cdot I)^{-1} X^T y$
3. Compute the regularized residuals:
   $r_i = \| y - X_i \cdot \hat{\rho}_i \|_2 / \| \hat{\rho}_i \|_2$
4. Output the identity of y as:
   $Identity(y) = \arg\min_i \{r_i\}$.

5 EXPERIMENTAL RESULTS

5.1 *Experimental data*

All the experimental data in this paper is from the wireless sensor network database (the Acoustic Vehicle Classification Dataset) of the SITEX02 experiment of Project SensIT at DARPA: *http://www.ecs.umass.edu/~mduarte/Software.html* [accessed 7 December 2018]. The database contains two audio files of military vehicles: an Assault Amphibian Vehicle (AAV) and a Dragon Wagon (DW).

5.2 *Description and experimental results*

For the experiment we select the audio signal with a frequency of 4,960 Hz (see Figure 1), and then use the FFT and MFCC algorithms to extract features.

First, we use the SRC algorithms to classify the signal using either an FFT or an MFCC eigenvector, and then we make a comparison of the two eigenvectors (see Figure 2). For the experiment, we used 90 samples. After extracting the features, we randomly selected n samples as training samples, and used the remainder as testing samples. Finally, we compared the recognition rate.

From Table 1, we can see that as the number of training samples is increased, the recognition rate of both eigenvectors increases slightly for each classification algorithm. In addition, we can see that the MFCC eigenvectors perform better than the FFT eigenvectors in classifying the AAV signal, and a little worse than the latter in classifying the DW signal. However, if we consider the overall performance, we can see that the MFCC eigenvectors outperform the FFT ones.

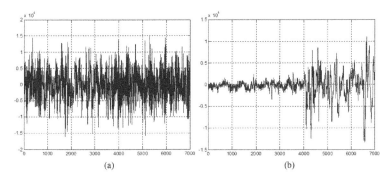

Figure 1. Time series of military vehicle sounds: (a) AAV; (b) DW.

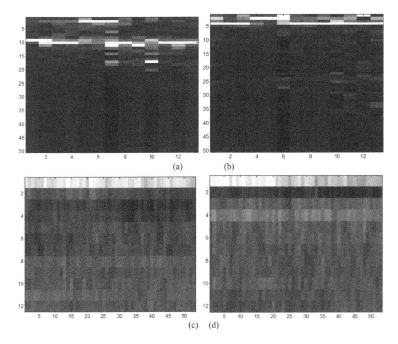

Figure 2. Eigenvectors of military vehicle sounds: (a) FFT eigenvector of AAV; (b) FFT eigenvector of DW; (c) MFCC eigenvector of AAV; (d) MFCC eigenvector of DW.

Table 1. Comparison of FFT and MFCC eigenvectors in signal classification.

		Training number		
Eigenvector	Vehicle category	20	40	60
FFT	AAV	75.14%	73.60%	75.33%
	DW	91.71%	95.60%	94.67%
MFCC	AAV	82.57%	85.47%	84.89%
	DW	91.14%	92.93%	93.11%

Thus, having extracted the MFCC eigenvectors, we used either the SRC or the CRC algorithm to classify the signal, and then made a comparison of the two. For the experiment, we used 90 samples. After extracting the features, we randomly selected n samples as training samples, and used the remainder as testing samples. Last, we compared the recognition rate and time cost of the two algorithms.

Table 2. Comparison of SRC and CRC algorithms in signal classification (recognition rate and compute time).

Classification algorithm	Vehicle category	Training number							
		20		30		40		50	
SRC	AAV	0.76	12.65 s	0.82	31.07 s	0.88	54.14 s	0.90	72.94 s
	DW	0.96		0.95		0.92		0.90	
CRC	AAV	0.81	0.04 s	0.80	0.06 s	0.78	0.06 s	0.80	0.07 s
	DW	0.91		0.93		0.90		0.85	

From Table 2, we can see that as the number of training samples is increased, the time cost of both algorithms increases, while the overall recognition rate increases slightly. Although the CRC algorithm produces a slightly lower recognition rate than the SRC algorithm, it costs much less in time terms.

6 CONCLUSION

In this paper, we briefly introduce the acoustic signal recognition system and its component parts. Then we introduce a new acoustic signal recognition algorithm, CRC, and make a detailed comparison with the SRC method, comparing recognition rate and time cost. We find that the CRC algorithm has almost the same recognition rate as SRC, but at a much lower cost in time. However, we have not investigated further the impact that different signals may have on the result. In ongoing research, we may explore the method prior to signal extraction. Because SRC is becoming increasingly popular in signal processing, the advantage of CRC in costing much less time should give it broader application in the future.

REFERENCES

Boucheron, L.E., De Leon, P.L. & Sandoval, S. (2012). Low bit-rate speech coding through quantization of mel-frequency cepstral coefficients. *IEEE Transactions on Audio, Speech, and Language Processing, 20*(2), 610–619.
Feng, Q. & Zhou, Y. (2016). Kernel combined sparse representation for disease recognition. *IEEE Transactions on Multimedia, 18*(10), 1956–1968.
Gang, N., Kai, W., Xizhi, F. & Naishu, C. (2010). Voice activity detection algorithm based on mel-scale frequency log-spectral energy difference in noise environment. In *Proceedings of Third International Conference on Information and Computing (ICIC 2010), Wuxi, Jiang Su, China, 4–6 June 2010* (vol. 4, pp. 212–215). Los Alamitos, CA: IEEE Computer Society. doi:10.1109/ICIC.2010.324
Kopparapu, S.K. & Laxminarayana, M. (2010). Choice of mel filter bank in computing MFCC of a resampled speech. In *10th International Conference on Information Sciences, Signal Processing and their Applications (ISSPA 2010), Kuala Lumpur, Malaysia, 10–13 May 2010* (pp. 121–124). New York: IEEE. doi:10.1109/ISSPA.2010.5605491.
Kua, J.M.K., Ambikairajah, E., Epps, J. & Togneri, R. (2011). Speaker verification using sparse representation classification. In *2011 IEEE International Conference on Acoustics, Speech, and Signal Processing: Proceedings* (pp. 4548–4551). New York: IEEE. doi:10.1109/ICASSP.2011.5947366
Larrain, T., Bernhard, J.S., Mery, D. & Bowyer, K.W. (2017). Face recognition using spare fingerprint classification algorithm. *IEEE Transactions on Information Forensics and Security, 12*(7), 1646–1657.
McLaughlin, N., Ming, J. & Crookes, D. (2017). Largest matching areas for illumination and occlusion robust face recognition. *IEEE Transactions on Cybernetics, 47*(3), 796–808.
Yang, M., Zhang, L., Feng, X. & Zhang, D. (2011). Fisher discrimination dictionary learning for sparse representation. In *Proceedings of 2011 International Conference on Computer Vision* (pp. 543–550). Washington, DC: IEEE Computer Society. doi:10.1109/ICCV.2011.6126286
Zhang, L., Yang, M. & Feng, X. (2011). Sparse representation or collaborative representation: Which helps face recognition? In *Proceedings of 2011 International Conference on Computer Vision* (pp. 471–478). Washington, DC: IEEE Computer Society. doi:10.1109/ICCV.2011.6126277

Automatic Control, Mechatronics and Industrial Engineering – He & Qing (Eds)
© 2019 Taylor & Francis Group, London, ISBN 978-1-138-60427-8

Structured assessment of ABET student outcomes 'a' and 'e' in mathematics for engineering students

S. Kadry
Department of Mathematics and Computer Science, Faculty of Science, Beirut Arab University, Lebanon

ABSTRACT: Student outcomes are statements that describe the attributes, skills and abilities that students should have by the time of graduation. For quality assurance evaluation, these outcomes must be assessed through the curriculum and extracurricular activities. In order to evaluate the level to which an outcome has been met, it is necessary to select some courses where the outcome is highly covered. Whether an outcome is 'highly' covered can be identified using Bloom's taxonomy levels, the number of hours of coverage, the weighting of its assessment and other parameters. Course goals must be linked to the student outcomes and defined in measurable terms. Using just the two ABET outcomes 'a' and 'e' as examples, this paper presents a structured approach to directly and indirectly assessing student outcomes in mathematics through a partial differential equation course and by defining appropriate measurable course goals. The strategy to assess the outcomes is also explained.

Keywords: assessment, outcomes, ABET, accreditation

1 INTRODUCTION

Academic accreditation is an official recognition and validation that a higher education institution is assessed positively or negatively against a set of standards and criteria. It is a sign of the quality of continuous assessment and improvement of their programs. There are two types of academic accreditation: institutional and specialized. The first of these accredits the institution as a whole and not a specific program within the institution, which is what the specialized accreditation does. This research study focuses on this second type of accreditation. One of the important factors of this accreditation is involving different stakeholders and the community (parents, employers, etc.) in the process of assessment. Accreditation benefits students, instructors and the community. Accreditation is evidence that an academic program that leads to a degree has met certain standards essential in producing graduates who are ready to enter their chosen professions. Students who graduate from accredited universities have access to enhanced opportunities in employment, licensure (e.g. Professional Engineer), registration and certification (e.g. IEEE society membership), graduate education and global mobility (e.g. student transfer acceptance, applications for graduate schools).

ABET (formerly the Accreditation Board of Engineering and Technology) is a nonprofit and non-governmental accreditation agency for academic programs in four disciplines: applied science, computing, engineering, and engineering technology. As of 22 November 2014, ABET accredits more than 3,400 academic programs at nearly 700 colleges and universities in 28 countries (USA and 27 other countries). ABET provides specialized, programmatic accreditation that evaluates an individual program of study, rather than evaluating the entire institution.

ABET accreditation processes and procedures are totally voluntary and achieved through peer review. It provides assurance that a college or university program meets the quality standards established by the profession for which the program prepares its students. In the USA, ABET is recognized by the Council for Higher Education Accreditation (CHEA), and by mutual agreement with many countries like Canada and Australia.

Recently, I developed a systematic process to assess the student outcomes in mathematics courses (Kadry, 2015a, 2015b). In this paper, I focus on just two outcomes as an example, and discuss the new approach I used to develop an assessment methodology in mathematics courses, conforming with ABET accreditation standards (ABET, 2018).

2 STUDENT OUTCOMES

ABET requires each program seeking accreditation to develop a clear set of student outcomes (referred to as Criterion 3), collect direct and indirect assessment data through several courses and surveys, determine the degree to which the outcomes are achieved, and use the results of the evaluation to improve the program. The ABET website has several documents related to assessment (ABET, 2018).

Criterion 3 (Student Outcomes) of the ABET Criteria is a set of characteristics that each program must have documented in relation to student outcomes that prepare graduates to accomplish the institutional program educational objectives. As mentioned, student outcomes describe the attributes, skills, and abilities that students should have upon graduating from the program. Listed below are the 11 student outcomes defined by ABET (normally referred to as 'a to k'):

a. An ability to apply knowledge of mathematics, science, and engineering;
b. An ability to design and conduct experiments, as well as to analyze and interpret data;
c. An ability to design a system, component, or process to meet desired needs within realistic constraints such as economic, environmental, social, political, ethical, health and safety, manufacturability, and sustainability;
d. An ability to function on multidisciplinary teams;
e. An ability to identify, formulate, and solve engineering problems;
f. An understanding of professional and ethical responsibility;
g. An ability to communicate effectively;
h. The broad education necessary to understand the impact of engineering solutions in a global, economic, environmental, and societal context;
i. Recognition of the need for, and an ability to engage in, lifelong learning;
j. A knowledge of contemporary issues;
k. An ability to use the techniques, skills, and modern engineering tools necessary for engineering practice.

Every core course in a curriculum includes a set of course goals; the goals are linked to the student outcomes (i.e. 'a to k'). At the end of a semester, instructors are expected to measure those outcomes using direct and indirect assessment for the courses they have taught.

3 ASSESSMENT TYPES

Broadly speaking, there are two types of assessment: formative and summative. Formative assessment observes student learning progressively during the learning process. For example, the instructor asks students at the end of each class for their feedback and, based on that, the instructor might adjust his/her teaching method of the topic in the next class. The assessment techniques for this type may include quizzes, minutes paper, very small survey, and so on.

Summative assessment evaluates student learning at the end of a chapter or topic. The assessment technique for this type may include exams, projects, presentations, and so on.

4 PERFORMANCE INDICATORS

A performance indicator is a statement that includes an action verb. This verb is usually taken from Bloom's taxonomy in order to be measurable. The idea here is to use the measurable verb in order to evaluate accurately and effectively whether a student outcome can be judged to have been achieved. Performance indicators are quantitative tools and are usually expressed as a rate, ratio or percentage. In ABET terminology (ABET, 2018), performance indicators are "effective assessments that use relevant direct, indirect, quantitative and qualitative measures as appropriate to the outcome being measured". However, with growing maturity in the assessment process, programs are expected to place greater emphasis on direct assessment of student learning, based on demonstrable student work, rather than indirect assessment, which is based on student, alumni, or employer surveys.

In this paper, I will focus solely on outcomes 'a' and 'e' as examples, and describe the assessment methodology. I will also use direct and indirect assessment of the outcome through evaluation of student work and survey. Clearly, outcomes 'a' and 'e' are broad statements and their assessment requires de composition into a number of simpler measurable aspects that allow one to determine the extent to which the outcome is met. In ABET literature, these measurable aspects are known as performance indicators. Performance indicators are written with action verbs such as define, demonstrate, discriminate, evaluate, and interpret. Performance indicators also spell out specific subject contents (Darandari & Murphy, 2013; Briedis, 2002; Keshavarz & Baghdarnia, 2007; Bloxham & Boyd, 2007). In this case, we select a partial differential equation course for assessment. The reason for selection of this course is explained below.

4.1 *Bloom's taxonomy in the math context*

Table 1 describes Bloom's taxonomy and related measurable verbs that can be applied in math assessment.

It is strongly recommended that the verbs in Table 1 are used while preparing the 'course goals' of mathematics courses so that the assessment of the outcomes is made more straightforward.

4.2 *Outcomes 'a' and 'e'*

In order to remove any ambiguity and understand more deeply the difference between these two outcomes, I shall break them down to different factors, so the assessment of each one becomes more accurate.

Outcome 'a': An ability to apply knowledge of mathematics, science, and engineering.

- a1: Apply knowledge of mathematics to distinguish between dependent and independent variables. Describe derivative and integral of functions. Select an appropriate model. Use analytical and numerical techniques to solve equations. Apply concepts of integral and differential calculus and linear algebra to solve problems.
- a2: Apply knowledge of science and engineering to describe the principal of physical, chemical or biological systems. Apply thermodynamic principles. Apply materials principles. Analyze data using statistics principles. Analyze models of systems.

Outcome 'e': An ability to identify, formulate, and solve engineering problems.

- e1: Identifies and understands all problems associated with current methods.
- e2: Uses knowledge to construct models of physical systems.
- e3: Identifies and analyzes potential solutions to engineering problem.
- e4: Assesses the effectiveness and accuracy of different approach.
- e5: Indicates how theory can be used in practice.
- e6: Select appropriate solutions to the engineering problem.

4.3 *Course goals*

An example of the course goals of a Partial Differential Equation (PDE) course might be:

- CG1: Model engineering systems using PDE(e)
- CG2: Solve linear first and second order PDE(a)

Table 1. The application of Bloom's taxonomy to math assessment.

Bloom's level	Application examples	Verbs
Remember	Student remembers how to: – Use number properties – Use functions' notation – Use basic formula of probability – Use basic formula of statistics – Use methods to solve equations	Use, Select, List, Locate
Understand	Student understands how to: – Convert between units – Interpret components of expressions – Interpret experiments – Use existing function to build new one	Estimate, Explain, Relate
Apply	Students applies: – Solve equations – Solve inequality – Perform polynomial operations – Draw functions – Perform operations of vectors – Perform operations of matrices	Solve, Perform, Draw
Analyze	Student analyzes: – Compare different solutions – Categorize differential equations – Differentiate different functions – Interpret graphical solution	Compare, Classify, Categorize, Interpret
Evaluate	Student evaluates: – Solve problems with different models – Interpret and predict result – Describe geometrically different models	Solve, Describe
Create	Student creates: – Proof of theorems – Model problem – Design problems	Design, Model

– CG3: Find Fourier series of function(a)
– ....
– ....

The letters contained in parentheses indicate the associated student outcome.

5 OUTCOMES ASSESSMENT

In order to assess outcomes 'a' and 'e', I will use direct and indirect assessment. For the direct assessment, I will include in the exam the problem described below and illustrated in Figure 1.

A thin rectangular plate coincides with the region defined by $2 \leq x \leq 8$, $1 \leq y \leq 5$. The left end and the bottom of the plate are insulated. The top of the plate is held at temperature zero, and the right end of the plate is held at temperature $f(y)=y$.

Q1: Model the problem using a partial differential equation.
Q2: Solve the partial differential equation obtained in Q1, using an appropriate technique.

As we can see, Q1 addresses CG1 and covers outcome 'e' (in particular, e2) and Q2 addresses CG2 and covers outcome 'a' (in particular, a1). Now, in order to decide if the outcome is achieved, we need to define the associated threshold. For instance, we can say that if 70% of students get 70 and above on Q1 then outcome 'e' is achieved in this course; otherwise, we need to identify and implement some improvements in relation to CG1 for the next semester.

For the indirect assessment, we use a survey, as illustrated in Table 2, that students and instructors are required to complete at the end of the term.

The outcome is achieved if the average of all students is '3' or above, for example, and the same for the instructor evaluation. At the end, an outcome is achieved in this course if all assessment, direct and indirect, is achieved.

We recommend that each outcome is assessed over three courses. Hence an outcome is achieved if it is achieved in at least two courses. Thus, in the example of Figure 2, outcome 'e' is achieved.

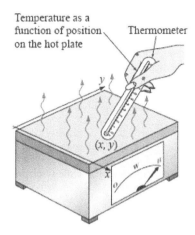

Figure 1. Steady-state temperatures in a rectangular plate.

Table 2. End-of-term survey for indirect assessment.

Please rate how well this class addressed each one of the following course learning outcomes or goals. (4 = Excellent; 3 = Good; 2 = Average; 1 = Poor; 0 = N/A)

Course goal	Student evaluation						
	4	3	2	1	0	Average	%
CG1: Model engineering systems using PDE(e)							
CG2: Solve linear first and second order PDE(a)							
......							

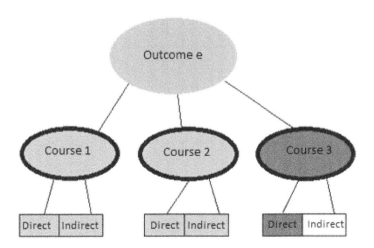

Figure 2. Outcome assessment regime.

6 CONCLUSION

In this paper, I explained the difference between the two ABET student outcomes 'a' and 'e', and how a faculty can assess these through direct and indirect assessment. The proposed strategy can be extended to cover all other outcomes.

REFERENCES

ABET. (2018). *Accreditation board for engineering and Technology*. Retrieved from *www.abet.org.*

Bloxham, S. & Boyd, P. (2007). *Developing effective assessment in higher education: A practical guide.* Maidenhead, UK: Open University Press.

Briedis, D. (2002). Developing effective assessment of student professional outcomes. *International Journal of Engineering Education, 18*(2), 208–216.

Darandari, E. & Murphy, A. (2013). Assessment of student learning. In L. Smith & A. Abouammoh (Eds.), *Higher education in Saudi Arabia: Achievements, challenges and opportunities* (pp. 61–71). London: Springer.

Kadry, S. (2015a). Systematic assessment of student outcomes in mathematics for engineering students. In *2015 IEEE Global Engineering Education Conference (EDUCON), Tallinn* (pp. 782–788). doi: 10.1109/EDUCON.2015.7096060.

Kadry, S. (2015b). Quality-assurance assessment of learning outcomes in mathematics. *International Journal of Quality Assurance in Engineering and Technology Education (IJQAETE), 4*(2), 37–48. doi:10.4018/IJQAETE.2015040104.

Keshavarz, M. & Baghdarnia, M. (2013). Assessment of student professional outcomes for continuous improvement. *Journal of Learning Design, 6*(2), 33–40.

Automatic Control, Mechatronics and Industrial Engineering – He & Qing (Eds)
© *2019 Taylor & Francis Group, London, ISBN 978-1-138-60427-8*

Research and application of voiceprint recognition based on a deep recurrent neural network

K. Luo & L. Fu

Key Laboratory of Dependable Service Computing in Cyber Physical Society, Ministry of Education, School of Big Data and Software Engineering, Chongqing University, Chongqing, China

ABSTRACT: Voiceprint recognition is one of the most popular biometric recognition technologies, which can recognize the identity of the speaker through the voice. Based on Convolutional Neural Network (CNN) and deep Recurrent Neural Network (RNN) voiceprint recognition program, the research of voiceprint recognition technology is called CDRNN. The CDRNN processes the speaker's original speech information through a series of processes and generates a two-dimensional spectrogram. The CNN is better than the advantage of processing the image to extract the personality characteristics of the speech signal from the spectrogram. These personality features are then inputted into the deep RNN. Voiceprint identification is then used to determine the identity of the speaker. The experimental results showed that the CDRNN can obtain better recognition accuracy than other schemes such as Gaussian Mixture Model-Universal Background Model (GMM-UBM).

Keywords: voiceprint recognition, deep recurrent neural network, convolutional neural network, spectrogram

1   INTRODUCTION

Compared with the traditional account password scheme, the identity authentication mechanism based on the biometrics (Jain et al., 2004) of people has a more secure and reliable advantage. Voiceprint recognition technology has a wide range of needs and prospects in the fields of finance, online transactions, and national defense (Furui, 1997).

There has been much research on voiceprint recognition technology. Early research on speaker recognition concentrated on feature parameter extraction and model matching. From the aspect of acoustic feature parameter extraction, the simulated auditory feature line of linear predictive coefficients, perceptual linear prediction coefficients (Hermansky, 1990) and mel-frequency cepstral coefficients (Vergin et al., 1999) have been proposed. For model matching, speech recognition technology is used in human voiceprint recognition such as dynamic time warping (Dutta, 2008), vector quantization (Gray, 1990), artificial neural network (Gardner & Dorling, 1998; Jain et al., 1996) and other technologies. The Gaussian Mixture Model (GMM) is one of the key methods of voiceprint recognition because of its simple and reliable performance and stable performance (Reynolds & Rose, 1995). Based on GMM, Reynolds proposed the Gaussian Mixture Model-Universal Background Model (GUM-UBM) to push the voiceprint recognition to practical applications (Reynolds et al., 2000).

With the development of deep learning technology, it achieved good results in the field of image processing and speech recognition (Abdel-Hamid et al., 2012; Schmidhuber, 2014; Simonyan & Zisserman, 2014) such as Palaz et al. (2015) who analyzed the Convolutional Neural Network (CNN), and for voice recognition, achieved good results. Inspired by this, some studies began to apply deep learning techniques to speaker recognition (Kanagasundaram et al., 2016; Phapatanaburi et al., 2016; Richardson et al., 2015). Richardson used Deep Neural Networks (DNN) for speaker recognition, and extracted a frame-level feature from speech signals by constructing an i-vector system based on bottleneck features. Phapatanaburi used GMM and DNN to identify dialects in the dialect environment with reverberation. Kanagasundaram (2016) used a singular phoneme, combined with DNN and simplified the Gaussian probability linear discriminant analysis to model a phrase's tone signal and to identify the speaker. Since speech information is a continuous context-correlated signal, and Recurrent Neural Networks (RNN) is good at processing sequence signals, Graves (Graves et al., 2013; Sak et al., 2014) introduced the identification of speakers through RNN. Wang et al., (2016) used CNN and RNN to complete multi-label image classification, Fan et al., (2016) used it for video-based emotional perception, and Jiang et al., (2016) used it for event detection of motion video and so on, but there is almost no work for voiceprint recognition.

CNN is good at image feature extraction, and an RNN network has advantages in timing modeling. Therefore, combining the advantages of CNN and RNN, this paper proposes a Voiceprint Recognition Mechanism (CDRNN) based on CNN and deep RNN, and applies CNN and RNN to voiceprint recognition. The CDRNN first converts the speaker's original speech into a spectrogram, and then uses the structural advantages of CNN to automatically extract the speaker's personality characteristics from the spectrogram; these personality features are inputted into the deep RNN to complete the classification. Achieve the recognition of the speaker's voiceprint.

2 CDRNN DESIGN

After generation of the spectrogram, this data will be inputted into the neural network for feature extraction and classification. This paper uses the CNN and deep RNN network to realize the feature extraction and classification of speech signals. CNN is particularly good at processing images, and the spectrogram is actually a two-dimensional grayscale image. The various properties of the image reflect various characteristics of the speaker's voice signal. Therefore, taking the spectrogram as input, the CNN network automatically extracts the personality features of the speech segment from the inputted two-dimensional grayscale spectrogram. CNN consists of multiple convolutional layers and pooling layers, where the convolutional layer can extract different features of the speech segment, and the pooling layer can perform translation, scaling or other deformation operations on the inputted two-dimensional grayscale image. After pooling, the same post-pooling features are still generated, thereby reducing the impact of spectrum changes.

The CNN structure part is composed of n convolution pooling units, as shown in Figure 1, where n needs to be set according to actual conditions.

The structure of a convolutional pooling unit is shown in Figure 2. The ReLU is the activation function and max pool is the pooling function. In order to enable the network to converge quickly, the training speed of the network is also accelerated by the batch normalization algorithm.

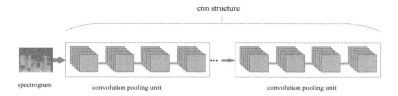

Figure 1. CNN network structure.

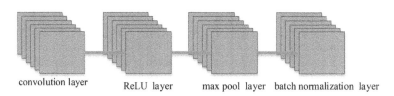

Figure 2. Convolution pooling unit.

For the convolution pooling unit, when the pooling layer performs the pooling operation, it only pools on the frequency and does not pool in time. This is mainly because pooling in time is likely to result in loss of timing information in the spectrogram. In addition, the number of convolutional pooling units and feature maps, the number of feature maps, the size of the convolution kernel, and the size of the step and even the size of the pooled area, also needs to be set experimentally based on specific problems and data sets.

2.1 Deep RNN network design

When the CNN network processes the two-dimensional gray image of the spectrogram, its output is used as an input to the deep RNN to perform further time series modeling. The deep RNN is constructed by superimposing the hidden layers of several RNNs, with the output of the previous hidden layer being the input of the next hidden layer. Compared to the neurons in the normal hidden layer, the neurons contained in the hidden layer of the deep RNN are connected.

2.1.1 Deep RNN input layer design

The CNN network processes the input spectrogram through n convolution pooling units. The processed output is a C-small spectrum with a size of $F \times T$, where C is the Number of Feature Maps, and F and T are the Height and Width of the output petmap, respectively. A sequence can be used to represent the output of the CNN network: $S = [S_1, S_1, \dots S_i, S_T]$; $1 \leq i \leq T$; and the element S_i in the sequence is a size $C \times F$ vector. CNN will output T vectors of size $C \times F$, there is a correspondence between them. CNN network output sequence S_i as an input at Time i for RNN. That is, the input of RNN, at Time i, is a vector of $C \times F$ dimension, and its step size is equal to T. Figure 3 shows the correspondence between the output sequence of the CNN and the RNN input.

2.1.2 RNN hidden layer

Deep RNN is stacked by a plurality of RNN, where in each layer of the input sequence as the output sequence of the next layer; this structure is shown in Figure 4. Compared with traditional

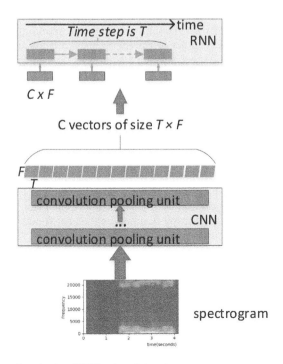

Figure 3.   CNN network output as RNN network input.

311

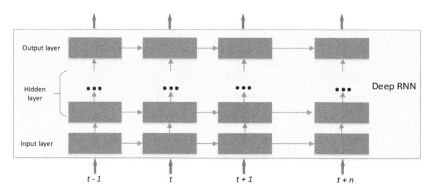

Figure 4.　Deep RNN network structure.

neural networks, deep RNN is characterized by a time feedback loop at each level. For practical problems, the hidden layer in deep RNN usually uses improved RNNs, such as LSTM-RNN or GRU-RNN, which solves the problem of a lack of long-term dependencies in the basic RNN. This allows the neural network to remember input data for a longer span of time. For deep RNN networks, when designing the structure of a hidden layer, the number of nodes in the hidden layer of the hidden layer needs to be considered. These two parameters also need to be set according to actual needs. In general, under the tendency of the same number of parameters, setting more layers can achieve better results than increasing the number of nodes per layer.

2.1.3　*Deep RNN output layer*
The output layer of deep RNN is classified using a softmax classifier, and the softmax classification makes the number of nodes in the output layer correspond to the number of speakers.

2.2　*Network model training*

The training of the CDRNN model adopts the method of supervised learning. First, all the data should be labeled, and then the data and the corresponding label are used as the training set. Assuming that the Number of Speech Signals to be Trained is K (generated by K speakers), the sequence of the spectrogram generated by the *i-th* speech signal is $S_i = (S_{i1}, S_{i2}, \ldots S_{im})$, where m is the Number of Spectrograms generated by the speech signal, and the *j-th* spectrogram S_{ij} corresponds to a Two-Dimensional Matrix. Giving it a tag value of $i-1$ means that all the spectrograms of the same speaker have the same tag, and this tag identifies the speaker's ID S_{ij}, and its label $i-1$ constitutes a Training Sample $(S_{ij}, i-1)$.

Before the training samples are trained, the sample data needs to be normalized. The data is scaled to a certain extent, mapped to a cell, thereby removing the unit limit of the data, and converted to a dimensionless number. At the same time, the data can be standardized to improve the convergence speed and accuracy of the model. In this paper, the min-max standardization mechanism commonly used in machine learning is used to normalize each pixel of a two-dimensional gray image. After data normalization, the pixel value interval is [0,1].

The CDRNN selects the cross-entropy function as the cost function, and uses the Error Back Propagation (BP) and Back Propagation Through Time (BPTT) algorithms to calculate the gradient and complete the training of the sample data.

2.3　*Network model identification*

The speaker produces a test speech signal that is pre-processed to generate n spectrograms, which are also normalized and then sequentially input into the CDRNN network model. The model will eventually output the ID of the speaker corresponding to each spectrogram. Obviously, n spectrograms will output the n IDs, and the ID with the most occurrences is considered to be the speaker ID of the test voice.

3 SIMULATION EXPERIMENT

3.1 *Experimental setup*

The experimental platform using Google's open source deep learning framework is Tensor-Flow. The sample data is trained on the TensorFlow platform, and the trained model, can be transplanted to the mobile phone. The latter can sample the speaker's voice and identify the voiceprint through the trained model. The machine that trains the sample data is configured as: Intel(R) Core™ i7-6700 CPU @ 3.40GHz8 core CPU and 2 NVIDIA GTX980Ti (6G) GPU. The memory size is 24G.

3.1.1 *Voice data set*

The speech data used in the experiment was collected from the real environment. 10 to 20 minutes of voice data was recorded for each of the 40 different students via a smartphone. Due to environmental factors, background noise data was inevitably included in the collected speech signals. Each student's speech data was divided into speech segments of 1s duration, with the first 80% of the data as the training data set and the last 20% of the data as the test data set. The recognition rate $= \frac{n_1}{n} \times 100\%$, of which n_1 is the Number of Correct Speech Segments and n is the Total Number of Speech Segments in the test data set.

3.1.2 *Spectral parameters*

When generating a spectrogram for each speech segment, the frame length is set to 512; 256 pixels are then obtained after generating the spectrogram, which corresponds to the height of the spectrogram. In fact, only the first 128 pixels are taken during the experiment because the frequency of the speech signal is generally in the range of 300–3000 Hz, and the signal outside the interval is a noise signal, which can be ignored. The other parameter frame shift is set to 160. Since the sampling frequency is 16 KHz, the speech segment of 1s duration will generate 16, 000 sample points. A 100 frames can be obtained, which means that the spectrogram width is 100 pixels.

3.1.3 *CNN structural parameters*

The parameters of the CNN, such as the number of convolution pooling units, the step size, the convolution kernel size, and the number of feature maps, are determined based on the parameters of the actual data set. After actual adjustment, the parameters of the CNN structure are set as follows:

a. The number of convolution pooling units is $n = 4$. The number of feature maps for the first pooled unit is set to 32, and the number of feature maps for the last three pooled units is set to 64.
b. The convolutional layer has a convolution kernel size of 5×5 and a step size of 1×1, and convolution is performed both in the frequency and time direction.
c. The size of the pooled area is set to 1×1. The step size is still 1×1, and the pooling is performed only in the frequency direction.

3.1.4 *Deep RNN structure parameters*

The two important parameters of deep RNN are the number of layers of the RNN and the number of nodes per layer. RNN more layers, the stronger the ability to identify the speaker ID. This means that a large number of layers of the RNN and the number of nodes per layer easier to over-fitting. RNN structures obtained by various different combinations of the two parameters, and tested recognition rate under different network structures. The number of layers at the highest recognition rate and the network structure corresponding to each layer of RNN selection nodes are used as Deep RNN parameters.

As shown in Figure 5, the number of layers of the RNN is 1, 3, 5, and 7, and the number of nodes per layer is 128, 256, and 512, so that 12 combinations can be obtained; corresponding to 12 network models. It was seen, with increasing number of RNN layers, that the recognition rate rose substantially. Similarly, when the number of RNN layers were not more than 5 (more nodes on each layer) the higher the recognition rate. However, when the number of RNN layers is 7, the number of nodes per layer is 512, and the recognition rate is lower than the recognition rate when the number of nodes per layer is 256. This shows that the more layers and the number of nodes per layer, the worse the recognition result is. The reason is that as the number of layers and the number of nodes per layer increase, the geometric number of parameters increases, and the size of the training set is limited, which easily leads to over-fitting. Based on the experimental results, the number of RNN layers is set to 7, and the number of nodes per layer is set to 256.

3.2 Experimental results

Firstly, the recognition accuracy of the four mechanisms based on CDRNN, GMM-UBM (Reynolds et al., 2000), DNN (Kanagasundaram et al., 2016) and GMM-DNN (Phapatanaburi et al., 2016) on this speech data set was compared. The results are shown in Figure 6. As the number of speakers increases, the recognition rate of the four mechanisms decreases, and the recognition rate of GMM-UBM decreases very rapidly. This is because the value of the key parameter mixture in GMM-UBM has a greater impact on the result. The recognition rate of CDRNN is slower, and the recognition rate of GMM-UBM is about 18% higher, especially when the number of speakers is large. In addition, CDRNN is about 6% higher than DNN and GMM-DNN, indicating that the back-end uses RNN to get better results than using DNN.

The CDRNN uses a front-end network model such as CNN+RNN. It compares performance with a deep network model that uses only RNN modeling, a front-end DNN and a back-end using RNN.

Figure 7 shows the recognition accuracy of the above three models when the number of RNN layers is 1, 3, and 5, and the number of nodes per layer is 256. It can be seen that with the increase in number of RNN layers, the recognition accuracy of the three network models has been

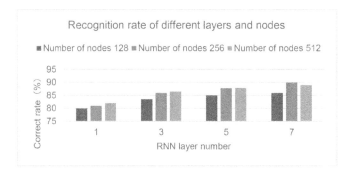

Figure 5.   Recognition rate of different network parameters in deep RNN.

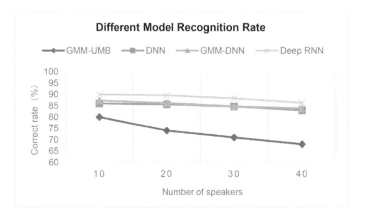

Figure 6.   Performance comparison of the four schemes.

improved to some extent, and the recognition accuracy of the RNN network model is the lowest. On the premise that the back-end uses the same RNN network, the recognition rate obtained by the front-end using the CNN network is higher than that of the front-end using the DNN network. Then the front-end of the network model is fixed as a CNN network, and the back-end is a DNN and RNN network, respectively. The number of layers is 1, 3, 5, and the number of nodes in each layer is 128, 256 and 512 respectively. The result is shown in the Figure 8.

As can be seen from Figure 8, the recognition rate also increases as the number of layers and the number of nodes per layer increases. Regardless of the number of layers in the back-end network of the two models and the number of nodes in each layer, when the parameters are the same, the recognition rate of the CNN+RNN network model is about 4% higher than that of CNN+DNN. This reflects the advantages of the CNN+RNN structure.

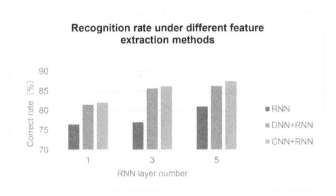

Figure 7. Recognition rate of RNN, DNN+RNN and CNN+RNN models.

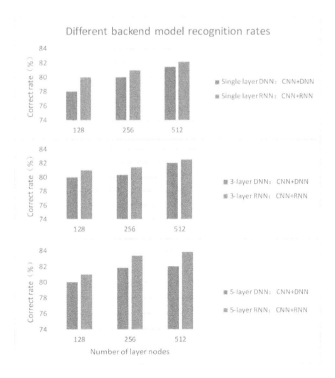

Figure 8. Recognition rate of CNN+DNN and CNN+RNN models.

4 CONCLUSION

Using CNN's ability to process images and RNN's ability to model time series data, this paper proposes a CDRNN model that combines the advantages of CNN and RNN and uses it for voiceprint recognition. Through the comparison of real voice data sets, using CDRNN for training and testing, the accuracy of voiceprint recognition is higher than other models.

REFERENCES

Abdel-Hamid, O., Mohamed, A., Jiang, H. (2012). Applying convolutional neural networks concepts to hybrid NN-HMM model for speech recognition. *Proceedings of IEEE International Conference on Acoustics, Speech and Signal Processing*, 4277–4280.

Dutta, T. (2008). Dynamic time warping based approach to text-dependent speaker identification using spectrograms. *Proceedings of the Congress on Image and Signal Processing*, New York, USA, 354–360.

Fan, Y., Lu, X., Li, D. (2016). Video-based emotion recognition using CNN-RNN and C3D hybrid networks. *Proceedings of the 18th ACM International Conference on Multimodal Interaction*, 445–450.

Furui, S. (1997). Recent advances in speaker recognition. *Pattern Recognition Letters*, *18*(9), 859–872.

Gardner, M.W. & Dorling, S. (1998). Artificial neural networks – a review of applications in the atmospheric sciences. *Atmospheric Environment*, *32*(14–15), 2627–2636.

Graves, A., Mohamed, A. & Hinton, G. (2013). Speech recognition with deep recurrent neural networks. *Proceedings of IEEE International Conference on Acoustics, Speech and Signal Processing*, Florence, Italy, 6645–6649.

Gray, R. (1990). Vector Quantization. *IEEE ASSP (Acoustics, Speech, and Signal Processing) Magazine, 1*(2), 75–100.

Hermansky, H. (1990). Perceptual linear predictive (PLP) analysis of speech. *Journal of the Acoustical Society of America*, *87*(4), 1738–52.

Jain, A., Mao, J. & Mohiuddin, K. (1996). Artificial neural networks: A tutorial. *Computer*, *29*(3), 31–44.

Jain, A, Ross, A. & Prabhakar, S. (2004). An introduction to biometric recognition. *IEEE Transactions on Circuits & Systems for Video Technology*, *14*(1), 4–20.

Jiang, H., Lu, Y. & Xue, J. (2016). Automatic soccer video event detection based on a deep neural network combined CNN and RNN. *Proceedings of the 28th IEEE International Conference on Tools with Artificial Intelligence*, 490–494.

Kanagasundaram, A., Dean, D., Sridharan, S. & Fookes, C. (2016). DNN based speaker recognition on short utterances. *Proceedings of Speaker & Language Recognition Workshop, 4–20.*

Palaz, D., Magimai, M. & Collobert, R. (2015). Analysis of CNN-based speech recognition system using raw speech as input. *Proceedings of International Speech*, 11–15.

Phapatanaburi, K., Wang, L. & Sakagami, R. (2016). Distant-talking accent recognition by combining GMM and DNN. *Multimedia Tools & Applications*, *75*(9), 5109–5124.

Reynolds, D. & Rose, R. (1995). Robust text-independent speaker identification using Gaussian mixture speaker models. *IEEE Transactions on Speech & Audio Processing*, *3*(1), 72–83.

Reynolds, D., Quatieri, T. & Dunn, R. (2000). Speaker verification using adapted Gaussian mixture models. *Digital Signal Processing*, *10*(1–3), 19–41.

Richardson, F., Reynolds, D. & Dehak, N. (2015). Deep neural network approaches to speaker and language recognition. *IEEE Signal Processing Letters*, *22*(10), 1671–1675.

Sak, H., Senior, A. & Beaufays, F. (2014). Long short-term memory based recurrent neural network architectures for large vocabulary speech recognition. *Computer Science*, *13*(8), 338–342.

TensorFlow. (2018, March 10, 2018). Retrieved from http://www.tensorflow.org.

Schmidhuber, J. (2014). Deep learning in neural networks: An overview. *Neural Networks*, *61*(3), 85–94.

Simon yan, K. & Zisserman, A. (2014). Very deep convolutional networks for large scale image recognition. *Computer Science*, *13*(2), 120–131.

Vergin, R., O'Shaughnessy, D. & Farhat, A. (1999). Generalized mel frequency cepstral coefficients for large-vocabulary speaker-independent continuous-speech recognition. *IEEE Transactions on Speech & Audio Processing*, *7*(5), 525–532.

Wang, J., Yang, Y., Mao, J. (2016). CNN-RNN: A unified framework for multi-label image classification. *Proceedings of IEEE Conference on Computer Vision and Pattern Recognition*, 2285–2294.

Automatic Control, Mechatronics and Industrial Engineering – He & Qing (Eds)
© *2019 Taylor & Francis Group, London, ISBN 978-1-138-60427-8*

Performance evaluation of TCP congestion control algorithms using a network simulator

A. Mohammad Ali
Arts and Technology University, Iraq

S. Kadry
Department of Mathematics and Computer Science, Faculty of Science, Beirut Arab University, Lebanon

ABSTRACT: The algorithms used in Transmission Control Protocol (TCP) congestion control are the main reason for the performance of a given TCP variant. There are different implementations among TCP: Tahoe, Reno, NewReno, SACK and Vegas. Simulation was used to evaluate these TCP congestion control algorithms from many aspects, such as effective resource utilization (throughput utilization and packet-dropping probability). A network simulator was used with different scenarios and factors to influence the performance of the TCP variants. Under congested network conditions, the more conservative algorithm of TCP Vegas achieves higher throughput and a lower dropped-packet rate than the more aggressive algorithms of the other TCP variants tested.

Keywords: TCP, congestions, TCP variants, Network Simulator

1 INTRODUCTION

The algorithms for Transmission Control Protocol (TCP) congestion control are the main reason for the performance of a TCP variant. There are different implementations among TCP: Tahoe, Reno, NewReno, SACK and Vegas. Simulation will be used to evaluate these TCP congestion control algorithms from many aspects, such as effective resource utilization (throughput utilization and packet-dropping probability) A network simulator with different scenarios and factors that affect the performance of a TCP variant will be used. Focus will be given to the effect of different queue-type algorithms, such as DropTail and RED, and different buffer sizes. Simulations to compare TCPs Tahoe, NewReno, Reno, SACK and Vegas in different situations will also be carried out. All the simulations use File Transfer Protocol (FTP) as a traffic source. The ratio of number of packets received in a given time will be used to indicate the throughput occupancy (cwnd/time = throughput). The capacity of the buffer and window sizes are the number of units in the packet. Our first evaluation will compare the impact of resource utilization to show the available network bandwidth as a performance metric (effective throughput) for each TCP variant and the percentage drop when each type of variant is running alone on a congested network. Our second evaluation answers the question of how frequently each of the TCP protocol variants drops packets when two different TCP agents are running together and sharing a bottle necked path.

2 THROUGHPUT RATE

Figure 1 depicts Topology 1, a simple simulation network. The circle (R) indicates a finite-buffer DropTail gateway, and the other circles (S and D) indicate sender and receiver hosts. The links are labeled with their bandwidth capacity and propagation delay. The parameters

Figure 1. Network Topology 1.

Table 1. Parameters of Topology 1.

No.	Fixed parameter	Value
1	TCP sender capacity	10 Mbps
2	Propagation delay	1 ms
3	TCP receiver capacity	1.5 Mbps
4	Queue type	DropTail
5	Traffic regenerator	FTP
6	Packet size	500
7	Buffer size	15
8	Window size	300

Table 2. Throughput rate (5 s).

Tahoe	NewReno	SACK	Reno	Vegas
Packets sent				
3257	4309	4422	3357	5855
Packets received				
3160	4202	4313	3270	5722
Packets dropped				
97	111	113	90	133
Drop percentage (%)				
2.97	2.57	2.55	2.68	2.27

used in the simulation are shown in Table 1, and were used in the simulation and an error model was introduced to simulate packet dropping and enable comparison of the effective throughput for each of these five congestion control algorithms.

Measurement of the performance of Tahoe, NewReno, Vegas, Reno and SACK protocols on Topology 1 was undertaken, the results of which are shown in Table 2. In terms of the throughput utilization of these TCP protocols, the simulation results over5 s show that in this scenario TCP Vegas achieves a higher throughput and lower dropped-packet percentage than the other TCP variants.

3 PACKET-DROPPING PROBABILITY

To investigate packet-dropping rates, the network configuration shown in Figure 2 (Topology 2) was used. R1 and R2 are finite-buffer switches and S1 to S4 are the end hosts. Connection 1 transfers packets from S1 to S3, while Connection 2 transfers from S2 to S4. The links are labeled with their capacities and propagation delays and Table 3 details all the parameters used in the topology.

In a series of simulations, Connections 1 and 2were each configured to a different TCP protocol variant, in order to obtain ten result combinations, as shown in Table 4. These two-sender-agent experiments are designed to test the dropping rates of TCPs Tahoe (Kadry & Al-Issa, 2015; Parvez et al., 2010), NewReno, Reno, SACK and Vegas when two of them are running together.

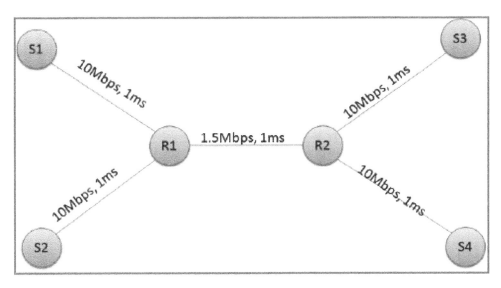

Figure 2.   Network Topology 2.

Table 3.   Parameters of Topology 2.

No.	Fixed parameter	Value
1	TCP sender capacity	10 Mbps
2	Propagation delay	1 ms
3	TCP receiver capacity	10 Mbps
4	Queue type	DropTail
5	Traffic regenerator	FTP
6	Packet size	500
7	Bottleneck	1.5 Mbps
8	Buffer size	10
9	Window size	300

Table 4.   Packet-dropping probability for protocol combinations.

Packet-dropping			
Tahoe/Vegas	28/6	NewReno/Reno	29/29
Vegas/Reno	3/26	Tahoe/SACK	30/28
Vegas/NewReno	3/27	SACK/NewReno	29/29
Vegas/SACK	3/27	SACK/Reno	29/29
Tahoe/NewReno	29/29	Tahoe/Reno	28/26

The results in Table 4 show the number of packets dropped by the TCP variants when two of them compete with each other. These results show that TCP Vegas has a lower rate of dropped packets than the other protocols (Afanasyev et al., 2010; Ibrahim et al., 2009), with the number varying between 3 and 6. TCP Vegas only loses a small number of packets because it uses a conservative algorithm that can enable it to detect and predict the congestion of an network before it happens. The other TCP protocols use aggressive algorithms to increase their window size until packet dropping occurs. When the SACK, NewReno, Reno and Tahoe protocols are competing, they share the bandwidth but at the expense of higher loss rates.

4 FAIRNESS BETWEEN THE SAME TCP AGENTS

This section explores the fairness between connections that have the same or different delay when identical variants of TCP run together on one bottlenecked link and share the bandwidth. An experiment on the effect of delay on TCPs Reno, SACK, NewReno, Tahoe and Vegas is carried out, together with a check of the behaviors of TCP variants in this situation.

In the simulation of network Topology 3 (illustrated in Figure 3), R1 and R2 are limited key buffers and S1 to S4 are the host ends. Connection 1 transfers packets from S2 to S4, and Connection 2 from S1 to S3. In each simulation, senders S1 and S2 are the same variants of TCP, such as two Tahoe, two NewReno, two Vegas, two SACK, and two Reno TCP agents, respectively. The links with capacity restriction (1.5 Mbps) and/or propagation delay are where the bandwidth or the router capacity generate a congestion state in the network. Other links operate at 10 Mbps and other simulation parameters are detailed in Table 5. The delay of the link connecting R2 and S3, which is denoted by X in Figure 3, was changed to show the effect of delay on the fairness between different connections; X had two values (1 msand30 ms).

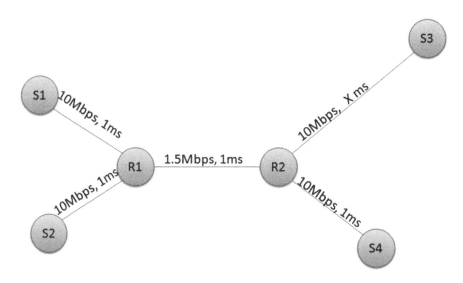

Figure 3.   Network Topology 3.

Table 5.   Parameter settings of Topology 3.

No.	Fixed parameter	Value
1	TCP sender capacity	10 Mbps
2	Router propagation delay	1 ms
3	TCP receiver capacity	10 Mbps
4	Queue type	DropTail, RED
5	Traffic regenerator	FTP
6	Packet size	500
7	Bottleneck	1.5 Mbps
8	Buffer size	15
9	Sender propagation delay	1 ms
10	Receiver propagation delay	1 ms; 30 ms
11	Window size	300

5 CONCLUSION

On links where there is traffic or congestion, TCP Vegas produces the highest throughput and lowest rate of packet drop of the TCP protocol variants tested (Vegas, Tahoe, NewReno, Reno and SACK). The performance of TCP Vegas is perfect when congestion appears and the number of dropped packets is between 3 and 6 when running with the other TCP protocols. When the SACK, NewReno, Reno and Tahoe protocols are in competition, they share the bandwidth almost half and half, but at the expense of higher loss rates.

REFERENCES

Afanasyev, A., Tilley, T., Reiher, P. & Kleinrock, L. (2010). Host-to-host congestion control for TCP. *IEEE Communications Surveys & Tutorials*, *12*(3), 304–342.

Ibrahim, M., Altman, E., Primet, P., Carofiglio, G. & Post, G. (2009). A simulation study of passive inference of TCP rate and detection of congestion. In *Proceedings of Fourth International ICST Conference on Performance Evaluation Methodologies and Tools (VALUETOOLS '09), Pisa, Italy* (article 8). doi:10.4108/ICST.VALUETOOLS2009.7639.

Kadry, S. & Al-Issa, A.E. (2015). Modeling and simulation of out-of-order impact in TCP protocol. *Journal of Advances in Computer Networks*, *3*(3), 220–224.

Parvez, N., Mahanti, A. & Williamson, C. (2010). An analytic throughput model for TCP NewReno. *IEEE/ACM Transactions on Networking*, *18*(2), 448–461.

Automatic Control, Mechatronics and Industrial Engineering – He & Qing (Eds)
© 2019 Taylor & Francis Group, London, ISBN 978-1-138-60427-8

A novel underwater image enhancement algorithm based on rough set

P.F. Shi, X.N. Fan, J.J. Ni & M. Li
College of IOT Engineering, Hohai University, Changzhou, China

D.W. Yang
Nanjing Hydraulic Research Institute, Nanjing, China

ABSTRACT: The underwater image enhancement is a challenging for the low signal-to-noise ratio and uneven illumination under water. To solve this problem, this paper proposes a novel image enhancement algorithm. First of all, the knowledge representation of the underwater image is constructed by rough set. Then the image is divided into optimal partition according to the approximate equivalence relation between upper and lower approximation. Finally, an improved affine shadow formation model is used to enhance each layer of the image based on the optimal partition. The simulation results verify the effectiveness of the algorithm.

1 INTRODUCTION

Due to the complexity of the underwater environment, the quality of the underwater images is usually unsatisfactory. In order to improve the image quality acquired from underwater, image enhancement is considered in this paper. The image enhancement is mainly divided into three categories (Wang W. et al. 2018). That is image enhancement based on additive noise model, statistical model and multiplicative noise model (Xu Y. et al. 2017). Such as Mask, Wallis, Curvelet, Contourlet, Retinex algorithm and so on. The Mask algorithm is commonly used in enhancing the SNR image. However, this algorithm needs to determine the optimal adaptive filter size (Shi C. et al. 2014). The Wallis algorithm is not suitable for a single image as it has to select the normal image and calculate the mean and variance (Tian J. et al. 2016). The Curvelet and Contourlet methods can effectively capture the outline of the image information, but they will result in a pseudo Gibbs phenomenon (Nason G. & Stevens K. 2015, Puranikmath S.S. & Vani K. 2016, Jai J.B.J. & Sudha G.F. 2014). The Retinex algorithm which simulates the characteristics of the human visual system is widely used in solving such problem (Chang H. et al. 2015). Image enhancement based on Gaussian mixture model nonlinear mapping and image enhancement based on fuzzy C-means clustering method are lack of the protection of texture details (Zhao F. et al. 2016, Meena Prakash R. & Shantha S.K.R. 2016). The above methods cannot be directly used in processing the images obtained from the complex underwater environment. Rough set is a classic theory in studying incomplete and uncertain data expression. Rough set can acquire knowledge from large data and generate some mathematical rules. Therefore, it is feasible to apply rough set theory for image information processing (Guo Y. et al. 2017).

This paper proposes a novel image enhancement algorithm based on the rough set. The knowledge representation of the underwater image is used to gain the optimal partition. Then the improved affine shadow formation model is used to enhance each layer of the crack image based on the optimal partition. The experimental results show the effectiveness of the proposed algorithm.

2 THE PROPOSED METHOD

2.1 *Modeling of image knowledge representation through rough set*

First of all, the image knowledge representation based on rough set is established. Here we define a four-tuple:

$$S = (U, R, V, f) \tag{1}$$

where U refers to the domain comprised by all the pixels of the image. R refers to the pixel gray and other attributes of the collection. V refers to the range value of the property R that is, the range of gray value with the same attribute in the image. f refers to the information function, that is, the mapping of pixels in the image to its attribute range.

$$g_{R_1}(x, y) = 1 / \left(\frac{X}{4} \cdot \frac{Y}{4} \right) \times \sum_{i=1}^{x/4} \sum_{j=1}^{y/4} K(i, j) \tag{2}$$

Secondly, background is extracted to get the light distribution of the image. Regard neighborhood image pixel gray value as knowledge R_1. X and Y are the width and height of the image, respectively. $K(i, j)$ is the neighborhood pixel value. Mask parameter filters the whole image through linear spatial mask. The value is calculated by Formula (2). The original underwater image and the result of background knowledge are shown in Figure 1.

Then, the brightness layers are calculated according to the upper and lower approximation (Phophalia A. & Mitra S.K. 2015). All pixels are divided into N sets. Each layer has the same brightness values. The results when N = 10 are shown in Figure 2.

2.2 *Obtaining the optimal layer and enhancing the image layer by layer*

After obtaining the brightness layers, the layer with over illumination or low illumination is repaired. Here we define $[\alpha \beta]$ $(0 < \alpha < \beta < 1)$ is the normal area. And $i < \alpha \cdot N$ is the dark area. $i > \beta \cdot N$ is the over illumination area. Then the repaired formula (3) is used for each layer.

Figure 1.   The original image and its background.

Figure 2.   Results of brightness layers calculated by upper and lower approximation when N = 10.

$$G(x,y)_{x_i}^{even} = \frac{\sigma^{normalarea}}{\sigma_i^{layer}}[G(x,y)_{x_i}^{layer} - \eta_i^{layer}] + \eta^{normalarea} \tag{3}$$

$G(x,y)_{x_i}^{even}$ is the result, $G(x,y)_{x_i}^{layer}$ is the value before equilibrium, $\sigma^{normalarea}$ $\eta^{normalarea}$ is standard deviation and mean of the normal area. The results of image enhancement for upper and lower approximation is shown in Figure 3. From Figure 3, we can see when the number of layers is small, there are some obvious boundaries among the layers. It shows that the division is not enough. On the contrary, when the number of layers is large, for example N = 10, the result is shown in Figure 4. Amount of calculation will be significantly increased. It means there is an optimal division of layers between 10 and 200, leading to a better enhancement performance with the less calculation time.

We can select the best N by calculating the approximate classification accuracy and system parameters importance degree in formula (4):

$$\begin{cases} \alpha_{R_1}(\pi(U)) = \dfrac{\sum\limits_{i=1}^{N}|\underline{R_1}(x_i)|}{\sum\limits_{i}^{N}|\overline{R_1}(x_i)|} \to 1 \\[4mm] \mathrm{sig}_{R_1}(X_i) = \dfrac{\sum\limits_{i=1}^{n}|U - bnR_1(X_i)|}{n|U|} \to 1 \end{cases} \tag{4}$$

Figure 3.   Results of image enhancement for upper and lower approximation respectively when N = 10.

Figure 4.   Results of image enhancement for upper and lower approximation respectively when N = 200.

Figure 5.   Result of image enhancement when N = 140.

where $\underline{U}$ is the pixels of the image; X_i is the *ith* layer consisting of pixels in the same gray value; $\overline{R_1}(X_i), \underline{R_1}(X_i)$ are the upper and lower approximation correspond to knowledge of R_1, respectively; $bnR_1(X_i)$ is the boundary of X_i. $\alpha_R(X)$ represents the percentage of correct decision. $\text{sig}_{R_1}(X_i)$ represents the accuracy of knowledge. When $\alpha_R(X)$ and $\text{sig}_{R_1}(X_i)$ converge to 1, knowledge R can accurately describe the set X. In this case, N is the optimal division.

For the original image in Figure 1, the optimal value of N is 140. When N = 140, the result is shown in Figure 5.

3 EXPERIMENTS

In this section, we show the performance of the proposed method by enhancing four typical images as shown in Figure 6.

In order to verify the performance improvement, we compare the proposed algorithm with other methods. Figure 7 shows a typical original image and the results of Mask (Shi C. et al. 2014), Wallis (Tian J. et al. 2016), Retinex (Tao F. et al. 2017) and the proposed method (PRO), respectively.

As can be seen from Figure 7. Mask algorithm has enhanced the texture of the shaded area in some degree but not completely. Excessive illumination area has been added too. Wallis algorithm has balanced the illumination in local area, but due to the division of blocks there are some clear dividing lines among the blocks. The Retinex algorithm has better gray balance effect. However, the gray-scale of the original image is changed and the image contrast is reduced. The proposed algorithm has the best gray balance effect, with the texture feature

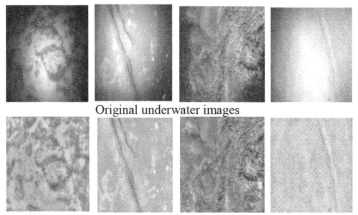

Original underwater images

The results of the proposed image enhancement algorithm

Figure 6.   Four typical images before and after enhancement.

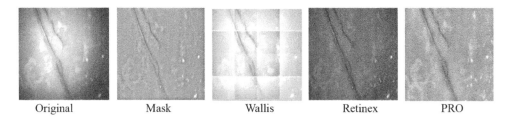

Original   Mask   Wallis   Retinex   PRO

Figure 7.   The proposed algorithm compared with different algorithms.

has been well preserved. In particular, the texture of the image boundary region is well maintained and enhanced.

4 CONCLUSION

In this paper, an underwater dam crack image enhancement algorithm based on Rough Set has been presented. The proposed algorithm can solve the low signal-to-noise ratio and uneven illumination problems of the image by upper/lower approximation. Besides, the proposed algorithm has the advantage of solving the optimal division for the set of pixels with approximate gray-scale. The experimental results show the effectiveness of the proposed algorithm.

ACKNOWLEDGMENTS

This work is financially supported by the Jiangsu Province Natural Science Foundation (No. BK20170305) and the National Natural Science Foundation of China (No. 61801169, No. 61573128, No. 61873086), and the Fundamental Research Funds of Nanjing Hydraulic Research Institute (Y717010 and Y717009).

REFERENCES

Chang H., Ng M.K., Wang W., et al. 2015. Retinex image enhancement via a learned dictionary[J]. Optical Engineering 54(1): 013107–013107.

Guo Y., Jiao L., Wang S., et al. 2017. Fuzzy Sparse Autoencoder Framework for Single Image Per Person Face Recognition[J]. IEEE Transactions on Cybernetics 48(8):2402–2415.

Jai J.B.J., Sudha G.F. 2014. Non-subsampled contourlet transform based image Denoising in ultrasound thyroid images using adaptive binary morphological operations[J]. Computer Vision Iet 8(6):718–728.

Meena Prakash R., Shantha S.K.R. 2016. Fuzzy C means integrated with spatial information and contrast enhancement for segmentation of MR brain images[J]. International Journal of Imaging Systems and Technology 26(2):116–123.

Nason G., Stevens K. 2015. Bayesian Wavelet Shrinkage of the Haar-Fisz Transformed Wavelet Periodogram[J]. PloS one 10(9): e0137662.

Phophalia A., Mitra S.K. 2015. Rough set based bilateral filter design for denoising brain MR images[J]. Applied Soft Computing 33:1–14.

Puranikmath S.S., Vani K. 2016. Enhancement of SAR images using curvelet with controlled shrinking technique[J]. Remote Sensing Letters 7(1):21–30.

Shi C., Liu F., Li L.L., et al. 2014. Pan-sharpening algorithm to remove thin cloud via mask dodging and nonsampled shift-invariant shearlet transform[J]. Journal of Applied Remote Sensing 8(1):083658.

Tao F., Yang X., Wu W., et al. 2017. Retinex-based image enhancement framework by using region covariance filter[J]. Soft Computing 22(4):1–22.

Tian J., Li X., Duan F., et al. 2016. An Efficient Seam Elimination Method for UAV Images Based on Wallis Dodging and Gaussian Distance Weight Enhancement[J]. Sensors 16(5):662.

Wang W., Xin B., Deng N., et al. 2018. Single Vision Based Identification of Yarn Hairiness Using Adaptive Threshold and Image Enhancement Method[J]. Measurement.

Xu Y., Wen J., Fei L., et al. 2017. Review of Video and Image Defogging Algorithms and Related Studies on Image Restoration and Enhancement[J]. IEEE Access 4:165–188.

Zhao F., Zhao J., Zhao W., et al. 2016. Gaussian mixture model-based gradient field reconstruction for infrared image detail enhancement and denoising[J]. Infrared Physics and Technology 76:408–414.

Automatic Control, Mechatronics and Industrial Engineering – He & Qing (Eds)
© 2019 Taylor & Francis Group, London, ISBN 978-1-138-60427-8

Multi-sensor data fusion for sign language recognition based on dynamic Bayesian networks and convolution neural networks

Y.D. Zhao, Q.K. Xiao & H. Wang
Department of Electronic Information Engineering, Xi'an Technological University,
Xi'an City, P.R. China

ABSTRACT: A new multi-sensor fusion framework is proposed, which is based on Convolution Neural Network (CNN) and Dynamic Bayesian Network (DBN) for Sign Language Recognition (SLR). In this framework, a Microsoft Kinect, which is a low-cost RGB-D sensor, is used as a tool for the Human-Computer Interaction (HCI). In our method, firstly, the color and depth videos are collected using Kinect, then all image sequences features are extracted out using the CNN. The color and depth feature sequences are input into the DBN as observation data. Based on the graph model fusion machine, the maximum hidden state probability is calculated as recognition results of dynamic isolated sign language. The dataset is tested using the existing SLR methods. Using the proposed DBN+CNN SLR framework, the highest recognition rate can reach 99.40%. The test results show that our approach is effective.

Keywords: sign language recognition, dynamic Bayesian network, convolution neural network, multi-sensor data

1 INTRODUCTION

Sign language always provides a natural interaction pattern between people with normal hearing and those with a hearing impairment. The Sign Language Recognition (SLR) system is designed to provide a powerful and reliable interface. As mentioned in (Chen et al., 2003; Pugeault et al., 2012; Marin et al., 2016), the SLR system is playing an increasingly significant role in gesture-driven Human-Computer Interaction (HCI) systems. Hence, many researchers try their efforts on automatic hand gesture recognition. Recently, some low-cost depth sensors have been developed for new methods of HCI. Kinect is an effective and inexpensive device for acquiring depth data, and it has been successfully used in SLR systems (Almeida et al., 2014; Elons et al., 2014).

Recently, many sign language and hand gesture recognition systems have also utilized 3D depth sensors, such as Kinect. (Lang et al., 2012) proposed an open source framework for isolated gesture recognition using Kinect. The 25 symbols of the German Sign Language (GSL) were discriminated by the HMM and the accuracy went up to 97%. Pugeault et al., (2012) developed an interactive American Sign Language (ASL) hand-spelling system using Kinect, which uses static symbols. They also use depth and color information to complete hand segmentation, and the hand characteristics are extracted using Gabor filtering. The letters are recognized using a random forest classifier. The authors add a dictionary to the system, allowing users to select letters that are not clearly defined. Pedersoli et al., (2014) use a Kinect sensor to develop a system that recognizes static and dynamic gestures. Its principle is to extract the angle characteristics among hand centers in continuous frames based on the motion trajectory of the 3D hand.

Although the HMM is extremely powerful for the modeling of time series data, it can only be used in a simple state space with discrete hidden variables. Assume that we have two hidden variables, which are essentially independent, but these procedures affect each other and are highly related to each other. The state space will be as large as the product of the dimensions of all hidden variables if we use a traditional HMM to express a model with a single hidden variable. Therefore, we need a lot of data to estimate the model parameters. In addition, the conventional HMM can be used if the conditions under the Markov model are satisfied. Brand et al., (1997) construct CHMM by coupling HMM, which means that the model is likely to have interactions between different morphological structures and processes of mutual influence. Such issues frequently appear at computer vision, speech or both. The CHMM has been regarded as a particular type of Dynamic Bayesian Network (DBN) (Chu, & Huang, 2007), because its structure can capture the dependent relationship of asynchronous and time modes between two different information channels. Brand et al., (1997) indicated that the CHMM was superior to traditional HMM in recognizing hand gestures, and its accuracy rate can reach 94.2%. A CHMM is regarded as multiple HMM collections, one of which is equivalent to a data flow. Nefian et al., (2002) simulate the audio and video sequences using an audio-visual word recognition system based on the double current CHMM. According to the report, the accuracy rate of word recognition can reach 98.14%, which is better than the basic HMM models. In this paper, we use a multi-sensor data fusion framework for sign language recognition using the CHMM.

2 PROPOSED SYSTEM

2.1 Overview of the system

In this paper, firstly, we use a multi-sensor device to get sign language information. As shown in Figure 1, Kinect is a good choice to obtain color and depth sign language sequence images at the same time. Secondly, based on the Convolution Neural Network (CNN), we can extract features of color and depth images. The CNN-based image features always provide excellent performance in many areas of computer vision, such as pattern recognition, image retrieval, video tracking and so on. Thirdly, the color and depth image feature sequences are input into the Dynamic Bayesian Network (DBN). Based on inference of graph model probability, the image sequences can be classified correctly.

2.2 Extract features using the CNN

The CNN has made successful achievements in computer vision and solved many problems of feed-forward neural networks. The convolution layer is used to find the local relationship of the input layer, and the role of pooling layers is to gradually reduce the input dimension. When the CNN is used for gesture recognition tasks, the image is as input of the network, at the same time, the network will output the image labels or category probability. In this paper, we use the CNN to extract features of color and depth images, before the fully-connected layer is connected to the softmax regression layer, the CNN output image features, as shown in Figure 2. There are ten layers in the CNN, including an input layer, three convolution layers, three nonlinear layers, two max pooling layers, one fully-connected layer and one classification output layer. The first convolution layer is computed using 8 different 3×3 kernels

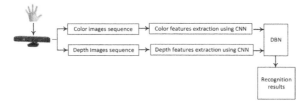

Figure 1. Multi-sensor data fusion based on DBN for sign language recognition.

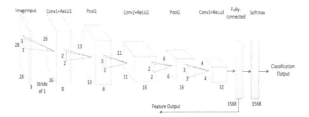

Figure 2.   Illustration of extracting image features using the CNN.

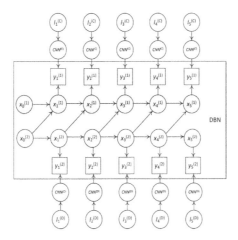

Figure 3.   The DBN model construction based on multi-sensor data fusion.

and the 8 feature maps are obtained, and the second convolution layer is computed using 16 different 3×3 kernels, and 16 feature maps are obtained. The convolution layers are used to find the local relationship of the input layer. The corresponding pooling layer is the largest pooling layer in the 2×2 neighborhood domain, and the original size of the feature map is obtained. Similarly, the third convolution layer has 32 different 3×3 kernels. The fifth pooling layer is connected to a fully-connected layer as the feature vector output.

2.3   Using DBN inference to classify sign language

2.3.1   Using DBN to represent sign language sequences
In this article, we use a revised Couple Hidden Markov Model (CHMM) for sign language sequential data modeling. The CHMM is a probability graph model, according to graph model theory, and the revised CHMM can be regarded as a kind of DBN (Marin et al., 2005).

As we all know, we often use the DBN to represent non-stationary random signals. Based on the DBN, we can analyze or calculate many non-stationary random signals. It is reasonable to use the DBN graph model to describe human body movement according to graph model and non-liner signal processing theory. As Figure 3 shows, we build a mixed-state DBN model to represent a motion sequence. Our motivation is summarized as follows: (1) first of all, for a given input sequence signal Y(C), similar as the HMM viterbi encode, we can use the CHMM principle to estimate main sensor hidden sequence X(C) and assistant sensor sequence X(D). (2) based on the X(D) and Y(D), in addition, another HMM is built, this means that there is a close relationship between the X(C) and the Y(D). Meanwhile, the Y(D) is as new-information to renew the prior estimation of X(C). Based on above analysis, we combine the two HMM together to build a mix-state DBN model, as shown in Figure 3.

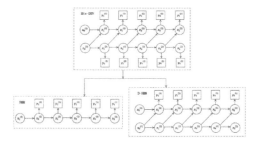

Figure 4.　The original DBN is divided into HMM and D-HMM.

2.3.2 *Inference*

In the DBN, the inference goal is to get the posterior probability of the hidden state. In order to achieve inference of the mixed-state DBN, we can first divide the DBN into two sub-graphs: D-HMM and HMM, as shown in Figure 4. The D-HMM (Depth-Hidden Markov Model) is constructed based on HMM, the depth state sequence is added into color-based state sequence, and the final color state would be decided by color-based observation and depth-based state. Another sub-graph is HMM.

The detail of graph model inference can be described as follows. Firstly, HMM inference is discussed.

The HMM includes 3 parameters:

$$\pi_0^D = P\left(x_0^D\right) = p_i^D, i = 1, \ldots, n \tag{1}$$

$$A^D = \left[a_{ij}^D\right]_{n \times n} = P\left(x_{t+1}^D \mid x_t^D\right) \tag{2}$$

$$\begin{aligned} B^D = P\left(y_t^D \mid x_t^D\right) = n\left(\mu_y^D, {\sum}_y^D\right)\left(y_t^D\right) \\ = \frac{1}{\sqrt{2\pi}{\sum}_y^D} \exp\left\{-\frac{1}{2}\left[\left(y_t^D - \mu_y^D\right)^T\right.\right. \\ \left.\left. {\sum}_y^D\left(y_t^D - \mu_y^D\right)\right]\right\} \end{aligned} \tag{3}$$

where the π_0^D is the prior probability distribution of state x_0^D, the n is the number of hidden state, the A^D is the state transaction matrix, the element a_{ij}^D denotes the transaction probability from state *i* to state *j*. The B^D is the conditional probability of observation y_t^D known state x_t^D.

The goal of HMM inference is to obtain optimal state estimation based on observation, according to Bayesian theory, and initial node probability is computed:

$$P(x_1^D) = P(x_1^D \mid x_0^D)P(x_0^D) \tag{4}$$

Then the state is updated by observation:

$$P(x_1^D \mid y_1^D) = \frac{P(y_1^D \mid x_1^D)P(x_1^D)}{P(y_1^D)} \tag{5}$$

In general, at time t, we have:

$$P(x_{t+1}^D) = P(x_{t+1}^D \mid x_t^D)P(x_t^D) \tag{6}$$

In the end, the state sequence is calculated and updated by observation sequence:

$$
\begin{aligned}
P(x_{1+t}^D \mid y_{1:1+t}^D) &= P(x_{1+t}^D \mid y_{1+t}^D, y_{1:t}^D) \\
&= P(x_{1+t}^D \mid y_{1+t}^D) \cdot P(x_{1+t}^D \mid y_{1:t}^D) \\
&= \alpha P(y_{1+t}^D \mid x_{1+t}^D) \cdot P(x_{1+t}^D \mid y_{1:t}^D) \\
&= \alpha P(y_{1+t}^D \mid x_{1+t}^D) P(x_{1+t}^D \mid x_t^D) P(x_t^D) \\
&\quad \sum_{x_t} P(x_{1+t}^D \mid x_t^D) P(x_{1+t}^D \mid y_{1:t}^D)
\end{aligned}
\tag{7}
$$

The D-HMM inference is similar to HMM inference, the 3 parameters are given firstly:

$$
\pi_0^C = P(x_0^C) = [p_i]_{1 \times n}
\tag{8}
$$

$$
A^C = P(x_{1+t}^D \mid x_t^C, x_t^D) = [a_{ij}^C]_{n \times m}
\tag{9}
$$

$$
\begin{aligned}
B^C &= P(y_t^C \mid x_t^C) = \mathrm{n}(\mu_y^C, \Sigma_y^C)(y_t^C) \\
&= \frac{1}{\sqrt{2\pi}\Sigma_y^C} \exp\left\{ -\frac{1}{2}\left[(y_t^C - \mu_y^C)^T \sum_y{}^C (y_t^C - \mu_y^C) \right] \right\}
\end{aligned}
\tag{10}
$$

where the π_0^C, A^C, B^C are prior probability distribution, state conditional transaction matrix, and observation matrix, respectively.

The next, similar to HMM inference, according to Bayesian theory, we can calculate initial state node probability:

$$
\begin{aligned}
P(x_1^C) &= P(x_1^C \mid x_0^C, x_0^D) P(x_0^C, x_0^D) \\
&= P(x_1^C \mid x_0^C, x_0^D) P(x_0^C) P(x_0^D)
\end{aligned}
\tag{11}
$$

And initial state value is updated by observation:

$$
\begin{aligned}
P(x_1^C \mid y_1^C) &= \frac{P(y_1^C \mid x_1^C) P(x_1^C)}{P(y_1^C)} \\
&= \frac{P(y_1^C \mid x_1^C) P(x_1^C \mid x_0^C, x_0^D) P(x_0^C) P(x_0^D)}{P(y_1^C)}
\end{aligned}
\tag{12}
$$

In general, at time t, the node state probability is calculated:

$$
\begin{aligned}
P(x_{1+t}^C) &= P(x_{1+t}^C \mid x_t^C, x_t^D) P(x_t^C, x_t^D) \\
&= P(x_{1+t}^C \mid x_t^C, x_t^D) P(x_t^C) P(x_t^D)
\end{aligned}
\tag{13}
$$

Finally, calculate the hidden state sequence by observing sequences:

$$
\begin{aligned}
P(x_{1+t}^C \mid y_{1:1+t}^C) &= P(x_{1+t}^C \mid y_{1+t}^C, y_{1:t}^C) \\
&= P(x_{1+t}^C \mid y_{1+t}^C) \cdot P(x_{1+t}^C \mid y_{1:t}^C) \\
&= \alpha P(y_{1+t}^C \mid x_{1+t}^C) \cdot P(x_{1+t}^C \mid y_{1:t}^C) \\
&= \alpha P(y_{1+t}^C \mid x_{1+t}^C) P(x_{1+t}^C) \cdot P(x_{1+t}^C \mid y_{1:t}^C) \\
&= \alpha P(y_{1+t}^C \mid x_{1+t}^C) P(x_{1+t}^C \mid x_t^C \mid x_t^C) P(x_t^C) \\
&\quad P(x_t^D) \cdot P(x_{1+t}^C \mid y_{1:t}^C) \\
&= \alpha P(y_{1+t}^C \mid x_{1+t}^C) P(x_{1+t}^C \mid x_t^C \mid x_t^C) P(x_t^C) \\
&\quad P(x_t^D) \sum_{x_t} P(x_{1+t}^C \mid x_t^C) P(x_t^C \mid y_{1:t}^C)
\end{aligned}
\tag{14}
$$

Based on eq.14, according to Bayesian theory [10], we have:

$$
\begin{aligned}
&\max_{x_1, \boldsymbol{n}, x_t} P(x_{1:1+t}^C \mid y_{1:1+t}^C) \\
&= \alpha P(y_{1+t}^C \mid y_t^C) \max_{x_t} P(x_{1+t}^C \mid x_t^C) \max_{x_1, \boldsymbol{n}, x_{t-1}} P(x_t^C \mid y_{1:t}^C)
\end{aligned}
\tag{15}
$$

In the end, the optimal classification prediction is:

$$(\hat{x}_{1:t}^C)^* = E[x_{1:t}^C \mid y_{1:t}^C]$$
$$= \int_{-\infty}^{\infty} x_{1:t}^C \cdot \left(\max_{x_1,\dots,x_{t+1}} P(x_{1:t}^C \mid y_{1:t}^C) \right) dx^C \tag{16}$$

Based on above mentioned graph model inference, according to maximum probability sequence theory, we can obtain multi-sensor data fusion sign language classification results.

3 RESULTS AND DISCUSSION

To test the proposed framework, we collected a sign language dataset. In this section, we firstly discuss the details of the sign language dataset. Next, we use DBN to show the recognition results.

3.1 *Dataset collection*

As shown in Figure 5, we have collected datasets of sign language based on the Chinese Sign Language (CSL), which is composed of ten dynamic symbols. All gestures are done by only one hand. All sign gestures were done by the school's ten different signers. Each signer repeats each signature gesture ten times, so that each signer creates a total of 100 signature gestures. Hence, we have collected 10,000 samples of symbolic words. The dataset includes various sign words. Datasets can be obtained online in future research (Nefian et al., 2002).

3.2 *Sign language recognition consequence*

We will divide the complete dataset into three parts, 60% of the total data is used for training, and the remaining data is used as the verification set and testing. We have tested our framework on the Core i5 CPU with 8 GB of RAM on the Microsoft Windows 10 operating system, and spent approximately 5 minutes on training the dataset.

In Figure 6, we show three kinds of sign language recognition results based on different recognition composition. 6,000 images are considered as training data, 1,000 images are regarded as validation data, 3,000 images are taken as testing data. We only use depth camera to get video sequence. In Figure 6(a), we use the CNN (D) + Softmax method to classify sign language. The confusion matrix of testing data recognition is shown, the accuracy is 91.6%. In Figure 6(b), we use the CNN(C) + Softmax method to classify sign language. In Figure 6(c), the confusion matrix of testing data recognition is shown, the accuracy is 95.4%, In Figure 6(a), we use the CNN(C) + CNN(D) + DBN method to classify sign language. In Figure 6(c), the confusion matrix of testing data recognition is shown, the accuracy is 99.2%.

To test the performance of DBN-based data fusion model, various recognition composition methods are compared. As shown Table 1, the overall recognition accuracy of the

Figure 5.   Illustration of collected sign language dataset, the color sequence is corresponding to depth sequence.

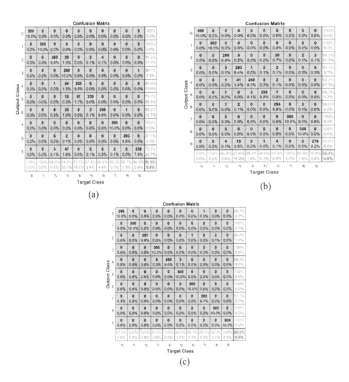

(a)

(b)

(c)

Figure 6. The confusion matrix of testing data recognition: (a) Sign language recognition results based on CNN(D) + Softmax; (b) Sign language recognition results based on CNN(D) + Softmax; (c) The confusion matrix of testing data recognition.

Table 1. Performance compares with various recognition approaches.

Various recognition composition	Recognition accuracy
CNN(D) + Softmax	91.2%
CNN(C) + Softmax	95.2%
CNN(C) + CNN(D) + DBN	99.4%

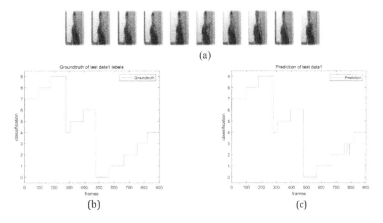

Figure 7. Testing experiment: (a) Illustration of video sequence; (b) Ground truth of image label; (c) Prediction labels of the DBN.

CNN(C) + CNN (D) + DBN composition is up to 99.40%. Compared with our method, the accuracy of the CNN (D) + Softmax composition is 91.2% and the accuracy of the CNN(C) + Softmax composition is 95.2%. The comparison results indicate that our method is perfect.

In Figure 7 some experiment results are illustrated, for example, in Figure 7(a), some key-frame images in video are shown. The video recognition results are shown in Figure 7(c), and if we compare labels from Figure 7(b) to Figure 7(c), there is little difference between prediction and truth, which means the results indicate that our proposed method is effectiveness.

4   CONCLUSIONS

In this article, we have proposed a SLR algorithm based on multi-sensor data fusion and deep learning. The method firstly built a mix-DBN to describe the SLR system, and in this model, the features are extracted using the CNN. Next, a fusion approach was proposed, based on graph model theory. Additionally we have tested the proposed algorithm on our SLR database. The comparison between the experimental results and the existing methods shows that this method is effective.

ACKNOWLEDGMENTS

This work is supported by the Nature Science Foundation of China (Nos. 60972095, 61271362, 61671362) and the Nature Science Basic Research Plan in Shaanxi Province of China (Nos. 2017 JM6041).

REFERENCES

Almeida, S.G.M., Guimarães, F.G. & Ramírez, J.A. (2014). *Feature extraction in Brazilian Sign Language Recognition based on phonological structure and using RGB-D sensors*. Pergamon Press, Inc.

Brand, M., Oliver, N. & Pentland, A. (1997). Coupled hidden Markov models for complex action recognition. Computer vision and pattern recognition. *Proceedings of the IEEE Computer Society Conference* (p. 994).

Chen, F.S., Fu, C.M. & Huang, C.L. (2003). Hand gesture recognition using a real-time tracking method and hidden markov models. *Image & Vision Computing, 21*(8), 745–758.

Chu, S.M., & Huang, T.S. (2007). Audio-Visual speech fusion using coupled Hidden Markov Models. Computer vision and pattern recognition. *Proceedings of the CVPR '07 IEEE Conference*, 2, (pp. 1–2).

Elons, A.S., Ahmed, M., Shedid, H. & Tolba, M.F. (2014). Arabic sign language recognition using leap motion sensor. *Proceedings of the International Conference on Computer Engineering & Systems*, 30, (pp. 368–373).

Lang, S., Block, M. & Rojas, R. (2012). Sign language recognition using Kinect. Artificial Intelligence and Soft Computing. Berlin, Germany: Springer.

Marin, G., Dominio, F. & Zanuttigh, P. (2015). Hand gesture recognition with leap motion and Kinect devices. *Proceedings of the IEEE International Conference on Image Processing* (pp.1565–1569).

Marin, G., Dominio, F. & Zanuttigh, P. (2016). *Hand gesture recognition with jointly calibrated Leap Motion and depth sensor*. Kluwer Academic Publishers.

Nefian, A.V., Liang, L., Pi, X. & Xiaoxiang, L. (2002). A coupled HMM for audio-visual speech recognition. *Proceedings of the IEEE International Conference on Acoustics, Speech, and Signal Processing*, 2, (pp. II-2013–II-2016).

Pedersoli, F., Benini, S., Adami, N. & Leonardi, R. (2014). Xkin: An open source framework for hand pose and gesture recognition using Kinect. *Visual Computer, 30*(10), 1107–1122.

Pugeault, N. & Bowden, R. (2012). Spelling it out: Real-time ASL fingerspelling recognition. *Proceedings of the IEEE International Conference on Computer Vision Workshops*, 28, (pp. 1114–1119).

Automatic Control, Mechatronics and Industrial Engineering – He & Qing (Eds)
© 2019 Taylor & Francis Group, London, ISBN 978-1-138-60427-8

Weakly supervised co-segmentation by neural attention

Y. Zhao, F. Zhang, Z.L. Zhang & X.H. Liang
State Key Laboratory of Virtual Reality Technology and Systems, School of Computer Science and Engineering, Beihang University, Beijing, China

ABSTRACT: Most existing co-segmentation algorithms mainly focus on using low-level features, failing to deal with complex scenes. In this paper, we propose a novel weakly supervised method that can learn high-level features of different images in an image group for accurate common object segmentation. Specifically, we introduce a top-down neural attention model to generate class-specific maps as aprior hint about possible foreground locations. Then, we apply a Fully Connected Conditional Random Field (FC-CRF) model for accurate boundary recovery. Our method only requires image-level classification tags as supervision, so it can significantly reduce the cost in producing training data. The comparison experiments on the public datasets iCoseg and MSRC demonstrate the superior performance of the proposed method.

Keywords: co-segmentation, convolutional neural network, neural attention

1 INTRODUCTION

With the advance in digital cameras, smart phones and photo-sharing websites, people are now likely to amass large numbers of images, which might contain common objects. In computer vision, image segmentation has been one of the most attractive topics in the last few decades.

Given a set of images which share an object from the same semantic category, co-segmentation acts to segment the common objects. Most co-segmentation methods assume that the common objects have the same low-level features, such as color and intensity. However, these methods may fail when the scene is complex or when the objects are in the same class but are not the same.

Due to the observation that the low-level features are not robust enough, in this paper we propose an approach for co-segmentation based on extracting high-level semantic cues. Our approach is to adopt a Convolutional Neural Network (CNN) to support automatic high-level feature extraction. CNN has been demonstrated to have superior performance in feature extraction for many computer vision areas (Krizhevsky et al., 2012; He et al., 2016; Ren et al., 2017), especially the ability to capture semantic meaning. Fully supervised learning is costly

Figure 1. Sample results from our method. (a) Original images from the image group "plane". (b) Ground truth. (c) Attention maps. (d) Final co-segmentation results after FC-CRF.

and inefficient because the ground truth data is manual pixel-wise annotation. Our approach of weak supervision only needs image-level tags as supervision. We adopt a top-down neural attention model CAM (Zhou et al., 2016) and it can map the class into an attention map which is able to capture the semantic information. Therefore, we obtain the class-specific attention map by mapping the common object class. Figure 1 shows the intermediate and final results of our method.

2 RELATED WORK

2.1 Co-segmentation

Since image co-segmentation was first put forward by Rother et al. (2006), it has received a lot of attention. They presented the MRF-based co-segmentation method, which segmented common objects through adding a foreground similarity constraint into traditional MRF-based segmentation methods. Hochbaum and Singh (2009) then proposed a method that rewards the foreground similarity, with the result that the energy function could be optimized by a graph-cuts algorithm. Chang et al. (2011) introduced a co-saliency prior and used it to construct the MRF data terms; the submodular energy function could also be efficiently optimized with graph cut. Collins et al. (2012) used Random Walker (RW) segmentation as the core segmentation algorithm, rather than the traditional MRF approach. Rubinstein et al. (2013) proposed to use dense correspondences between images to capture the sparsity and visual variability of the common objects, so it ignores the noise that might be salient within their own images but did not commonly occur in others. For each image, Li et al. (2016) proposed a multi-search strategy to extract each target individually, and an adaptive decision criterion was raised to give each candidate a reliable judgment automatically. Most of these co-segmentation methods assume the common objects contain similar colors. The fixed features are used to measure the foreground similarity for co-segmentation, which produces unsuccessful co-segmentation results when different common objects contain different types of low-level features.

2.2 Top-down neural attention of CNNs

Top-down factors such as expectations, goals and knowledge can influence the co-segmentation result. Various methods have been proposed for modelling the top-down neural attention. Simonyan et al. (2014) applied a CNN to visualize relevant image regions for a certain class by activated hidden neurons. They made use of a back-propagation process to extract the attention maps. Cao et al. (2015) proposed feedback CNN architecture for capturing the top-down attention mechanism that could successfully identify task-relevant regions. Zhou et al. (2016) developed a mapping method by using the global average pooling layer in CNNs which has remarkable localization ability despite only being trained on image-level labels.

3 MODEL

Given a set of M images $\left\{I^i\right\}_{i=1}^M$ for co-segmentation, the foreground of I^i is the area R^i containing an instance of the common object.

Figure 2 illustrates the procedure for our weakly supervised co-segmentation method. We establish two training sets which contain the same image classes as that in the iCoseg dataset and the MSRC dataset. Then we train a ResNet-50 (He et al., 2016) for classification with only image-level tags. The Softmax loss function is represented in the following equation:

$$loss = -\sum_{i=1}^{K} y_i \log f(z_i) = -\log f(z_k) \qquad (1)$$

338

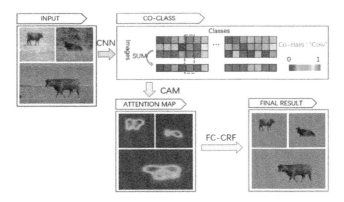

Figure 2. The flowchart of the proposed method.

$$f(z_k) = \frac{e^{z_k}}{\sum\limits_{i=1}^{K} e^{z_i}} \tag{2}$$

where K is the number of classes, $y_i = \{0,1\}$ represents the existence of the corresponding class label, and k is the neural unit whose class label is 1.

The class of the input image group is identified by the predicted probability of each image. We can extract the top-down attention maps through the remarkable localization ability of CAM. Due to the pooling, the size of the feature maps is small and the attention maps are vague and blurred after up-sampling. Therefore, we apply a fully connected CRF model to refine the results.

3.1 *Identifying the co-class*

The co-class is the class of the image group where each image contains the same category of object. Although the ResNet-50 achieves a high classification accuracy, a small number of images still cannot be classified correctly. Determining the image class only by individual image is therefore not reliable enough. To tackle the problem, we exploit the classification result of related images to address the limitation of a single image. We identify the co-class of the input image group by summing the predicted probabilities corresponding to the classification labels of each image, which is formulated as follows:

$$c = argmax_m \sum_n P_{n,m} \tag{3}$$

where c is the class of the image group, and $P_{n,m}$ is the probability that the n-th image is classified into the m-th class. The class having the maximum sum of probabilities is regarded as the co-class for the image group.

3.2 *Extracting attention map*

The attention map for a particular category indicates the discriminative image regions used by the CNN to identify that category. We extract our attention map by CAM, which is illustrated in Figure 3.

We perform global average pooling instead of max pooling on the convolutional feature maps and use those as features for a fully-connected layer that produces the Softmax output layer. The importance of the image regions can be represented by the weights of the output layer. Global average pooling outputs the spatial average of the feature map of each unit at the last convolutional layer. The class-specific attention map, which represents high-level

339

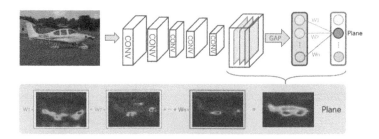

Figure 3. Architecture overview of extracting attention map.

Figure 4. Examples of the neural attention of the input images.

semantic information, is obtained by computing a weighted sum of the feature maps of the last convolutional layer. We define A^c as the attention map for co-class c, and it can be obtained as follows:

$$A^c = \sum_{k=1}^{K} w_k^c F_k \qquad (4)$$

where w_k^c is the weight corresponding to co-class c for unit k, and F_k is the k-th feature map of the last convolutional layer. In Figure 4, we show some results of CAM and we observe that discriminative regions of each image for co-class are highlighted.

3.3 Refining by fully connected CRF

As a classical probabilistic graphical model, Fully Connected Conditional Random Field (FC-CRF) (Krahenbuhl & Koltun, 2011) considers both node priors and consistency between nodes. Since the attention map obtained in the previous subsection is vague, we apply an FC-CRF model to both enhance spatial coherence and refine boundary accuracy. Every pixel is regarded as a node, and every node is connected to every other node. The energy function is showed as follows:

$$E(\mathbf{x}) = \sum_i \theta_i(x_i) + \sum_{i,j} \theta_{i,j}(x_i, x_j) \qquad (5)$$

where $\mathbf{x}$ is the label assignment on pixels. We use as unary potential $\theta_i(x_i) = -logA_{x_i}$, where A_{x_i} is the attention value at pixel i in the attention map. The pairwise term of Equation 5 is:

$$\theta_{i,j}\left(x_i,x_j\right) = \mu\left(x_i,x_j\right)\left[w_1\exp\left(-\frac{\left|p_i-p_j\right|^2}{2\sigma_\alpha^2} - \frac{\left|I_i-I_j\right|^2}{2\sigma_\beta^2}\right) + w_2\exp\left(-\frac{\left|p_i-p_j\right|^2}{2\sigma_\gamma^2}\right)\right] \qquad (6)$$

where $\mu(x_i,x_j)=1$ if $x_i \neq x_j$, and 0 otherwise, which means the model penalizes nodes with different labels. The following two Gaussian kernels extract different features from pixel i and j. The first bilateral kernel depends on both pixel positions (denoted as p) and RGB color (denoted as I), and the second kernel only depends on pixel positions. The hyper parameters σ_α, σ_β and σ_γ control the scale of Gaussian kernels.

4 RESULTS AND ANALYSIS

In this section, we evaluate the performance of the proposed method on two public datasets iCoseg (Batra et al., 2010) and MSRC (Winn et al., 2005). The MSRC and iCoseg datasets are widely used in co-segmentation which contain pixel-wise, level, manually labelled ground truth.

To quantitatively evaluate the performance of our proposed method against the state-of-the-art co-segmentation methods, we adopt two widely used evaluation metrics: IoU (Intersection over Union) score and mean error rate. The quantitative comparison result will be presented in subsection 4.2.

In the experiments, we use the ResNet-50 trained over ImageNet dataset (Krizhevsky et al., 2012) as a basic CNN, and fine-tune it with the training dataset which has the same class as iCoseg or MSRC; notice that we treat the players in white and red as two different classes. We implemented the proposed method with Caffe (Jia et al., 2014) and the experiments are conducted on a PC with an Intel Xeon CPU and a TITAN X GPU. For fine-tuned details, the learning rate is 0.001 and the momentum is 0.9. As for hyper-parameters of fully connected CRF, we use the following parameters: $w_1 = w_2 = 1$, $\sigma_\alpha = 40$, $\sigma_\beta = 10$, $\sigma_\gamma = 3$.

4.1 Subjective results

For subjective comparison, we apply our method on two benchmark datasets, iCoseg and MSRC, to further test the performance. The co-segmentation results of ten classes are shown in Figure 5; the left block is from iCoseg and the other is from MSRC. For each image class, five original images and the co-segmentation results are presented.

The images in the same group contain considerable changes in viewpoints, outlines and illumination. In some more challenging image groups, the background is complex and the target objects are diverse.

The co-segmentation result will be more robust if the high-level semantic information is included. The examples in groups 'dog' and 'cat' indicate that our proposed method can effectively locate the co-segmentation objects even when the target objects vary significantly, such as by color, shape, and texture. In relation to groups 'face' and 'car', whose background is very complicated, our proposed method has a strong ability to suppress the complex background, benefitting from the semantic information.

4.2 Quantitative results

We evaluate the proposed co-segmentation method by the mean error rate and IoU score. The error rate is defined as the ratio of the number of wrongly segmented pixels to the total number of pixels, meaning that a small error rate refers to a successful segmentation and the mean error rate over all images is used to evaluate the performance of an image group.

In Tables 1 and 2, we compare our method with the existing co-segmentation approaches: Li et al. (2016), Joulin et al. (2010), Joulin et al. (2012), Vicente et al. (2011), Meng et al. (2012) and

341

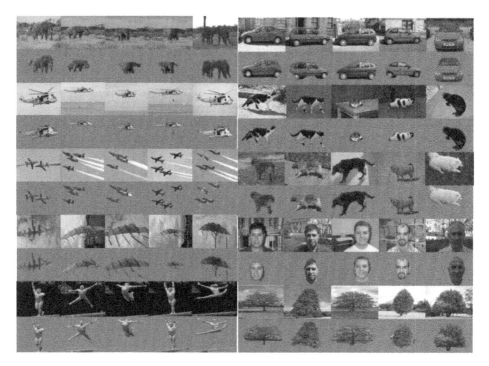

Figure 5. Examples of the input images and the co-segmentation results. The left block is the iCoseg dataset and the right block is the MSRC dataset.

Table 1. Segmentation performance between our approach and existing methods in terms of mean error rate (%, iCoseg); the best results are shown in **boldface**.

iCoseg (error %)	Ours	Li et al. (2016)	Vicente et al. (2011)	Meng et al. (2012)	Joulin et al. (2010)	Joulin et al. (2012)
Alaskan	12.26	**8.36**	10.00	11.97	23.69	23.19
Stonehenge	10.93	**2.42**	36.70	13.96	8.29	10.10
Farrari	11.26	**8.28**	10.10	8.52	35.06	34.35
Taj Mahal	10.62	12.01	8.90	**4.49**	28.16	31.90
Pyramids	**8.00**	10.94	39.49	18.70	42.73	43.25
Panda	15.16	23.09	**7.30**	21.58	13.13	9.85
Helicopter	1.15	1.21	2.73	**0.86**	5.06	47.13
Airshows	4.38	1.53	4.15	**0.46**	8.32	8.10
Cheetah	**11.71**	12.18	20.49	12.75	33.52	41.00
Gymnastics	**3.57**	4.99	8.30	3.88	86.14	43.01
Skating	**1.09**	1.13	2.77	2.62	37.14	41.84
Balloon	**2.25**	5.08	9.90	8.78	20.15	28.76
Statue	7.93	6.32	6.20	10.46	**4.07**	4.77
Christ	12.7	10.69	16.84	13.11	**5.14**	6.41
Windmill	**5.72**	8.95	37.96	10.98	66.80	48.15
Kite	**2.69**	3.31	9.70	4.90	18.66	13.77
Kite Panda	8.38	**6.10**	9.80	18.82	21.88	18.37
Average	7.93	**7.75**	14.03	11.27	26.39	26.43

Wang et al. (2014). Since there is no unified evaluation measurement, we adopt a different measurement for iCoseg (mean error rate) and MSRC (Intersection-over-Union score) in order to use as many publicly available evaluation results as possible for fair comparison. It can be seen that our proposed method achieves promising performance; it achieves the lowest error rates for many

Table 2. Segmentation performance between our approach and existing methods in terms of IoU (%, MSRC); the best results are shown in **boldface**.

MSRC (IoU)	Ours	Li et al. (2016)	Wang et al. (2014)	Meng et al. (2012)	Joulin et al. (2010)	Joulin et al. (2012)
Plane	**0.68**	0.48	0.52	0.46	0.22	0.33
Car	**0.74**	0.63	0.73	0.47	0.59	0.60
Cat	**0.69**	0.58	0.66	0.57	0.30	0.32
Dog	**0.70**	0.64	0.56	0.50	0.41	0.42
Cow	**0.73**	**0.73**	0.68	0.69	0.45	0.53
Average	**0.72**	0.61	0.63	0.59	0.39	0.44

image groups. The other comparison methods also achieve good performance in the iCoseg dataset, since the common objects contain similar colors in the iCoseg dataset. For MSRC dataset, the IoU rates over all classes are 0.61, 0.63, 0.59, 0.39, 0.44 and 0.72 for the methods in Li et al. (2016), Wang et al. (2014), Meng et al. (2012), Joulin et al. (2010), Joulin et al. (2012) and our proposed method, respectively. The IoU score is increased by our proposed method.

4.3 *Discussion*

In our experiment, we use ResNet-50 as the basic CNN, and the attention map generated from ResNet-50 is satisfactory. To further improve the result, enhancing the performance of the attention map in both precision and recall is significant.

In this paper, common targets are mainly defined as objects of the same category. However, in the iCoseg dataset, an image group of football players exists; players in red clothes or white are more likely to be viewed as the real common targets since they appear in every single image and are highly correlated with each other. Accordingly, we treat football players in red and white clothes as two different classes. To some extent, the problem was solved. Thus, integrating with high-level features in our method will give a better solution.

5 CONCLUSION

In this paper, we discover the top-down neural attention and introduce high-level semantic features into co-segmentation by CAM. We propose a weakly supervised co-segmentation method by class-specific attention and FC-CRF. The attention map has the ability to highlight discriminative object regions across a set of relevant images without requiring pixel-level or any bounding box annotations. In our evaluation on the challenging iCoseg dataset and the MSRC dataset, we demonstrate the effectiveness of our method qualitatively and quantitatively.

If the attention map can cover all the implied common targets, it will improve the efficiency of the proposed method. In the future, we will extend our proposed learning method with more discriminative ability for co-segmentation.

ACKNOWLEDGMENTS

This work is supported by the funds of National Key R&D Program of China (2017YFB1002702) and National Natural Science Foundation of China (No. 61572058).

REFERENCES

Batra, D., Kowdle, A., Parikh, D., Luo, J. & Chen, T. (2010). iCoseg: Interactive co-segmentation with intelligent scribble guidance [C]. *Computer Vision and Pattern Recognition*, 3169–3176.

Cao, C., Liu, X., Yang, Y., Yu, Y., Wang, J., Wang, Z. & Huang, T.S. (2015). Look and think twice: Capturing top-down visual attention with feedback convolutional neural networks [C]. *International Conference on Computer Vision*, 2956–2964.

Chang, K., Liu, T. & Lai, S. (2011). From co-saliency to co-segmentation: An efficient and fully unsupervised energy minimization model [C]. *Computer Vision and Pattern Recognition*, 2129–2136.

Collins, M.D., Xu, J., Grady, L. & Singh, V. (2012). Random walks based multi-image segmentation: Quasiconvexity results and GPU-based solutions [C]. *Computer Vision and Pattern Recognition*, 1656–1663.

He, K., Zhang, X., Ren, S. & Sun, J. (2016). Deep residual learning for image recognition [C]. *Computer Vision and Pattern Recognition*, 770–778.

Hochbaum, D.S., & Singh, V. (2009). An efficient algorithm for Co-segmentation [C]. *International Conference on Computer Vision*, 269–276.

Jia, Y., Shelhamer, E., Donahue, J., Karayev, S., Long, J., Girshick, R.B. & Darrell, T. (2014). Caffe: Convolutional Architecture for Fast Feature Embedding. *ACM Multimedia*, 675–678.

Joulin, A., Bach, F.R. & Ponce, J. (2010). Discriminative clustering for image co-segmentation [C]. *Computer Vision and Pattern Recognition*, 1943–1950.

Joulin, A., Bach, F.R. & Ponce, J. (2012). Multi-class cosegmentation [C]. *Computer Vision and Pattern Recognition*, 542–549.

Krahenbuhl, P. & Koltun, V. (2011). Efficient inference in fully connected CRFs with Gaussian edge potentials [C]. *Neural Information Processing Systems*, 109–117.

Krizhevsky, A., Sutskever, I. & Hinton, G.E. (2012). ImageNet classification with deep convolutional neural networks [C]. *Neural Information Processing Systems*, 1097–1105.

Li, K., Zhang, J. & Tao, W. (2016). Unsupervised co-Segmentation for indefinite number of common foreground objects. *IEEE Transactions on Image Processing*, *25*(4), 1898–1909.

Meng, F., Li, H., Liu, G. & Ngan, K.N. (2012). Object co-Segmentation based on shortest path algorithm and saliency model. *IEEE Transactions on Multimedia*, *14*(5), 1429–1441.

Ren, S., He, K., Girshick, R.B. & Sun, J. (2017). Faster R-CNN: Towards real-time object detection with region proposal networks. *IEEE Transactions on Pattern Analysis and Machine Intelligence*, *39*(6), 1137–1149.

Rother, C., Minka, T.P., Blake, A. & Kolmogorov, V. (2006). Cosegmentation of image pairs by histogram matching—Incorporating a global constraint into MRFs [C]. *Computer Vision and Pattern Recognition*, 993–1000.

Rubinstein, M., Joulin, A., Kopf, J. & Liu, C. (2013). Unsupervised joint object discovery and segmentation in internet images [C]. *Computer Vision and Pattern Recognition*, 1939–1946.

Simonyan, K., Vedaldi, A. & Zisserman, A. (2014). Deep inside convolutional networks: Visualising image classification models and saliency maps [C]. *International Conference on Learning Representations*.

Vicente, S., Rother, C. & Kolmogorov, V. (2011). Object cosegmentation [C]. *Computer Vision and Pattern Recognition*, 2217–2224.

Wang, F., Huang, Q., Ovsjanikov, M. & Guibas, L.J. (2014). Unsupervised multi-class joint image segmentation [C]. *Computer Vision and Pattern Recognition*, 3142–3149.

Winn, J., Criminisi, A. & Minka, T.P. (2005). Object categorization by learned universal visual dictionary [C]. *International Conference on Computer Vision*, 1800–1807.

Zhou, B., Khosla, A., Lapedriza, A. (2016). Learning deep features for discriminative localization [C]. *Computer Vision and Pattern Recognition*, 2921–2929.

Automatic Control, Mechatronics and Industrial Engineering – He & Qing (Eds)
© 2019 Taylor & Francis Group, London, ISBN 978-1-138-60427-8

Author index

Printed and bound by CPI Group (UK) Ltd, Croydon, CR0 4YY

24/10/2024

01778293-0005